www.KnowledgePublications.com

Movable Insulation

www.KnowledgePublications.com

A Guide to Reducing
Heating and Cooling
Losses Through the
Windows in Your Home

Movable Insulation

by William K. Langdon

Illustrations by William Graham

Rodale Press  Emmaus, Pa.

Book designed by Kim E. Morrow

Printed in the United States of America on recycled paper, containing a high percentage of de-inked fiber.

Artwork accompanying the Introduction reprinted with permission of John Schade, Associate Professor, School of Architecture and Urban Planning, University of Wisconsin—Milwaukee.

Library of Congress Cataloging in Publication Data

Langdon, William K
Movable insulation.

Bibliography: p.
Includes index.
1. Dwellings—Insulation. 2. Windows. I. Title.
TH1715.L34 693.8'32 80-11300
ISBN 0-87857-298-8 hardcover
ISBN 0-87857-310-0 paperback

2 4 6 8 10 9 7 5 3 1 hardcover
2 4 6 8 10 9 7 5 3 1 paperback

For Peter, Kathi, and Pat

Contents

Preface

In 1974, I seriously began to explore how windows affect the use of energy in buildings. For several decades, glass had been indiscriminately used in buildings without regard to climate or orientation. In response to rising fuel costs, many were advocating drastic reduction in window areas to save energy. My attention, however, focused on the positive effects received from carefully planned windows in energy-efficient building designs.

Windows have a lot to offer a building or home aesthetically, in the way of vision to the outside. The freedom that modern architecture had attained at this time with glass—opening entire wall areas to exterior views and sunlight—was a genuine improvement in the quality of modern life, and it seemed unlikely that many people would revert back to dimly lit rooms with minimal window areas. The daylight which entered through windows not only had aesthetic qualities but energy benefits as well. Electric lighting could be reduced and windows that faced south were providers of solar heat during the winter.

Instead of arguing about the pros and cons of windows, some architects and designers took a more enlightened approach and examined the proper design and placement of windows. Windows are very beneficial if properly located. The heat gained during the day by south-facing windows far exceeds their losses at night. Good building design during this period not only took into account the proper location of windows, but included proper siting and proper arrangement of interior spaces to allow the maximum benefit from windows in each room.

Windows of all orientations, regardless of whether or not they gained heat during the day, still lost significant quantities of heat during the night. To combat these nighttime heat losses, Steve Baer and other pioneers in the field developed shutter designs. Inspired by their work, I included, at this time, thermal curtains as part of the sun-tempered buildings I was designing. These window insulation designs and principles were presented in a section that I wrote for Eugene Eccli's *Low Cost, Energy-Efficient Shelter for the Owner and Builder* (Emmaus, Pennsylvania: Rodale, 1976).

Things have changed significantly in house design and construction since 1974. Many of the energy-efficient construction practices advocated in Eccli's book have become the norm, rather than the exception, as fuel prices have taken a sharp upward turn that has not yet ended. The use of south-facing windows for solar heating has expanded into a compelling passive solar movement that now challenges the imagination of nearly all architects and designers. Passive solar homes are under construction from coast to coast, and existing homes are receiving energy retrofitting everywhere, many with passive solar features.

In spite of the rapid progress to make new and existing homes more energy-efficient, the development of window insulation has proceeded at a snail's pace. Between 1974 and 1978, there was little new activity in this field. For a handful of movable insulation and design enthusiasts, this period was a time to work patiently behind the scenes, sketching ideas and developing prototypes as money and time permitted.

I realized that things were beginning to change in this field during the winter of 1978, when I received a draft edition of William Shurcliff's *Thermal Shutters and Shades*. Having waited for over two years after the release of Eccli's book for a more detailed guide to window insulation, I studied Shurcliff's book with a great deal of enthusiasm. Like so much of his work, this publication boldly plunged into an entirely new territory and brilliantly illuminated the character of its terrain. While reading this survey, I realized that the wheels of progress were still slowly turning and that a well-researched book on window insulation, including construction details and information from people who have built and lived with such systems, would be very useful to both designers and lay people. It was shortly thereafter that I decided to research and write this book. With Shurcliff's draft survey as a crude map of this strange territory, I began what turned out to be a long and exhausting journey.

After several months on the road, visiting and interviewing people who had built and installed movable insulation systems in their own homes, and more than a year at the desk writing and researching, the book is now finished. While I am pleased at the amount of information that I have been able to uncover, the book no doubt has some errors, and essential information in certain areas is still incomplete. The greatest dilemma in writing this state-of-the-art piece has been the critical need for solid and complete information amid an embryonic and rapidly developing field. I have had firsthand experiences with roughly one-half of the designs in Part II, or one-quarter of those in the entire book. The remaining designs were contributed by others active in passive solar and energy-efficient design in houses. If time and money would have allowed, I would have liked to construct all the systems I described in order to verify that they are free of design errors. However, this could have taken several years to complete, and in the meantime, a great deal of useful information would have been withheld.

Having acknowledged that this book is not without flaws, I'll release it to you. My research is now your research; use it as a stepping-stone to better window designs. As you experiment with the concepts and details here, you, too, will be helping to expand the hands-on knowledge that is vital to this growing field. I hope that you enjoy *Movable Insulation* and that it helps to make your next winter a warmer one.

Bill Langdon
September 1979

Acknowledgments

I would like to thank Bill Shurcliff for generously sharing his knowledge with me and for encouraging me to continue the research of this book when the task at hand seemed insurmountable.

Second, I would like to thank Judy Thiele and Lawrence Doxsey for their help in researching and writing. Without them, the completion of this project would not have been possible. Judy's talents in typing, corresponding, and editing copy produced alone what would have normally required two or three staff positions, but her greatest contribution was her ability to organize, catalog, and file a continual influx of information. As a meticulous researcher, Lawrence Doxsey spent many months charting correspondence, studying product literature, computing tables, and preparing the examples in the appendices. He also prepared original sketches for many of the illustrations.

The next line of thanks is extended to Bill Graham, Carol Stoner, and the Rodale Press Photography Department. I was very pleased to be able to enlist Bill Graham's graphic talents to lucidly illustrate the many designs and ideas presented in this book. Carol Stoner of Rodale Press was a real pleasure to work with on this project. She always was able to remain supportive even at times when constructive criticism was required. Rodale's Photography Department was also quite helpful as this project neared completion, supplying me with a number of key photographs.

Several other people also deserve to be recognized for their help in researching the designs for this book. Those who have helped directly in lengthy phone conversations and correspondence include: Nancy Korda, who has developed a number of designs at The Center for Community Technology; Denise A. Guerin, an interior designer who has researched the use of fabrics in window energy controls; Fuller Moore, an architect who is developing a greenhouse shade with magnetic edge seals; Clare Moorhead, who has helped to thermally improve the design of curtains;

and Tim Maloney with his ingenious, thermal garage door designs. A more complete list of credits for inventors and designers is outlined in Appendix V.

Finally, I would like to thank several friends, both near and far, whose warm encouragement and advice have helped to see this project into completion. They include Barbara Putnam, Jack Parks, Joy Lovoy, Susan Nichols, Victor V. Olgyay, Alyce Smith, and Dr. Robert Cole.

Concerning Patents

Patents give special rights to inventors to exclusively produce and sell their inventions, usually for a period of 17 years. Many of the designs in this book are covered by patents; however, it is beyond the scope of this book to legally interpret and identify all of them. Some of these patents are footnoted or mentioned in the text when the author had specific knowledge about them and in general, the author has attempted to identify the inventors of each design in this book. The reader should recognize that the appearance of a design or idea in *Movable Insulation* does not in any way convey a release of patent rights held by the inventors.

Introduction

If your home is a typical one, it loses 25 to 30% of its heat through its windows. If you have particularly large window areas, this heat loss may be 50% or more, even if they are of insulating glass and face south! In this book you will find numerous movable insulation systems that will cut the heat loss through these windows in half.

All types of window glass are poor insulators. A double-glazed or "insulating glass" window conducts 10 times more heat from your home than a well-insulated wall of the same area. For only single-glazed windows, this rate of heat loss can be doubled once again. The ideal glazing material for your windows would be as transparent as glass but insulate as well as your walls. Unfortunately, no such material exists. In the wintertime, windows benefit us only during the 10 hours of daylight; during the other 14 hours of darkness, insulating curtains, shutters, or shades can be drawn over exposed glass areas, bringing their heat-retaining value up to that of our walls.

Insulation has traditionally been a fixed item placed permanently inside walls, ceilings, and floors of a building to trap heat and create an effective thermal envelope. Insulation is most effective when this envelope is continuous and complete, with no gaps or weak spots. Windows penetrate this envelope and allow heat to escape. *Movable* insulation covers these gaps in the building's thermal envelope at night. Because this insulation is movable and not stationary, it can be removed during the daytime to allow light and solar heat to enter.

Many of the movable insulation systems in this book are first-generation items, created by people who are building and living in energy-responsive environments. Many can be home-fabricated and are fully diagrammed and described in Parts II through VI. Systems which are ready-made and available as off-the-shelf products are also included in each chapter. As you read this book, you will learn how much you can save by adding thermal protection to your windows.

Homeowners will probably be the first to use the concepts in this book because they have complete freedom to change the design of their own windows. However, this book is not just to aid the homeowner. The information here can also be useful in commercial and institutional buildings where abundant use of single glazing is causing a colossal loss of heat. This book should also aid those architects and engineers who are designing buildings which will respond to natural energy cycles. Window insulation speaks a new design language that is still foreign to many architects, and this guide helps to reveal some of the essential design details and criteria needed to apply such insulation. A method of assessing the economic return from a window insulation system in Appendix II and the listings of products, hardware, and components in Appendix III may be of particular interest to design professionals.

All practical solar designs begin with an energy-efficient design. Movable insulation reduces heating loads, thereby reducing the collector area required. In passive solar heating systems, movable insulation is often an integral component, helping heat-storage walls to retain their heat for longer periods of time. In cold climates, movable insulation enables the solar greenhouse to maintain growing temperatures without a supplemental source of heat. By employing movable insulation to enhance solar water heating systems, pumps, collector panels, and heat exchangers can sometimes be eliminated altogether!

Much can be done to reduce these heat losses in your present home by only slightly modifying your existing window drapes. And those who are willing to depart from this conventional window treatment will find many imaginative systems described and detailed in this book which reduce these losses further. The nuts and bolts of these systems, the dollars you can save, and their means of construction and operation are fully examined in the following chapters. Beyond the heat you save from these systems, there are additional benefits. Solar heat gain can be enhanced and controlled, privacy can be insured, and the spaces where these systems are employed take on new depths of relevance to their surrounding environs.

New possibilities for human shelter spring from the advent of these dynamic thermal valves. Shelter is an extended microclimatic adaptation of space for humans and other living things. As our shelters begin to pulse and breathe in response to changes in the outside environment, to enhance the metabolic processes of living systems within, the very shelter takes on the biotic or lifelike functions of the life it contains. From this metabolic responsiveness is emerging the true "organic" architecture of the twentieth century.

Part I

Movable Insulation— Why Is It Needed Today?

Today's homes are better insulated than they were 80 years ago but they also generally have much larger glass areas than a home built at the turn of the century. Even with double glazing, today's windows are liable for an increasing share of the home heating budget.

Part I of this book examines how heat is lost through windows and explores options available to reduce these losses. If you are planning to install movable insulation in your home, reading Part I is strongly recommended to help you make a wise investment. Unfortunately, the flow of heat is difficult to block, and Chapters 2 and 3 require careful study by anyone who is not already familiar with the subject. Some readers may elect to breeze through these two chapters because of their technical content and read them more thoroughly at a later date. Chapter 4, however, contains essential guidelines for the design of any window insulation system and should be read carefully before beginning Parts II through VI.

Chapter 1

The Energy-Responsive Dwelling, Past to Present

Window openings have always been the most vulnerable part of a building's shell. The exterior walls and roof of a building form an envelope to protect the occupants from wind, rain, and excessive sunlight. The resistance of this envelope to heat transfer allows a comfortable temperature to be maintained inside the building. The walls and roof of a building can be constructed to accommodate insulation of whatever thickness is needed, limited only by cost, space requirements, and the structural support they provide. Windows, on the other hand, are inherently vulnerable to heat losses and gains because they don't have the benefits of such insulation.

Windows interrupt the protective envelope of a shelter by exchanging both heating and cooling energy with the outside. But a dwelling without windows would be a very dismal place to live—dark, stuffy, and vaultlike. Windows provide natural light, a view to the outside, and ventilation. Windows and other glass areas are also very important exchangers of thermal energy. Glass allows radiant heat from the sun to enter a building. The sun's rays can be a liability to cooling in the summer but are a real bonus to heating during most days in the winter. However, windows are poor thermal insulators, and much of the winter heat gained through them during the day is lost on cold nights and very cloudy days unless adequate measures are taken to keep this heat inside.

Effective windows are energy control points in the protective envelope, opening to a beneficial environment and closing to a hostile one. The amount of insulation used in walls and ceilings has been greatly increased during the past decade, in both new and existing homes, to isolate the interior from temperatures on the exterior. Windows cannot simply isolate us from the exterior because their main function is to allow communication or exchange with the outside. However, with movable insulation, windows can respond to the outside environment, sometimes allowing energy to flow into the dwelling and at other times preventing an energy transfer between the inside and the outside.

A History of the Residential Window

In dwellings of the Middle Ages, the notion of a window opening and closing in response to outside weather conditions preceded the use of glass. Although large, stained-glass windows adorned the Gothic cathedrals constructed during this era, the dwellings of the village peasants usually contained only simple holes in the walls, without glass, to allow daylight and fresh air to enter. These bare openings were small, keeping drafts to a minimum, and had little view to the outside. Wooden shutters were closed over them at night and during storms. Without glass, there was no way to allow sunlight to enter and screen out the winter wind at the same time. Shutters were the only means to control the energy flow through the window.

Although glass windows were used in some buildings dating back to the Roman Empire, even small glass windows were rare in the home of the common person until the eighteenth century when production methods improved. Previously, glass was made by reheating a globe of blown glass on an iron rod or "pontil" and then spinning the rod until the globe flared out into a disk (see Figure 1-1). The disk was thicker at the center than the edges and was uneven with concentric ridges. Small panes were cut from the outer regions of the disk where the glass was most uniform, and these were then tediously leaded together to make the windows of the day.

Figure 1-1: Spinning a pontil to make glass.

Figure 1-2: Casement window.

Figure 1-3: Double-hung window.

As glass was introduced in homes in northern Europe, it was the accepted practice to put it only in the upper portion of the window and to leave bare openings with shutters below. Glass was later introduced into the shutter itself and gradually evolved into the casement window as we know it today (see Figure 1-2).

The double-hung window (see Figure 1-3), widespread in the United States today, came into common use in England during the late seventeenth and eighteenth centuries. Instead of swinging to the side like hinged casement-window shutters, two sashes slid vertically, up and down, in separate wooden channels.

The nineteenth century saw tremendous improvements in the methods of glass production, increasing the size of individual panes and decreasing the number of panes per window. A window that contained 4 panes of glass in 1830 would have required 18 to 24 panes 100 years earlier.

By the twentieth century glass could be made to essentially any size desired. The only limitations to its use were generated by structural considerations (which were usually respected) and by thermal considerations (which were usually ignored). The structural limitations diminished as technology improved. The development of heavy plate glass, followed by tempered glass, allowed the construction of very large panes that could withstand heavier wind loads. At the same time the steel skeleton frame helped to free buildings of heavy, masonry walls. "Modern architecture" as we know it was born using several times the glass area on the exterior walls of buildings as was used previously.

After World War II, the window designs in many new homes were influenced by the "window wall" concept and by the influx of new window types and materials. Large, sealed, double-glazed units, which could be fixed between structural wooden posts or mullions, formed the residential window wall. Sealed double-glazed glass had insulating properties superior to single-glazed glass with none of the cleaning problems of storm windows. Glass was used indiscriminately in homes to capture the visual features of a site with no regard to climate or solar orientation. Heating fuels were cheap then. Heating offset winter heat losses and large air conditioning (cooling) systems were added to remove summer heat gains.

Since the 1950s, the large fixed-glass sash has become a common element in the American home. Two additional types of operating sashes have become popular since World War II—the sliding window sash and awning or hopper window sashes (see Figure 1-4). With larger window glass areas, it is no longer necessary to open the entire window for ventilation. Awning or hopper windows are commonly used above or below a fixed-glass section to allow adequate ventilation above or below it. Horizontal sliding windows allow larger glass panels to be used as operating sashes than the vertical, sliding double-hung sashes. This type of window is often extended all the way down to the floor, becoming the "patio door" which allows access to outdoor living spaces.

With the development of the glass window into its modern-day version, the functional window shutter declined in use. Exterior shutters as shown in Photo 1-1 were an

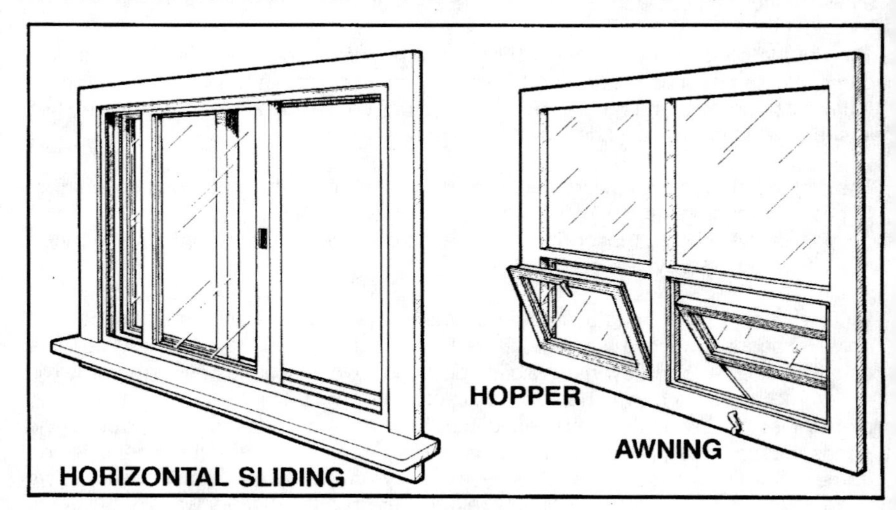

Figure 1-4: Modern residential windows.

Photo 1-1: Exterior shutters on an old Pennsylvania house.

integral part of the homes constructed in northern states until the mid-1800s. These shutters were opened and closed daily during the winter to help seal the dwelling against the wind and cold. Louvered interior shutters were also common in warmer climates to screen out intense summer heat. When central heating and cheap fossil fuels such as coal came into use, the New England shutter became merely an inoperable ornament.

Many years later, air conditioning was introduced and sun-shading shutters also began to disappear. Housing styles that were originally indigenous and climatically responsive to a particular region spread across the country and lost much of their regional indentity and purpose. Georgian colonials were constructed in New England, New England saltboxes in Kansas, and prairie or ranch homes were constructed everywhere. The original function of items such as window shutters was lost as these energy-responsive building components became merely ornaments in look-alike housing developments across the American continent.

The importance of operable window shutters has been recently revived, because of rising fuel costs and a new interest in energy conservation. Windows are becoming widely recognized as very important sources of the sun's renewable energy but also as great contributors to winter heat loss. Shutters are appearing in homes today in new forms to enhance our ability to live in a balanced and non-energy-intensive manner. For over a decade now Steve Baer at Zomeworks in Albuquerque, New Mexico, has been pioneering in finding new ways that shutters can help direct sunlight to sustain a warm and comfortable interior in a home during the winter. Charlie Wing at Cornerstones, a house-building school in Brunswick, Maine, has recently developed a low-cost, home-built shutter that is becoming popular in New England. The interest and developmental work in window insulation is growing rapidly not only among designers, but among builders and homeowners as well. On the exterior of homes in sunny New Mexico and the interior of homes in coastal Maine, the tradition of the window shutter is being revived and updated, and the new versions have become a viable means of controlling the flow of energy through windows.

Windows and the Energy-Responsive Dwelling

Window heat loss was not as significant a problem to people settling in the United States 200 years ago as it is today. The homes built then had no insulation. Air infiltration was by far the biggest heating problem. A home built in those days would often lose five to eight times as much heat per winter as a well-insulated home of comparable size built today.

Windows were generally small in the old days due to the expense of glass, and they were often covered with shutters to reduce the air infiltration around leaky sashes. These small glass windows were a real luxury at the time because they allowed light to enter a home while to some degree kept out cold winds.

The pie graphs in Figure 1-5 show four homes comparable in size and space built over the past 120 years. Each of the four homes contains a typical amount of insulation and window protection for its period in time. Two general trends can be observed from this scenario. Each newer house shows a reduction in total heat losses through the building envelope due to continually improved insulation standards and the reduction of air infiltration through cracks. House D built in 1978 uses less than one-fifth the energy of House A built in 1858! However, the percentage of heat lost through window glazing in each subsequent house increased as a result of reduction in heat losses in other parts of the building's envelope.

Caution should be exercised in assuming that any house is typical for a given period in time. General trends in construction and insulating practices can be observed but a survey of the homes in any given period in time will reveal vast differences in the distribution of heat losses among the walls, ceilings, windows, and air leaks. The energy retrofitting of homes today generally brings their insulation standards up to

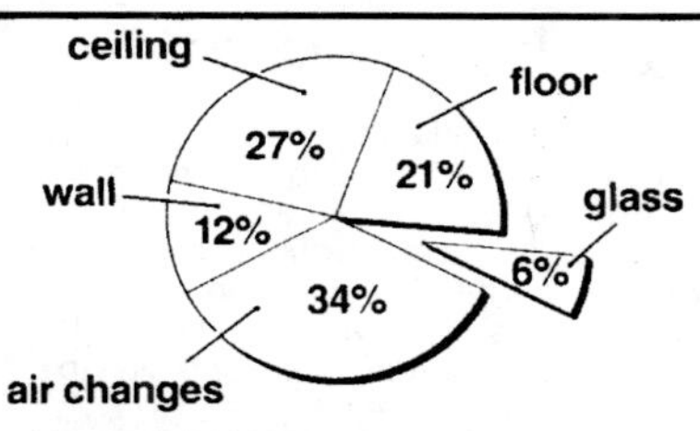

House A (1858) has no insulation. The windows are small and equivalent to only 9% of the floor area (121 square feet). The house is full of cracks and it has an open fireplace so four air changes per hour are common. Even though the windows in this house are single glazed, they account for only 6% of the heat losses.

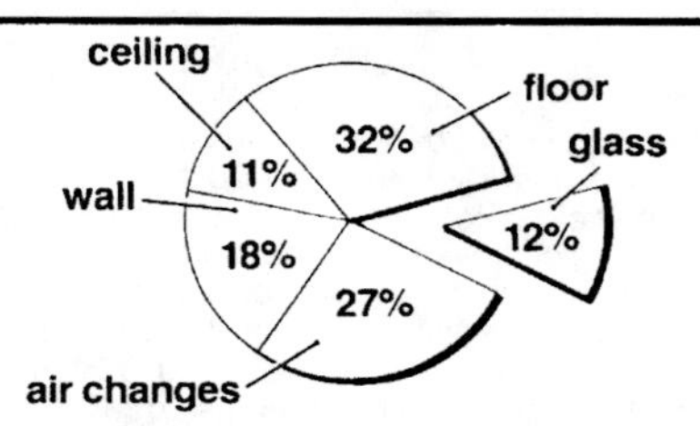

House B (1928) is similar to House A except that 2 inches of insulation have been added to the attic and tighter construction has reduced the air changes to two per hour. A few windows have also been enlarged to make a total glass area of 145 square feet or 11% of the floor area. Still using single glass, the windows in this home lose through them 12% of the house's heat.

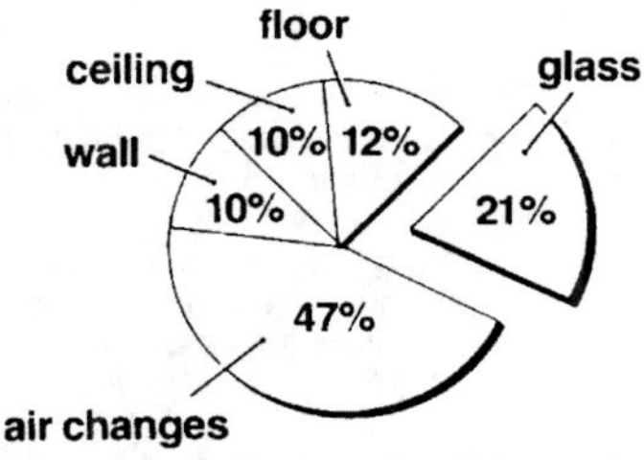

House C (1958) was built in the post-World War II era. Glass had become much lower in cost and was available in large double-glazed sheets. The window areas of this home are 19% of the floor area or a total of 242 square feet. Three inches of insulation were used in the walls, 6 inches in the ceiling, and 3 inches in the floor, and air changes have been reduced to 1.5 per hour. Even though all the glass in this house is double glazed, the percentage of heat lost through the windows has jumped to 21% of the total losses. The window heat losses here are similar to losses in houses built 50 years earlier, when single glazing was used, since the reduction caused by double glazing has been offset by the additional window area.

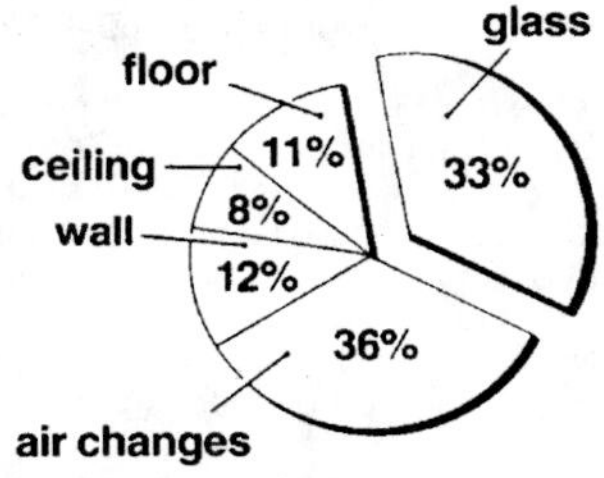

House D (1978) was built during the diminishing fuel reserves era of the late 1970s. With the housing market getting serious about energy conservation, the air changes have now been reduced to 0.75 per hour. Six inches of insulation are used in the floor, 12 inches in the ceiling, and 5 inches in the walls. The 242 square feet of double-glazed glass—the same as in House C—causes 33% of this home's heat to be lost through windows.

Figure 1-5: A scenario—four homes over the last 120 years.

WINDOW AREAS AND THE BUILDING CODE

The purpose of building codes is to protect the safety and general well-being of the public against shoddy and unsafe building practices. Residential window areas were not an area of particular concern in building codes until after World War II when rapid construction of housing on government-subsidized loans raised a genuine concern about needed window areas for daylighting and ventilation. As a result, most residential codes adopted "light and ventilation" requirements for the main or "inhabitable" rooms in a residence. A glass or window area equal to no less than 10% of the floor area was required in each inhabitable room. Openable window areas equal to 5% of the floor area were also required.

The rising costs of fuels and impending shortages recently brought about concern for the energy wasted in buildings with oversize glass areas. Window areas were deemed energy losers regardless of orientation, and some states naively instituted prescriptive codes with a flat, maximum area requirement for windows. As the energy panic subsided, a more enlightened viewpoint gradually emerged with regard to windows, and it was recognized that windows that are properly oriented can provide more energy than they lose. These codes were then amended to allow larger glass areas if the builder could prove that the home loses no more energy than permitted by the code maxima. The problem with this type of code is that it makes it easy for the builder to construct a mediocre home. But to construct a quality home with large, carefully planned glass areas facing south, the builder is required to go through a lot of rigamarole.

One of the most sensible codes for regulating window areas in residences has been instituted by city officials of Davis, California. This code defines what are called "unearned and earned" glass areas. The builder begins with an allotment of "unearned" glass area not to exceed 12½% of the floor area for single-glazed windows. If the window is double-glazed instead of single, the allotment is increased 40%, allowing the glass area to total 17½% of the floor area. In other words, a home that is allowed 100 square feet of single-glazed windows can have 140 square feet if the windows are double-glazed. The builder "earns" this extra glass area by meeting energy conservation standards.

Thermal shutters, which are the most effective way to control window energy transactions, also allow the most freedom in the design of glass areas. When shutters are used, the area of the allowable glass can be multiplied by the R value of the shutter. A 10-square-foot, single-glazed window area can be increased to 50 square feet with an R-5 shutter.

South-facing glass also allows the builder to "earn" extra glass area up to a point where the calculated mass of the building is no longer able to absorb the heat entering through this glass. The amount of mass in the home is weighed when assessing the amount of south-facing glass area that can be earned, which allows house designs with greater mass to incorporate more south-facing glass.

The Davis Code is very firm about limiting the glass areas that are unshaded from direct sun during the summer to 3% of the floor area of a home. None of the other design concessions for additional glass areas allow an increase of unshaded area. The proper shading of glass areas is an issue that most building codes do not address. Unshaded glass gives a disastrous boost to summer afternoon, electrical peak loads in areas where electric air conditioners provide cooling. Marshall Hunt is a Davis resident and one of the principal authors of the new code. In emphasizing the importance of shading he states, "Shading the homes in Davis can make the difference as to whether or not a nuclear power plant is built in the area." Each unshaded home adds 2 kilowatts to the afternoon peak load. Window shading, yes. Nuclear power plants, no!

those in House C or House D. As insulation is added to these homes, the heat losses through windows become more significant. Sooner or later windows emerge as the most promising and cost-effective place to put your insulation dollars.

It is important when retrofitting an existing house, to do weathertightening projects in order of priority, determined by their overall energy effectiveness. Caulking and weather-stripping to reduce wall, door, and window air-infiltration losses and adding insulation to the ceiling are usually the best weatherization tasks to do first. In homes with crawl spaces, insulating the floor would be the next step. Insulating walls can be difficult and expensive in existing homes. In cases where there is a choice between insulating walls and adding storm windows or movable insulation to single-glazed windows, the window retrofit option is the better choice. However, if storm windows already exist, it is usually more beneficial to insulate the walls before investing in movable window insulation.

Reducing heat losses from the home is like whittling a stick to a sharp point. A well-rounded approach is recommended, reducing it on all sides of the profile; otherwise you will just flatten one side.

Chapter 2

Window Heat Losses and Gains

Heat is never stationary. A kinetic form of energy, heat constantly moves toward materials and spaces having cooler temperatures, whereupon its energy dissipates. Even if it isn't moved by fans or blowers, heat devises its own means of travel, and we depend upon the clever placement of thermal insulation to help trap heat by providing resistance to its movement.

Moving a warm substance away from a heat source will increase the rate of heat dispersion. For example, the cooling system of an automobile engine pumps water through the hot engine block so that the water can collect heat and carry it to the radiator where it dissipates into the cooler air outside. When an exterior door is left open on a cold winter day the heat inside quickly travels to the cooler temperatures outside.

Two kinds of heat losses occur through your windows—conductive losses and air infiltration losses (see Figure 2-1). Conductive window heat losses occur as heat moves to window glass and passes through it to the outside. Movable insulation can be added to a window to slow down the rate at which this heat escapes. However, unless cracks around the window are sealed very tightly, not only will heat conduct through the glass, it will also be carried to the outside by the air moving around the window sash. The heat losses from warm air escaping to the outside around a window sash are called window air infiltration losses.

Window Air Infiltration Losses

Infiltration losses are difficult to measure precisely, yet they can be very significant. A single-glazed window that is loose and poorly fitted can lose twice as much heat through the air cracks (or through infiltration) than through the glass area itself (or through conduction) (see Figures 2-1a and 2-1b). The amount of heat lost by air

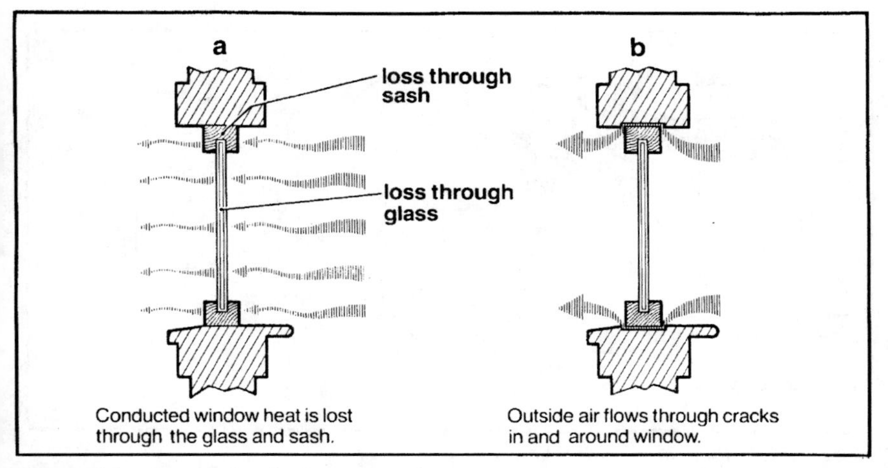

Figure 2-1: Two kinds of window heat losses.

leakage depends not only on the size of the cracks around the sash and frame, but also on the outside wind speed and direction. Table AIV-1 in Appendix IV shows the number of cubic feet of air flow per linear foot of crack for several different types of sash treatments and wind speeds. However, unless you are a serious researcher in shelter technology, all you need to know about air infiltration is that it robs a great deal of heat from your home. Where there is a crack, you are wasting heat and dollars! Caulking and weather stripping can be done by just about anyone and are invariably the most cost-effective forms of home energy retrofit.

Movable window insulation reduces the problem of conductive losses through window glass but it often does little to reduce air infiltration around the sash and frame unless it seals very tightly to the window frame. Rigid shutters with good edge seals can be effective barriers to air infiltration, and shades and curtains are generally successful at stopping through-the-wall air flow. But it is generally far easier and more effective to seal against air infiltration by caulking or weather-stripping cracks in the window frame and sash than to try to form a tighter barrier with some form of movable insulation.

Stopping Window Air Infiltration—New Products and Methods

Figure 2-2 shows the two main routes of air infiltration in a jamb section of a double-hung window. One route for air flow is the space between the window casing and the

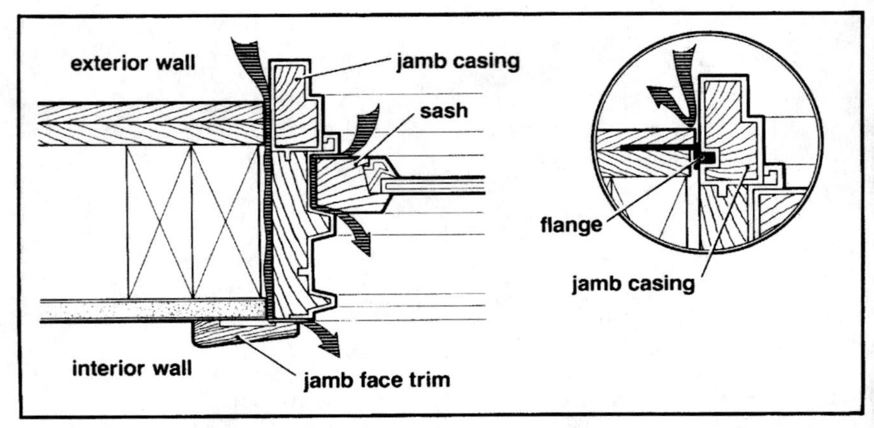

Figure 2-2: Route of air infiltration.

wall studs framing the window. The other is the clearances and cracks between the window sash and the casing or trim.

Flanges

The first problem (air flowing between the window casing and wall framing) is more serious in older homes than in new construction. Vinyl-clad windows marketed by Andersen and Pella for the past 10 years have a flange that can be nailed and caulked directly to the exterior strutting. The flange seals tightly to the wall studs or sheathing and completely stops air flow through this space. And one convenience is that siding can be applied directly over the flange without using exterior trim.

Foam Caulking

A new foam caulking material is now being used in homes to fill cracks that were mostly left unfilled before. While it was once difficult to fill cracks more than ¼ inch wide with the traditional caulkings, Great Stuff, marketed by Insta-Foam Products, Inc., Joliet, IL 60435, can be used to fill cracks of just about any size. This material (a urethane foam in an aerosol can) is also produced by several other companies and is available at most building supply outlets. A small can will lay a 1-inch bead 35 linear feet. Containers holding enough to fill 550 and 1,650 linear feet are also available.

It is becoming standard practice in new construction to fill the space between rough frame and window casings with this foam caulking. However, existing windows that leak around the frames should first be caulked thoroughly with a latex, butyl, or silicone caulking along the outside of the trim. If this caulking does not stop drafts at the edges of a window, the face trim on the inside can be removed and the gaps can then be filled with foam caulking.

If you have old double-hung windows on your house, the channels for window counterweights may admit drafts (see Figure 2-3). The slots where the sash cords run over the pulley and down to the lower sash allow air to exit through the weight channel and drain heat from your dwelling. A temporary solution is to put tape over these holes during the winter. Another measure is to caulk around the outside trim. You can also remove the interior face trim and caulk carefully from the inside (urethane foam caulking shown in the illustration), making sure you leave some room here to allow the window counterweights to drop down freely. If the sash cords are broken, you may simply want to remove them and fill this space with foam or insulation. Metal spring plates available from Wright Products, Inc., Rice Lake, WI 54868, can be mounted in old sash channels to hold the double-hung window sash open by friction. If these plates cause too much friction and make your window difficult to open, they can be taken out and flattened. This can be done by placing them on a concrete walk and pounding them with a hammer.

If spring metal weather strippings are installed in the sash channels with the proper amount of friction, they will eliminate the need for weighted sash cords. To apply

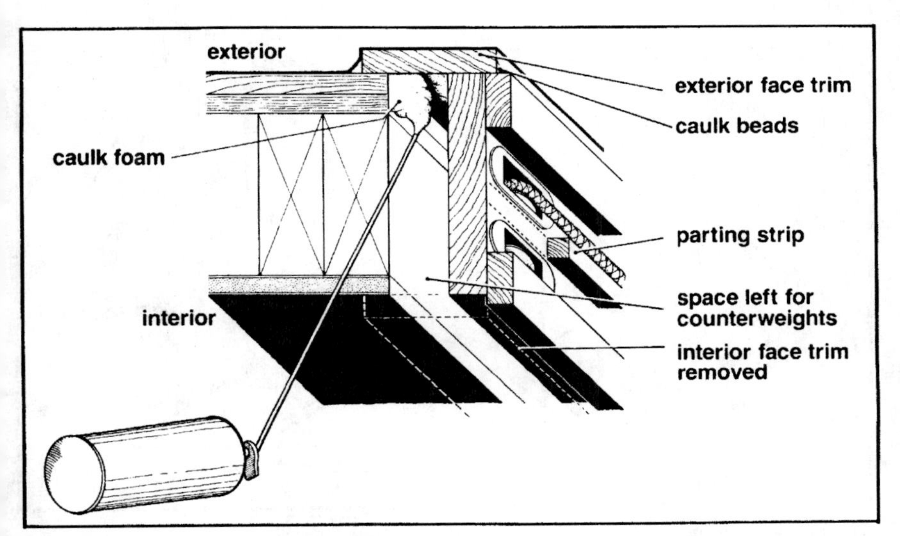

Figure 2-3: Sealing counterweighted sashes.

spring metal weather stripping along these channels, gently remove the sash stop and parting strips so that the window sashes can be taken out. Then tack the spring bronze or aluminum weather stripping into the sash channels and flare it out with a screwdriver to create the proper amount of friction when the window is reinstalled. Then remove the counterweights, fill the counter channels with insulation, and fill the holes leading into them with foam caulking, to eliminate a major source of heat loss through your windows.

If the upper window sash on double-hung windows in your home is stuck or rarely is opened during the summer, you may want to seal it permanently in place with a fine bead of silicone caulking. This upper sash is often of little use on north-facing win-

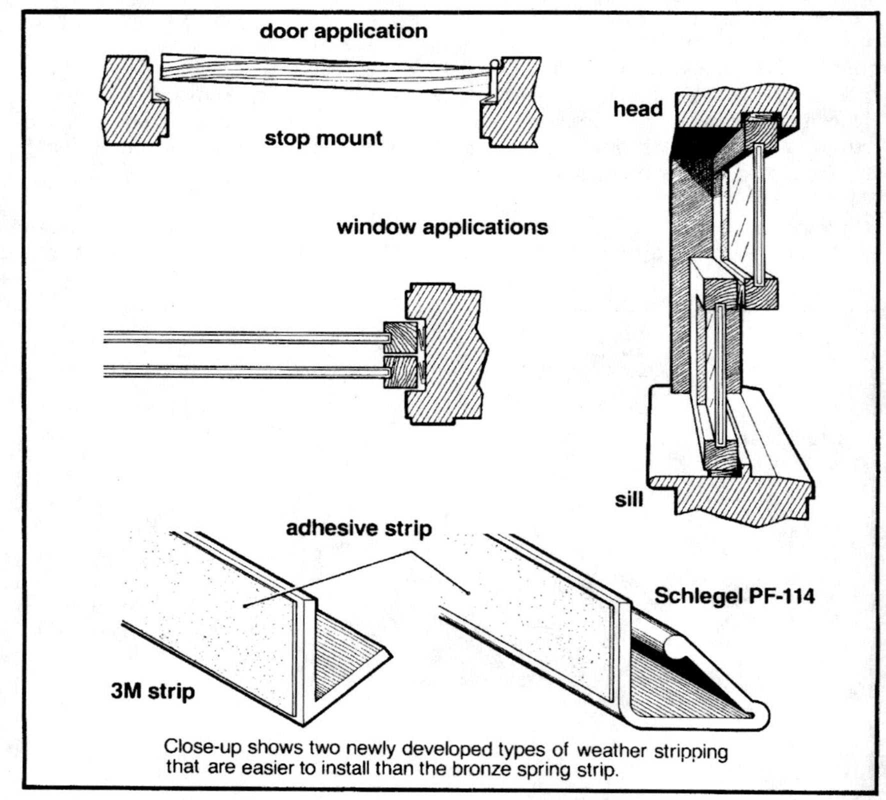

Close-up shows two newly developed types of weather stripping that are easier to install than the bronze spring strip.

Figure 2-4: Spring plastic weather strippings.

dows. However, on south-facing windows you may prefer to keep this upper sash in operation since lowering it during the summer is a very effective way to ventilate behind a shade or curtain.

Another solution to repairing old sash channels is to replace them with a single unit called Quaker Window Channels, made by Quaker Manufacturing, Sharon Hill, PA 19079. These channels form a tight fit on the sides of your window sashes and are very effective in stopping air infiltration. However, they are made of aluminum and since aluminum conducts heat readily, these channels increase the conductive losses through your window jamb.

The second major route for air flow—the cracks around operating window sashes—can be reduced drastically with weather stripping. Several new types of weather stripping have emerged which are superior in some ways to older methods. The traditional bronze spring strip is a very effective and durable weather stripping; however, it is expensive and must be tacked into some tight locations where it is difficult to maneuver. In addition, tacks must be spaced every couple of inches or air will leak behind the strips. Plastic spring strips that operate on the same principle are now available and are applied with a pressure-sensitive adhesive backing (see Figure 2-4). These weather strippings can be quickly and easily installed without special tools or skills. Unlike foam weather strippings, most of which deteriorate rapidly, these new plastic weather strippings promise to have a much longer life.

Schlegel Corporation makes a series of unique top-quality weather strippings. Their line includes Polyflex, a flexible, plastic sealing strip with an adhesive backing for inside and outside doors. Polyflex can also be used in some window applications. Fin-Seal, a fiber pile strip with a plastic center, is very effective on the bottom of swinging doors. Schlegel also has a series of foam strips in various sizes that are coated with a tough plastic skin and are available with one self-adhesive side. For information on these products, write: Schlegel Corporation, P.O. Box 23118, Rochester, NY 14692. Attention: CWO Department.

Stanley Hardware carries a line of weather strippings similar to the Schlegel products mentioned above. For a copy of their catalog, write: Stanley Hardware, Division of The Stanley Works, New Britain, CT 06050.

3M Company manufactures a light, versatile, polypropylene spring strip which comes in a roll and is called Scotch Brand Weatherstrip 2743. This strip is scored along its center so that it forms a V when folded. One leg of this V has an aggressive adhesive and paper backing. The V strip is suitable for both sliding and compression applications and is great for sealing cracks around old window sashes. If you can't find this product locally, write: 3M Company, Energy Control Products, Industrial Tape Division, 3M Center, Saint Paul, MN 55101.

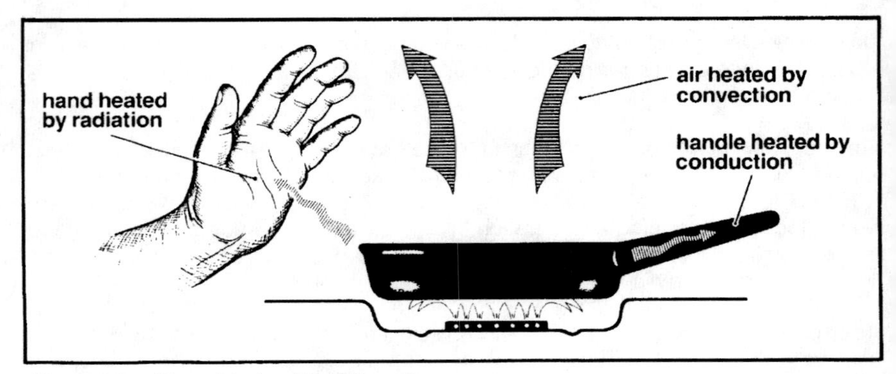

Figure 2-5: Three kinds of heat transfer.

Weather stripping can be tricky business and the products mentioned above can make it easier. For more complete instructions on applying some of the older forms of weather stripping, see *From the Walls In* by Charles Wing* and *Reader's Digest Complete Do-It-Yourself Manual.**

Conduction, Convection, and Radiation

With the problem of air leakage around windows taken care of, we can now turn our attention to reducing losses through the glass area and sash—the subject of the remainder of this book. All the heat passing *through* windows is *conducted* through the glass even though it reaches the glass and is released to the outside in several different ways. To understand how heat loss can be reduced with movable insulation or additional layers of glazing, we must take a careful look at the three basic modes of heat transfer. Just as water in rivers flows downward by gravity until it reaches an ocean or sea, heat always tends to dissipate according to several basic laws of physics. Warm objects or bodies invariably lose their heat to cooler ones around them. Heat moves by conduction, convection, and radiation (see Figure 2-5).

Conduction occurs in stationary objects or materials. Heat energy is transferred by increasing the vibration of molecules of the object or material and thus raising its temperature. If you have ever left a frying pan on an open fire while camping and gotten a burn from grabbing the hot handle, you have painfully experienced the conduction of heat.

*See Appendix V, Section 1 for book description and complete bibliographical information.

The ability of different materials to conduct heat varies tremendously. Table 2-1 shows the rate of heat conductance of several common materials. Dense materials, particularly metals, are excellent conductors. Aluminum conducts heat 1,775 times as quickly as wood. Keep that in mind next time you buy window sashes.* Glass is also a good conductor of heat and if it were not for thin, stagnant air films on each side of the glass pane (and aluminum window sashes), the amount of heat conducted through your windows would be much greater. Air is a poor conductor of heat when still. Materials that have many small pockets where air is trapped make good insulators for this very reason.

Table 2-1: The Rate of Heat Conductance of Several Common Materials

(Used In and Around Windows in Btu.-in./ft.^{2}/hr.-°F.)

Material	Rate
Still air	0.2±
Aluminum	1,450
Glass	5.5
Steel	310
Wood	0.8
Wool	0.26
Polystyrene	0.2±

One of nature's best insulators is goose down feathers. Tiny fibers in the feathers act to break up air movement. Several man-made fiberfills used in quilts and sleeping bags are based on this same principle. Urethane foam, one of the best man-made insulations available, breaks up air convection with a series of tiny air-filled cells or chambers. Walls of urethane foam cells are similar in principle to corkboard, except they are thinner and therefore conduct even less heat.

To decrease the amount of heat conducted by your windows at night, you can fix materials in front of window openings (see Figure 2-6) that have a low thermal conductance. Any good insulating material works well as do several air spaces with stagnant air films.

*This is noteworthy in terms of heat losses but a little misleading. Both aluminum and glass provide virtually no resistance to heat flow but the air films on each side of these materials inhibit the movement of heat into and out of them. Ten percent of the area of wooden sashes can be deducted from the effective glass area when computing heat losses, while aluminum sash areas simply count as if they are glass areas. If the aluminum sash contains a thermal break, it is computed like a wooden sash. In some aluminum window frames currently being manufactured, the thermal break is an epoxy gasket separating the interior and exterior frame sections.

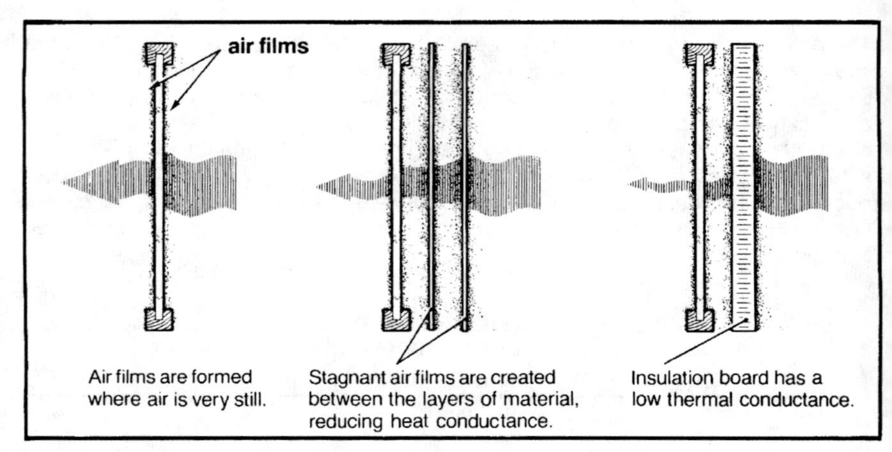

Figure 2-6: Reducing the conductance of heat through glass.

Convection is a motion of heat, occurring only in liquids and gases. Here the molecules are not stationary but move about freely. Warmer molecules move more quickly and tend to expand the liquid or gas as they increase in volume. They become lighter and rise upward while the cooler ones sink downward.

The winds outside are caused by air convection as is the draft up your chimney. There are more subtle forms of convection that cannot readily be detected, such as the convection in the space between a storm window and the inside sash. Some

Figure 2-7: Inside and outside air films.

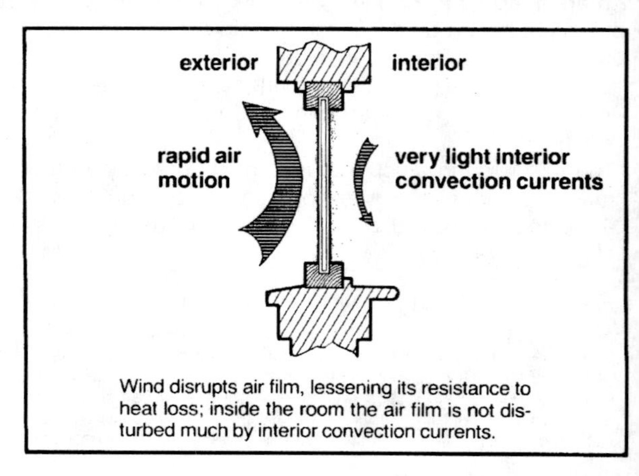

other examples are convection of cool air from a window down to the floor of a room, convection behind shades or drapes, and convection above a warm stove or heater. Wherever there are warm and cold surfaces in a room, there is convection.

Heat generally will not move closer than ⅛ inch or so to most smooth surfaces when it is transferred by convection. Air films or layers of air form along these surfaces where the air is very still. Still air is a poor conductor of heat so the air layers on both sides of window glass give it some resistance to heat flow (see Figure 2-7). In relatively still room air, convection currents do not disturb the air films very much. However, when the wind is blowing outside, the rapid motion of the air disrupts the air film outside the glass and much of it is removed, leaving an even thinner layer of still air with little resistance to heat transfer.

Since air *convects* heat readily but *conducts* heat poorly, air spaces are most effective when they have convection inhibitors. Fiber and foam insulations are essentially material filled with air pockets to block or inhibit air convection.

Window insulation is affected by convection currents. If an insulation panel, layer of glass, or some other material is placed in front of your window glass, there will always be small convection currents inside the air space (see Figure 2-8a). However, as long as the air space is sealed from the room or from another air space, the insulating value of each layer is maintained. Should air be allowed to convect from the room to directly behind your added insulating layer, the heat resistance of this material and accompanying air films will be short-circuited by convection currents and your win-

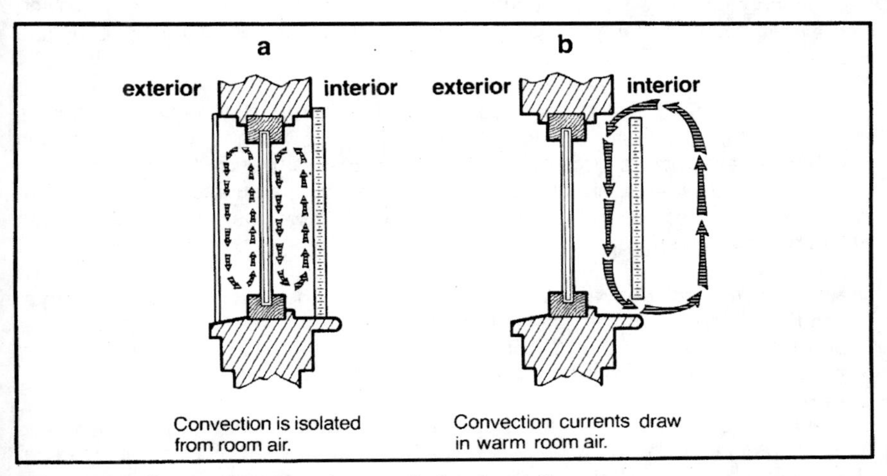

Figure 2-8: Convection currents and window insulation.

dow insulation will be much less effective (see Figure 2-8b). For this reason, edge seals are very important and are emphasized throughout this book.

Radiation is one of the most important ways heat moves from room surfaces to window glass and on to the outdoors. With single glazing this is only a two-step process. Heat radiates from room objects to window glass, is absorbed by and conducted through the glass, and then radiated to the cooler outdoors. Each layer of glass or opaque material added to your window provides one more barrier to radiant transfer to the outdoors.

Unlike conduction and convection which are readily understood by most homeowners, radiant heat lurks mysteriously in the shadows, sometimes visible in the glowing coals of a fireplace or stove, but more often invisible to the eye. Cameras with special film can be used to "see" radiant heat losses from a building, but this service is expensive and not yet widely available. We receive radiant energy from the sun; about 50% of it is visible light, and most of the rest is infrared radiation. Infrared radiation is the same type of heat as that given off by a wood stove or hot electric stove burner which has been turned off and has cooled past the point of a visible glow. Infrared radiation travels at the speed of light and originates in materials that are warmer than other materials and spaces around them.

All objects or "bodies" give off or "emit" radiant energy, but hotter ones emit more heat than cooler ones, following the principle stated earlier that heat always dissipates. You only need to concern yourself with the net flow of radiant energy. The reason the stove or radiant heat source makes you warm is that you are receiving more radiant energy than your body is giving off, resulting in a net transfer of heat to your body or a *net gain*.

When radiant energy strikes an object and is absorbed by it, the energy becomes heat. However, the heat flow doesn't stop here. The warming of an object can increase its temperature to the point where it too radiates heat to other objects. If our eyes could see low-temperature radiant heat, everything around us would be glowing constantly. People, plants, and animals would have the highly visible auras described by eastern mystics.

Radiant energy includes radio waves, X rays, and a whole range of waves characterized by different wavelengths. A brief description of these wavelengths can be found in Appendix IV, Section 2, but for a basic understanding of the heat losses and gains from windows in your house, the following should suffice.

There are three bands of wavelengths that are important. One range is the visible band of radiant waves. It accounts for 51% of the energy we receive from the sun. The second band is called the near-visible infrared, which accounts for most of the

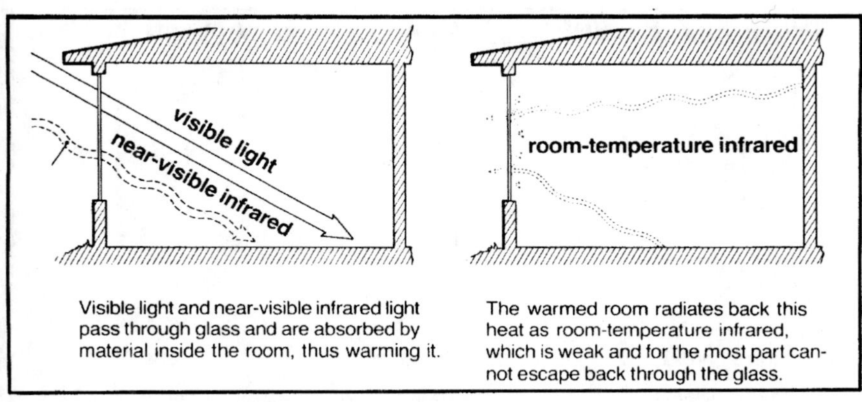

Visible light and near-visible infrared light pass through glass and are absorbed by material inside the room, thus warming it.

The warmed room radiates back this heat as room-temperature infrared, which is weak and for the most part cannot escape back through the glass.

Figure 2-9: The greenhouse effect.

remaining (45%) solar energy we receive. Window glass is transparent to both visible light and near-visible infrared energy, so window glass transmits nearly 90% of the solar energy that strikes its surfaces. The third band of wavelengths, called long-wave or room-temperature infrared, is given off not by the sun but by objects at room temperature, including everything from warm walls and furniture to hot radiators. This type of radiation cannot penetrate glass and is absorbed by it, creating the so-called "greenhouse effect" on windows that receive direct sunlight.

The "greenhouse effect" can be explained as the following process: Light comes through the glass as visible light and near-visible infrared, and is absorbed by materials inside. When these materials heat up and reradiate this heat (see Figure 2-9), their radiant waves are too weak to penetrate the glass. Therefore, most of this heat remains trapped inside the room.

To experience more directly how these principles apply to heat transfer from a room to your windows, you can experiment with a radiant heat source, a piece of thin polyethylene, a small sheet of glass, and a sheet of aluminum foil. A good, radiant heat source is a woodstove, a hot bed of coals in your fireplace, or the side of a big hot kettle on your stove. If you position your face near the radiant heat source, you will feel heat warming your skin.

Take a piece of polyethylene from a plastic bag, sandwich wrap, or the film sold for plastic storm windows. Place this material between you and the heat source. You will still feel the radiant heat since polyethylene is transparent to low-temperature infrared, as is air. Next, take a small pane of glass (one from a picture frame works just fine) and place it between you and the heat source. You will no longer feel the heat on

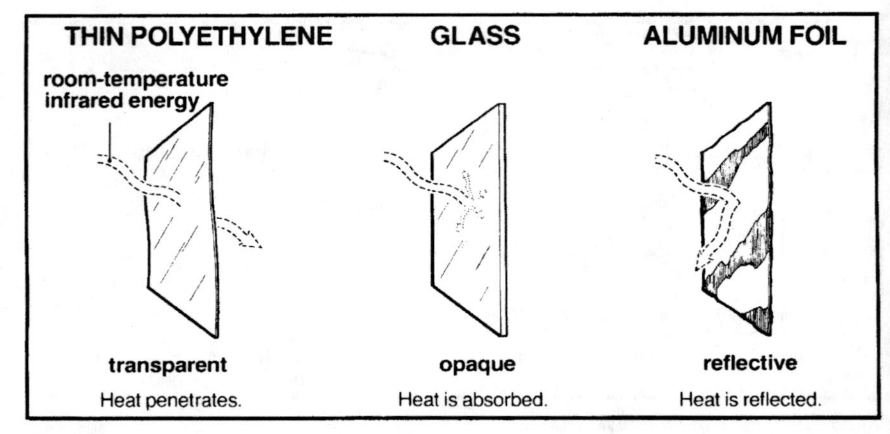

Figure 2-10: Three radiant properties.

your face because glass is opaque to low-temperature infrared and is absorbing the heat you were feeling before. Even if your heat source is glowing coals, most of the heat will still be in the low-temperature infrared range and will be shielded by the glass, although some radiant heat will go through the glass (see Figure 2-10).

Hold the glass near the heat source for a while and you will feel it rise in temperature. As it gets hotter, the glass will radiate more heat from its surface. If you place this glass close to your face, you may be able to feel a slight increase in the heat it is radiating. The same thing happens at your windows with heat from room surfaces. Heat radiates to the window glass, is absorbed, warms the glass slightly, and is reradiated to the outside or to the next pane of glass.

Next, take a piece of aluminum foil and hold it between your face and the heat source. Like the glass, it will stop heat from reaching you. However, unlike glass, the aluminum foil is not absorbing but reflecting heat back to the source. If you place your hand between the foil and the heat source you will feel the radiant heat on both sides of your hand—from the heat source on one side and reflection off the foil on the other.

Polished metallic surfaces not only reflect radiant energy but also do not radiate it very well. For this reason it is easy to burn your finger when trying to feel whether a polished-chrome steam iron is hot. You can place your finger very near the iron without sensing any radiant heat, and when you finally touch it you are surprised at how hot it is! A polished-chrome woodstove would not only be poor at throwing heat, but would be very dangerous.

To limit radiant heat transfer, a material must at least stop the radiant waves by absorbing them. Window glass serves as an opaque barrier to radiant heat transfer. However, radiant transfer is more effectively stopped if the heat is reflected back to its source. When adding window insulation, foil surfaces that face into the room's air space increase the heat resistance of a panel, shade, or curtain.

It makes little difference in overall heat transfer if the foil is on the warm or cold side of the air space between a window and interior insulation. If the foil is on the cold side of an air space, it, of course, reflects back the heat coming from the warm side. Foil-faced surfaces are also poor at giving off or emitting radiation (as described in the example with the steam iron). If a foil-faced liner is added to a curtain where the foil faces the glass, it gives off very little radiant energy and thereby limits the radiant transfer across the air space to the glass.

R Values

The term, R value, is common to most energy-conscious homeowners since R values are printed on the packages of nearly all insulation materials. The R value refers to the effective thermal resistance of a material itself, but it does not take into account air infiltration around the edges of the material once installed. Be aware that R values on products are usually derived from tests conducted under optimal conditions. Insulation used in buildings is sometimes compressed or collects moisture, or has a foil surface placed against another material. All of these conditions lower the R value and should be taken into account.

R values of all the materials used can be simply added to obtain the total resistance

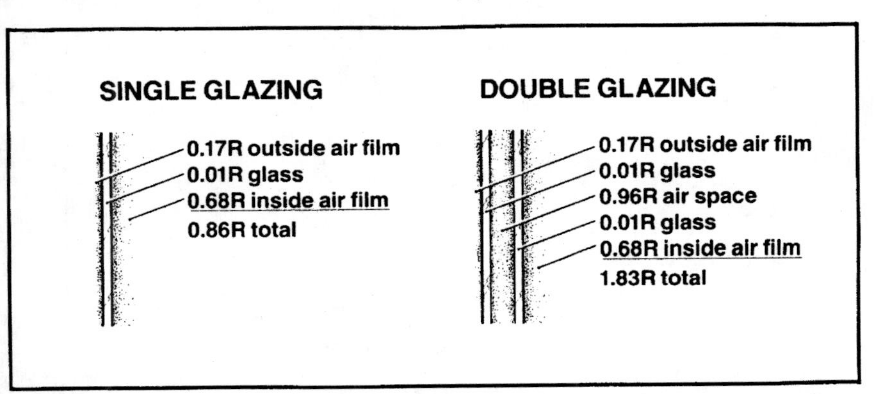

Figure 2-11: R values of single and double glazing.

to heat transfer through a window, wall, or other building section. For instance, in a 2-by-6 frame wall with R-19 fiberglass batts, the heat resistance from the exterior siding, interior paneling, and air films is small compared to the insulation provided by the fiberglass. But they do add up to an R value of about 3 that is added to the fiberglass's R value of 19 to yield a total R of about 22. The R value of a ⅛-inch sheet of glass, however, is only 0.01; therefore, the resistance provided by air films with glass is critical in reducing heat losses. Figure 2-11 shows how R values can be tallied for single and double glazing. The R value of the outside air film in a 15-mile-per-hour wind is 0.17. For the still air inside a room, R is 0.68.

An ordinary, vertical, ¾-inch air space has an R value of about 1. However, if a bright, metallic foil surface faces this air space, the R value increases, ranging between 2.36 and 3.48, according to how reflective or polished the surface is. If the layer that faces the room is reflective, an R of 1.35 can be substituted for the inside air film of 0.68.

The width of an air space also affects its rate of heat flow. Figure 2-12 shows a curve of the R value provided by an air space between two vertical surfaces. The R value falls off rapidly when the air space is any measure thinner than 3/16 inch because the thicknesses of the insulating air films on each surface are reduced. The R value increases up to about ¾ inch where air convection in this space begins to offset any

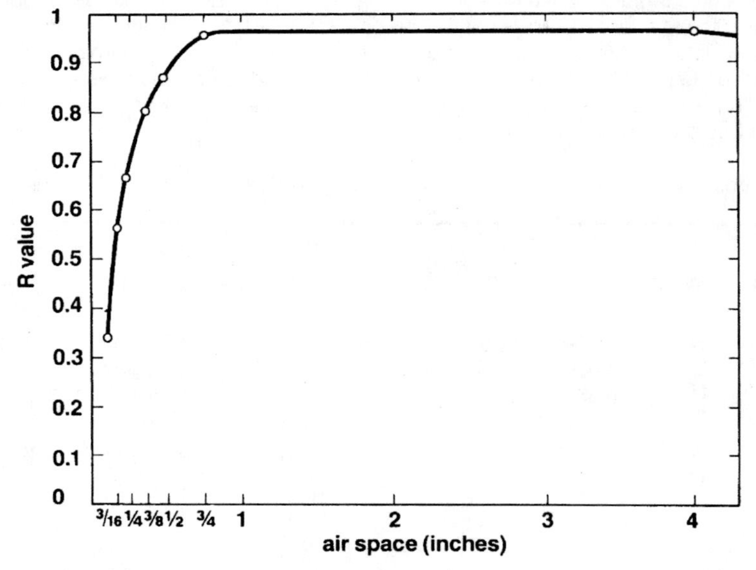

Figure 2-12: ***R value of a vertical air space.***

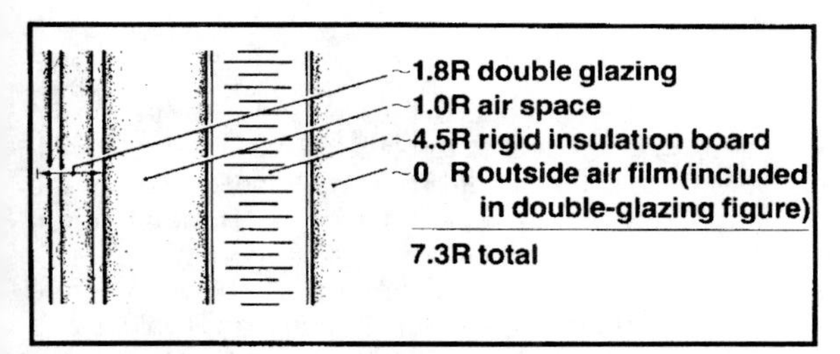

Figure 2-13: Tallying R values (disregarding edge losses).

increase in distance between the surfaces. The curve remains fairly level in this range with R value near 1 until the space exceeds 4 inches. Beyond 4 inches, convection currents disturb the air films and cause the R value of the air space to decrease. Because air spaces smaller than ½ inch have a reduced thermal resistance, the standard, ⅝-inch-thick sealed, double-glazed unit (two layers of $3/16$-inch glass and a ¼-inch air space) has a total R value of only 1.55, while a storm sash-window combination with a 3-inch air space has an R of 1.83. To simplify tallying approximate R values of window insulation, 0.9 can be used for the resistance of a single-glazed window and 1.8 for a double-glazed window.

If a foam insulation panel is added to the window opening and tightly sealed around the edges, the R value for this panel and additional air space can be added to the R value of the glass as shown in Figure 2-13. This gives the total R value for the foam panel plus the glass. Note that another inside air film does not have to be added in this case as it was already included in the R of 1.8 for the glass.

A great deal of caution should be used in applying this method of calculation to movable window insulation. It can be accurate for a very tight and well-sealed, thick, opaque shutter, but does not work well for thin, plastic membranes such as window shades. This method of calculating window insulation holds true in theory if the following conditions are met:

1. The insulating material does not let air flow through it or around its edges.

2. The insulating material does not transmit room-temperature infrared radiation. Most thick, opaque materials are also opaque to infrared radiation, but many thin, plastic films which are opaque to the eye are quite transparent to infrared radiation.

Standard window shades or curtains rarely meet both of these conditions adequately.

U Values

Mathematically speaking, U values are the reciprocal of R values, and they define the rate at which a window or wall section loses heat to the outside, apart from air infiltration losses. A U value therefore equals $1/R_T$ where R_T is the total R value for the section. A U value times the area of a wall (square feet) times the temperature difference (°F.) between inside and outside times the length of time (hours) in consideration equals the heat lost through a wall or window in Btu.'s as shown in Figure 2-14. The wall or window section that performs best thermally has high R values and low U values. Thus the window in Figure 2-14 with an R of 7.3 has a U of 1 ÷ 7.3 or 0.14, which means the window loses 0.14 Btu. per hour per square foot of window area for every degree (Fahrenheit) of temperature difference between inside and outside. Say for example the window area totaled 15 square feet, and for a 24-hour period the outside temperature was 20°F., while inside the temperature was maintained at 65°F. The total heat loss through the window would be:

$$U\text{-}0.14 \times 15 \text{ ft.}^2 \times 45°\text{F. temperature difference} \times 24 \text{ hours} = 2{,}268 \text{ Btu.'s}$$

And that's *with* movable insulation in place. Take away that insulation and you subtract R-5.5 from R-7.3, leaving a total R of 1.8 or a U of 0.56. The heat loss for the same, but uninsulated, window under the same conditions would be 9,072 Btu.'s, if it received no direct sunlight.

Window Losses and Gains

Windows not only lose heat to the chilly outdoors but can gain heat, even on cold, sunny days. South-facing, double-glazed windows collect more heat per square foot than practically any solar flat plate collector made! This is true even over a 24-hour period including the nighttime losses of heat through the window.

Because the surfaces of a solar collector are at a temperature of from 120° to 140°F., they constantly radiate heat back to the outside. As a result, it is only able to collect and retain about 40% of the solar heat available to it on a sunny day even though it has the advantage of not losing heat to the outside during the night. The solar collector must also "wake up" or warm up each morning before it starts work, and it gets caught snoozing when the sun goes behind a cloud. On the other hand, the south-facing window starts admitting heat as soon as the sun rises, and it keeps on working until it can no longer "see" the sun. It is such an earnest and thrifty character compared to flat plate collectors that it can afford to spend a little heat at night and still come out ahead.

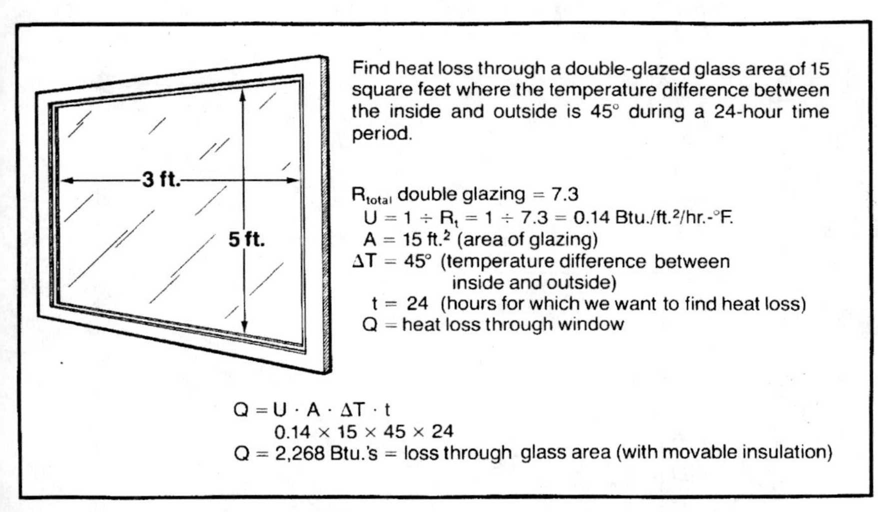

Figure 2-14: ***Example — computing heat loss through a window.***

Figure 2-15 shows the net heat transfer through several types of windows facing south, east, west, and north throughout the heating season in Madison, Wisconsin. The line in the center is the break-even line. Points below this line indicate that the glass is losing more heat than it is gaining in a given month. Those portions of the curves above the line indicate net gains of energy.

South-facing windows are the most beneficial. A window assembly facing south is a net provider of heat, except for single glazing which is a net loser of heat, November through February.

East-facing or west-facing windows all lose heat in Madison during December and January, although with double glazing and R-5 night insulation, these losses are small. East and west windows are most beneficial during the springtime when the days are longer and much more sun can shine on them during early morning and late afternoon.

North-facing windows are invariably heat losers with single-glazed units dipping way below the others, as one would predict. However, before you entirely write off north-facing windows, note that this graph shows only heat transactions and does not include the energy benefits from natural lighting nor any aesthetic considerations.

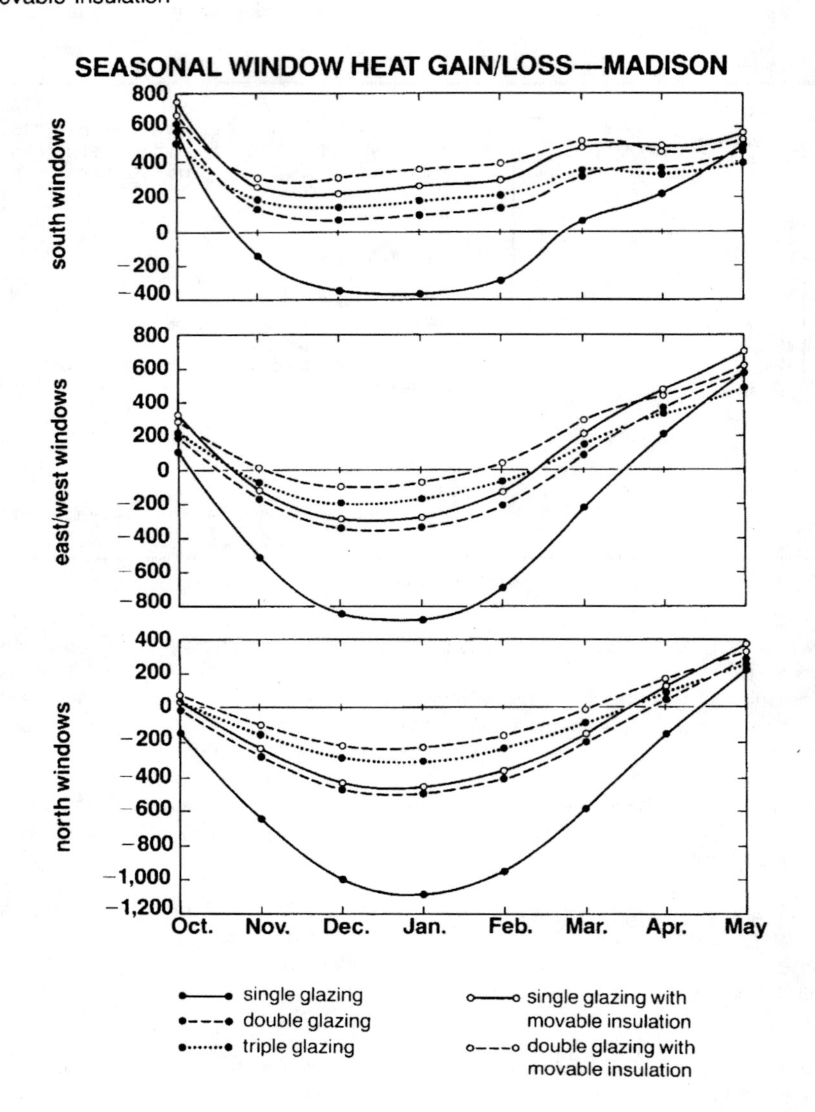

Figure 2-15: The net energy flow through various window assemblies (Madison, Wisconsin, for windows facing south, east, west, and north).

The heat lost and gained through windows varies not only according to the window orientation and season, but is also determined to a great extent by the climate or location of the building. Madison, Wisconsin, was chosen here because it has very cold and sunny winters. Graphs for several other cities are shown in Figure 3-1.

Generally a homeowner adds window insulation incrementally to a home, a few windows at a time, even in extreme northern climates.

In such locales it is often quite beneficial to protect all the windows of the home with night insulation. In more moderate climates, you will probably want to select only some of your windows to insulate. Even if you plan to protect all your windows, you may want to proceed incrementally, doing the most crucial ones first.

The priority windows are the largest ones because the size of the window is a more important factor than its orientation. Even though south-facing windows are usually net gainers of energy through the winter, they lose as much heat at night as a north-facing window of comparable size. Thus, more energy is often saved by adding movable insulation to a large south-facing window than to a small north-facing window even though northern windows show the poorest overall performance.

Most people think of using movable insulation for cutting heat loss in winter, and while this is certainly its major function, it's also effective in keeping heat out in summer, particularly from west-facing windows. The movable insulation systems described in the chapters that follow help to control these unwanted heat gains and shade the interior from unwanted sunlight as well. Special ways to shade unwanted summer heat are presented in Chapter 11.

In general, the window can be a tremendous asset to your dwelling, generating year-round comfort with a minimum of auxiliary energy. The window is an entry point for both light and heat, providing solar heat during the winter, ventilation during the summer, and natural light during all four seasons. However, the window is also a very vulnerable spot in your home's exterior shell, and controls are needed to regulate the heat and light that enter and exit through windows. Caulking and weather stripping, movable insulation and added glazing are the primary year-round controls for regulating window energy transactions.

Chapter 3

Enhanced Glazing Systems

The perfect glazing material would be clearer and more transparent than ordinary glass and would insulate as well as your walls. A single pane of window glass has a thermal resistance of about 1, a double pane of 2, triple pane of 3, etc. To achieve an R value of 5, which still allows four times the heat loss of a well-insulated wall, you need five layers of clear, ordinary glass. A window with this many layers of glass is not only expensive but transmits less than half the light and heat it receives from the sun (see Table 3-1). Clear vision through it is obscured by the many layers of glass and there are eight surfaces inside the window to collect dust and trap moisture. In any south-facing window the heat saved by such a multiglazed glass panel is more than nullified by a reduction in useful heat gained. Even for a north-facing window, the cost, loss of light, and other problems invariably rule out using four or five layers of window glass.

Table 3-1: Light Transmittance and Thermal Conductance of Multiple Layers of Glass Separated by a ½-Inch Air Space

Layers of ⅛-in. Glass	% Light Transmittance	Thermal Resistance* (R)	Thermal Conductance (U)
1	86	0.88	1.13
2	74	1.75	0.57
3	64	2.64	0.38
4	55	3.53	0.28
5	47	4.42	0.23

*Fifteen-mile-per-hour winds.

Double Glazing

Which is better, adding another glazing to single-glazed windows, or providing them with movable insulation? Technically speaking, single glazing with movable insulation results in less overall heat loss, but to make a practical decision other things must be considered.

Figure 3-1 shows a comparison of the net gains and losses during January for five different window assemblies in five cities with distinctly different climates. The five window assemblies are single glazing, double glazing, triple glazing, single glazing with an R-5 shutter used at night, and double glazing with the same shutter.

Single glazing with an R-5 nighttime shutter outperforms bare double glazing in every case. On south-facing windows in sunny or partly sunny climates such as Albuquerque, Madison, Boston, and Atlanta, the benefits of single glazing with an R-5 shutter over bare double glazing are quite pronounced, particularly in Atlanta and Albuquerque, where the single glazing with night insulation outperforms other assemblies.

Even though Figure 3-1 shows a better performance from the single glazing-shutter combination, there are several additional factors that must be taken into consideration. These include the following:

1. Moisture problems are more serious on single glazing with night insulation than on double glazing. A single layer of glass will become very cold when night insulation is added on the interior side of the glass. And moisture seals at the edges of the insulation must be very tight, particularly in cold, humid climates, to prevent the buildup of moisture on the glass.

2. Figure 3-1 assumes that you close the window insulation every night. Unless you religiously use the movable insulation 14 hours per day, the performance of your system may be less than that of double glazing. Certainly an exception can be made here for north-facing windows where you leave the insulation in place for days on end—the performance of your system will then exceed that shown on the graph.

3. Finding an R-5 window insulation shutter that is convenient to operate, meets these stringent moisture protection requirements, and is lower in cost than the addition of storm windows, can be difficult.

Cost will be one of the prime factors when making a decision between single glazing with night insulation and double glazing without it. There are no clear guidelines here since prices are constantly changing, but pricing each option available to you will be

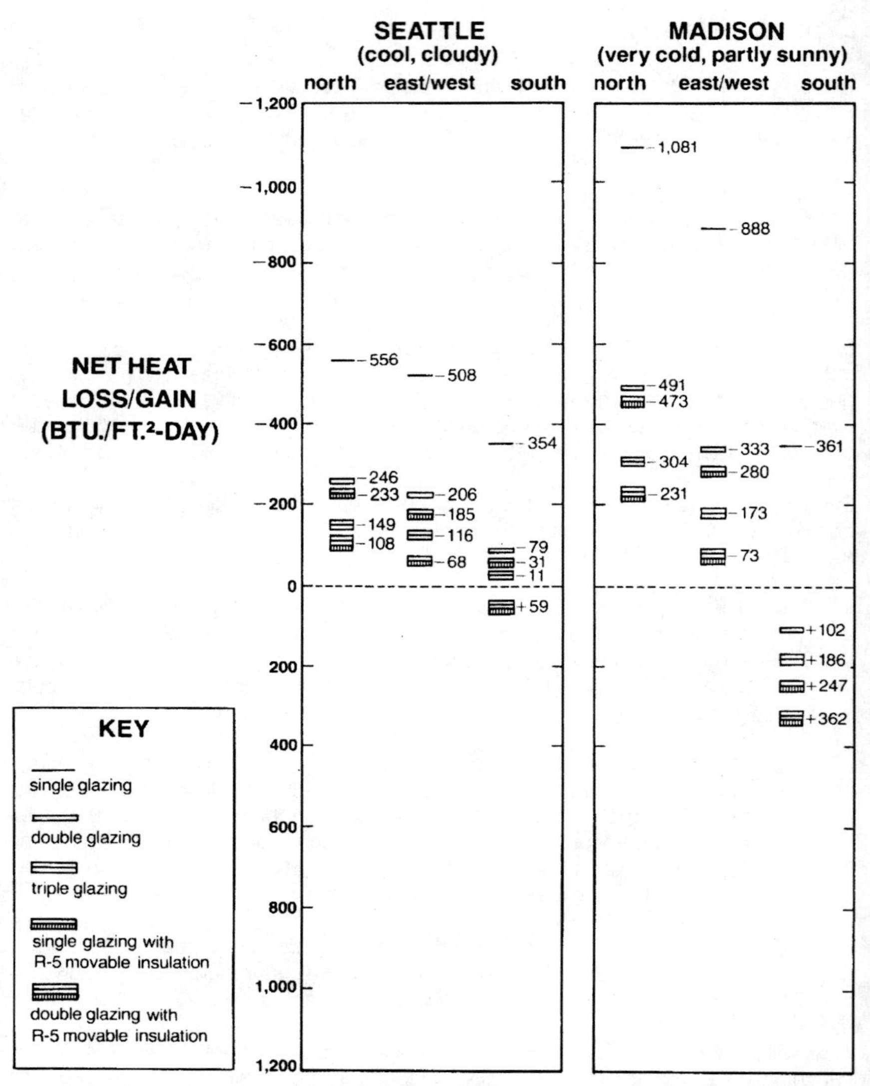

Figure 3-1: Net window gains and losses for five cities in January (here and on facing page).

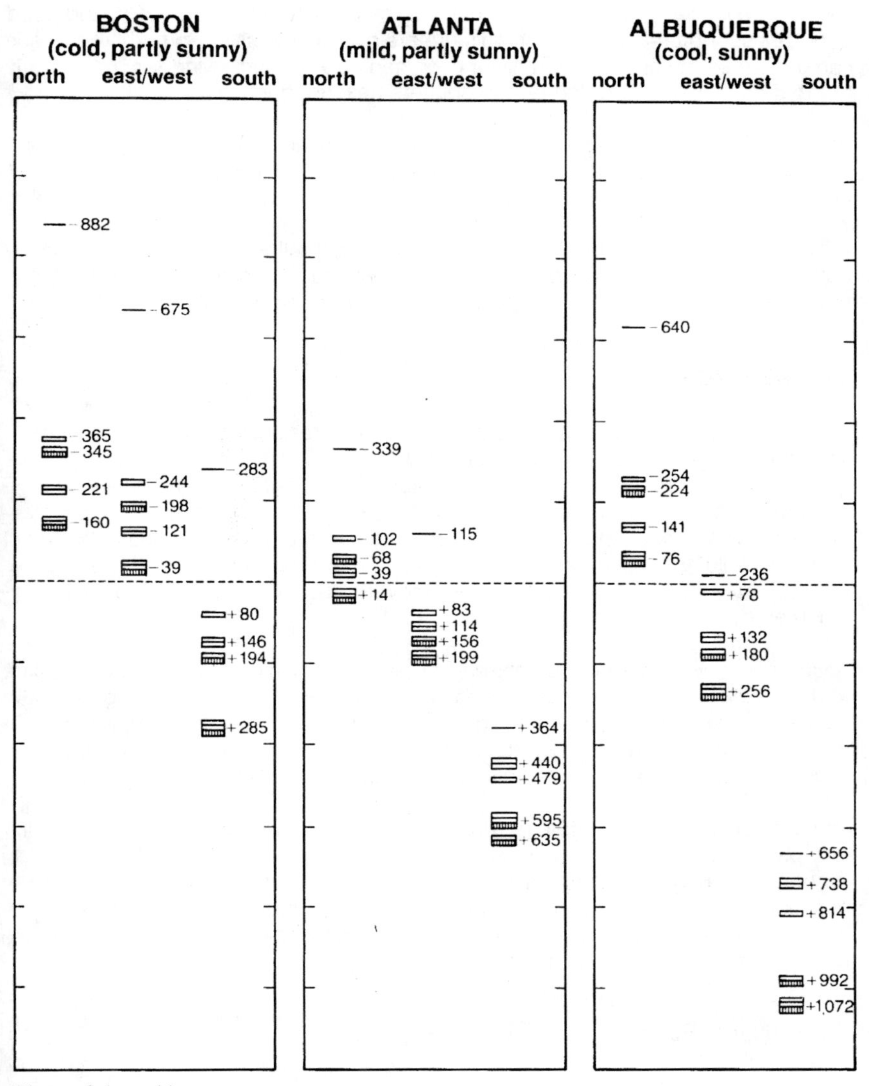

(Figure 3-1 cont.)

helpful. Storm windows generally run $2 to $4 per square foot of window. Many R-2 or R-3 window insulation products cost less than this, but systems with an R-5 and good edge seals for protection against moisture are more expensive. Home-built designs can usually be constructed at a lower cost than adding storm windows, particularly minimal-cost designs like the pop-in shutters described in Chapter 5.

A combination of double glazing and movable insulation is the optimal window insulation assembly, as shown in Figure 3-1. Adding double glazing first minimizes moisture condensation on your windowsill and helps insure the longevity of your existing window sashes. Adding window insulation first yields slightly greater energy savings, provided it is used properly. On many windows you will eventually want to employ both a second layer of glazing *and* night insulation, so the type of system you choose now should allow for the future addition of the other component.

Triple Glazing

If going from single to double glass doubles thermal resistance and cuts heat loss in half, why not go to triple glazing instead of using movable insulation with double glazing? As you go from double to triple glazing, you reduce the heat loss through the glass by only 33%, while the addition of movable insulation to bare double glazing reduces the heat loss up to 50%. In addition, triple glazing results in a light transmission of 64%, compared to a 74% light transmission through double glazing. Figure 3-1 shows that double glazing with movable insulation outperforms triple glazing in every situation.

South-facing windows must be allowed to take full advantage of the available sun, and triple glazing absorbs or shades so much solar heat that single glazing with night insulation is superior to triple glazing in most climates. On south-facing windows in Atlanta and Albuquerque the performance of bare triple glazing is especially poor, exceeded by every other assembly except bare single glazing.

The differences between triple glazing and double glazing with night insulation are less pronounced on east-, west-, and north-facing windows. On small, north-facing windows you may want to substitute triple glazing for night insulation for reasons of convenience. Another option for north-facing windows in rooms that are frequently vacant is to leave window insulation in place most of the time. It will increase the benefits beyond those shown in Figure 3-1.

Moisture Between Glazings

As was mentioned earlier, moisture is often a problem on the interior side of cold single-glazed glass. The condensation here is not a severe visual problem since it

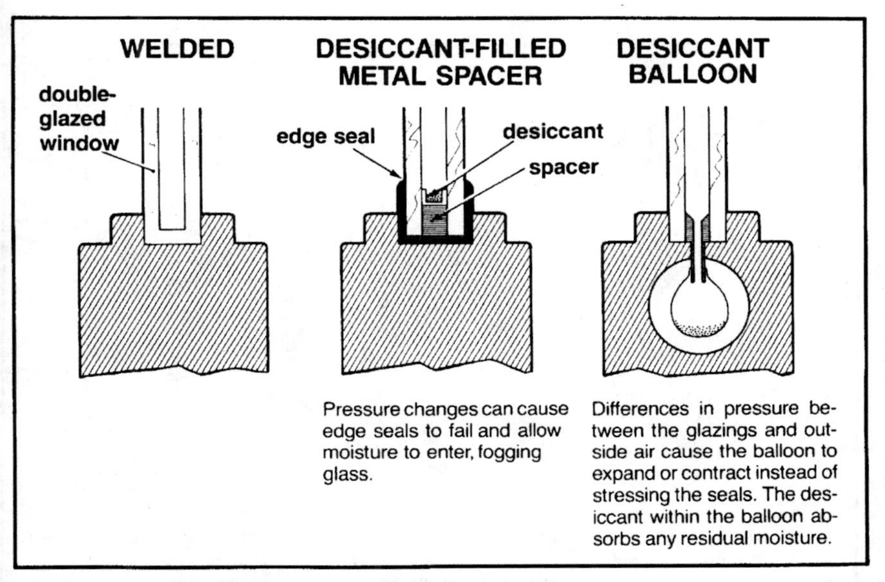

Figure 3-2: Three methods of sealing double glazing.

usually evaporates during the daytime as the window glass warms up, and if necessary, it can be easily wiped off. However, if moisture is allowed to condense repeatedly, it will eventually cause your window sash and the woodwork below it to rot and deteriorate.

With double glazing, moisture condensation on the interior side of a glass surface is usually not a problem because this layer of glass is near room temperature. However, moisture between the layers of glazing can be a serious problem. The space between glazings is usually inaccessible and moisture can condense here and block vision through the window for long periods of time. The double-glazed windows now commonly used are designed to prevent moisture buildup in one of two ways. They either permanently seal the inner space, preventing moisture from entering, as with "insulating glass" panels, or they breathe the moisture to the outside at a faster rate than it enters, as with storm windows.

The factory-sealed double-glazed panel that carries the generic term, "insulating glass" (remember, however, to compare its R value of 1.8 to an insulated wall with an R value of 15 to 20 to understand how "insulating" it really is) is either welded at the edges with glass or made up with desiccant-filled metal spacers (to absorb moisture) that are bonded at the glass edge with special sealants (see Figure 3-2). Many small, sealed double-glazed window sashes manufactured today have welded

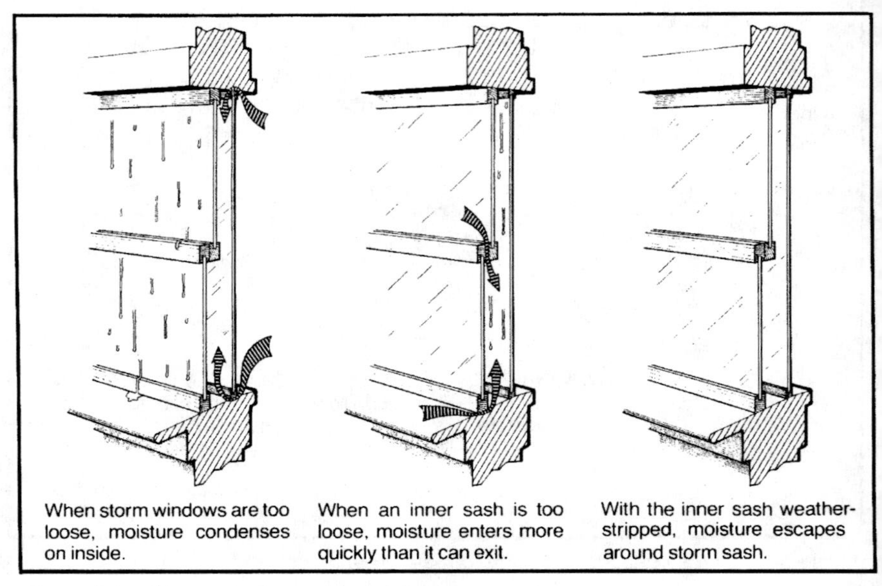

When storm windows are too loose, moisture condenses on inside.

When an inner sash is too loose, moisture enters more quickly than it can exit.

With the inner sash weather-stripped, moisture escapes around storm sash.

Figure 3-3: Storm windows and moisture.

edges. Since they are made entirely of glass, moisture will never penetrate the seal unless the glass is broken. The larger, double-glazed panels used in patio doors are made of tempered glass, using the spacer-sealant method of construction, because tempered glass cannot be welded. Sometimes the edge seal fails, causing a fogging problem. Once moisture gets in it will never leave, so if a fogging problem does occur, usually all you can do is replace the panel or live with moisture blocking part of your view. For those so inclined, however, the glazing unit can be taken apart, dried, rebuilt, and resealed.

A new, insulating glass system has been suggested that uses an expandable desiccant balloon* in the door or window frame to prevent fogging and equalize the pressure. (Differences in pressure between the glazings and outside air cause the panels to pop and allow moisture to enter.) The panes of glass are joined with the spacer-sealant technique similar to that used with tempered patio-door glass. Hid-

*Patented by Ralph K. Day; U.S. Patent Nos. 3,932,971 and 4,065,894. This system is not yet commercially available.

den in the window frame between the glazings is a cavity where an expandable balloon is connected to the air space by a plastic tube. As the pressure outside the window drops, the balloon expands instead of stressing the seals, and when the pressure outside rises, the balloon contracts. A small amount of desiccant is placed in this balloon to absorb any residual moisture.

Storm windows employ another approach to eliminate condensation. These windows normally have weep holes at the bottom and cracks around the edges so they can breathe moisture to the outside. However, if they fit too loosely they are ineffective in reducing heat loss. Moisture condensing on the interior side of the inner window glass indicates that too much outside air is circulating through the air space and cooling the inside glass. If this occurs, the windows should be weather-stripped.

A much more common problem is moisture on the storm sash between glazings (see Figure 3-3). This occurs because storm windows are often added to poorly weather-stripped windows. Moisture always moves from the inside to the outside during winter, so if the cracks are larger around the window sash than at the storm sash, moisture collects in this space. The solution to the problem lies in thoroughly weather-stripping your inside window sash.

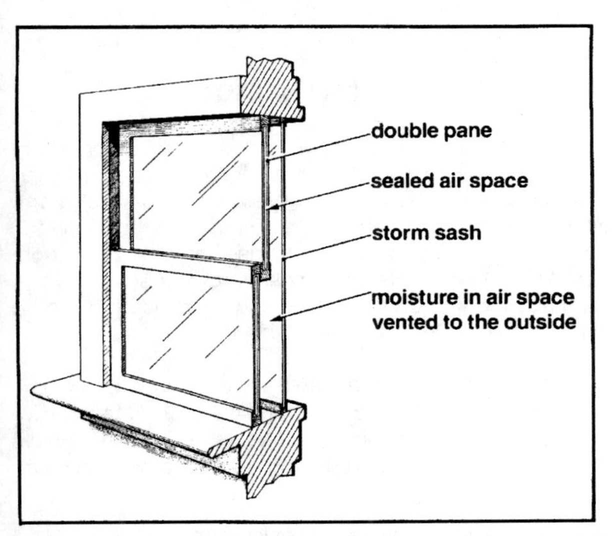

Figure 3-4: Double-pane sash with storm sash creates three layers of glazing.

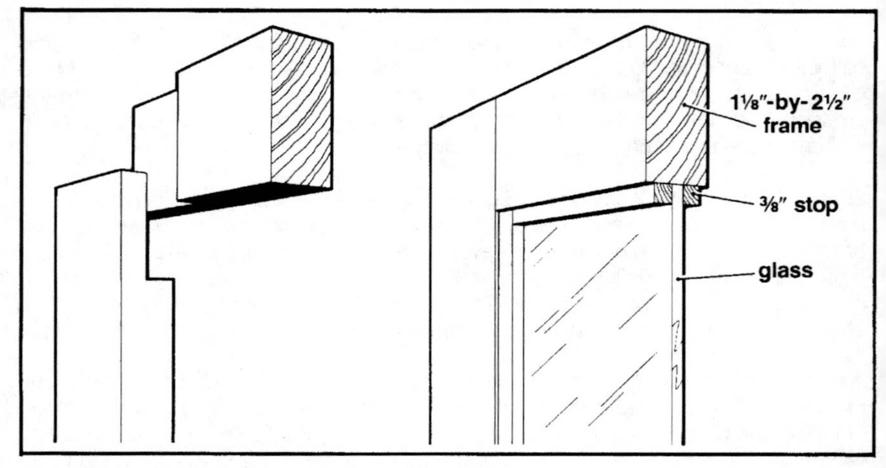

Figure 3-5: Owner-built storm window construction.

Many manufacturers are now offering double-glazed welded panels in their regular window sashes with an optional storm sash. This is an excellent way to achieve triple glazing without any moisture problem (see Figure 3-4).

Adding Layers of Glass to Existing Windows

The most common way to add additional glazing to windows is by adding storm windows. The standard exterior-type storm window is designed for double-hung windows but a type for sliding windows is also available. New, ready-to-install storm windows can be purchased for $2 to $4 per square foot and will pay for themselves in energy savings in 5 to 10 years. Storm windows can also be assembled at home from glass or rigid plastic and sash channels available in most hardware stores. These cost substantially less than ready-mades. Complete assembly instructions can be found in the *Reader's Digest Complete Do-It-Yourself Manual*.*

Wooden-sash storm windows have a number of advantages over ones with aluminum sashes, including a much lower rate of heat loss through the sash. One route for obtaining wooden-sash storm windows is to shop at garage sales. Be sure to have the sizes of your windows carefully measured. Even if the glass is broken in them,

*See Appendix V, Section 1 for book description and complete bibliographical information.

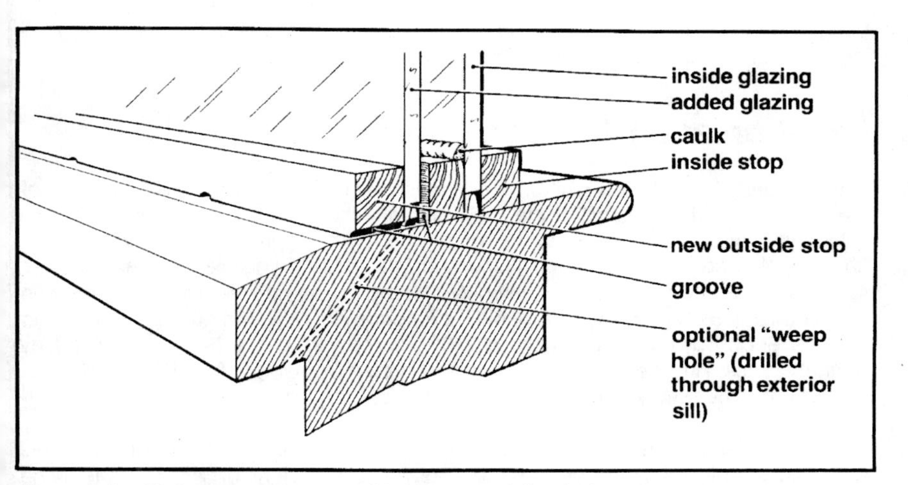

Figure 3-6: Moisture venting—multiple layers of fixed glass.

replacing the glass in a good frame is easier than making a frame from scratch. If a frame is slightly too large, you can also trim the frame down on a table saw after you remove the glass. (Be sure to wear goggles and check that nails or bits of metal aren't in the path of the saw blade.)

Figure 3-5 shows how you can make wooden-sash storm windows from five pieces of 5/4-by-3 nominal clear pine (which measures to be 1⅛ by 2½ inches, actual dimensions). Cut the ends so that they form a lap joint. Glue them with resorcinol or "red glue" and clamp them until the glue dries. Then tack ⅜-by-⅜-inch stops all the way around on the inside of the frame. Mount into the frame a pane of glass that is cut to dimensions ¼ inch smaller than the frame opening and apply another ⅜-inch overlapping stop to hold the glass in place. Caulk the glass with a bead of silicone and prime the frame with an exterior house paint, and you will have a top-quality storm window that will last many years.

If your window sash is the casement or awning type that swings out, an exterior storm window will block the operation of the sash. Interior-glass types are similar to exterior storm windows and can be installed wherever needed.

With fixed single glass, you can add additional glazing as shown in Figure 3-6. Caulk the inside sash with silicone to prevent any moisture entering from the room side. As you add an additional layer of glass, cut a couple of grooves through the bottom of the outside stop, or in the top of the sill. This allows moisture that collects in between the glazings to exit. An additional groove is required on the stop between the glazings to allow continuous moisture flow to the outside.

Low-Iron and High-Iron Glass

Low-iron glass, also "water white," is exceptionally clear and transmits 90 to 91% of the light it receives, instead of the 86% transmitted through regular glass. (These values vary with the thickness of the sheet and manufacturer so check the specifications.) You pay a premium for this kind of glass, but if you plan to maximize the solar gain from a window, it may well be worth the extra price. Two layers of low-iron glass transmit 81% of the solar radiation, compared to 74% with ordinary glass, and three layers of low-iron glass transmit 73%, compared to the 64% transmitted by ordinary glass. And because low-iron glass is clearer, using it for triple glazing will give you the visibility you get out of ordinary double glazing.

A high-iron glass is not recommended except in tropical and semitropical areas where sunshading is a must year-round. Despite its inefficiencies in most parts of the United States, this "heat-absorbing glass" has been in use for about 20 years, mostly in commercial buildings. It shades about half the solar radiation that strikes it, but ordinary shades or venetian blinds are just as effective in shading summer heat gains. The exterior-shade products listed in Chapter 12 are even more effective. Not only does high-iron-content glass have little practical value in controlling summer heat gains, but it also drastically reduces winter solar gains and should therefore be avoided. Mylar- or foil-faced liners on shades or curtains have been known to cause heat-absorbing glass to heat up and crack—another reason to avoid its use.

Plastic Glazings

Plastic glazings are becoming more popular today due to the ease of handling and installing them and because many of them are lower in cost than glass. Most plastics degrade in sunlight—some very slowly and some rapidly. Table 3-2 lists a number of plastic glazings, their costs, performances, and their manufacturers. Two plastics can look identical but behave very differently in sunlight, so compare specifications carefully.

Clear Glazing Films

Polyethylene film is widely used for plastic storm windows. Costing less than 3¢ per square foot when bought by the roll, it usually runs more than twice as much when purchased in plastic storm-window kits. Even if you buy it at this higher price and use it only one season, it pays for itself several times over in cutting heat losses.

Table 3-2: Comparison of Common Plastic Glazings*

Generic Type/ Trade Name	% Solar Transmission	Cost	Estimated Life (yrs.)	% Infrared Transmission	Company
Polyethylene film (4 mil)	90–93	Very low	1–2	72	Numerous manufacturers
Polyvinyl chloride film (6 mil)	90–95	Low-med.	4–5	15	Numerous manufacturers
FEP fluorocarbon film (Teflon—1 mil)	96	Low	....	58	Plastic & Resins Dept. Du Pont Co. Wilmington, DE 19898
Polyvinyl fluoride film (Tedlar—4 mil)	95	Low	....	....	Plastic & Resins Dept. Du Pont Co. Wilmington, DE 19898
Polyester film (3M—4 mil)	96	Low	15 (as inner glazing)	....	3M Co. Bldg. 219-1, 3M Center Saint Paul, MN 55101
Acrylic polyester film (Flexiguard 7410—7 mil)	89	Low	10	9.5	3M Co. Bldg. 223-2, 3M Center Saint Paul, MN 55101
Semirigid fiberglass reinforced polyester (FRP), (Sunlite), Premium II	85	Low-med.	20	5	Kalwall Corp. 1111 Candia Rd. Manchester, NH 03103
Rigid acrylic sheet (Lucite)	92	Med.	....	....	Plastic & Resins Dept. Du Pont Co. Wilmington, DE 19898
(Swedcast 300)	93	Med.	....	....	Swedlow, Inc. 7350 Empire Dr. Florence, KY 41022
Rigid acrylic sheet (Acrylite SDP)	93	Med.	25	0	CY/RO Ind. Wayne, NJ 07470
(Plexiglas G.)	91	Med.	10–20	....	Rohm & Haas Co. Independence Mall W Philadelphia, PA 19105
Acrylic fortified polyester, reinforced with fiberglass (FRP) (Filon with Tedlar)	86	Med.	....	....	Filon Div., Vistron Corp. 12333 Van Ness Ave. Hawthorne, CA 90250
Rigid polycarbonate sheet (Tuffak-Twinwall)	89	Med.-high	5–7	0	Rohm & Haas Co. Independence Mall W Philadelphia, PA 19105
Glass	86–92	Med.-high	100+	0	Numerous manufacturers

*Sources: "Summary of Glazing Properties" from the MASEA Glazing Conference, *The Solar Age Resource Book* (New York, Everest House, 1979), and several articles. Infrared transmission data was not entirely consistent from one source to the next. Manufacturers' tests and specifications should be consulted for the most accurate and current information on these properties.

For more permanent applications, you should consider other, clearer plastic films that cost more but perform better for a longer period of time. Not only is polyethylene less pleasant to look through than clearer plastic films, it deteriorates rapidly in sunlight. It is also highly transparent to the room-temperature radiant energy that you are trying to retain. Unlike glass, which absorbs nearly 100% of this radiation, 4-mil polyethylene absorbs only 28%, which means that it loses more heat than glass.

Clear vinyl is superior to polyethylene in every category except cost. It will last over five years if cared for, while polyethylene will, at best, last two years. Vinyl is much clearer and transmits far less of the room-temperature infrared radiation back to the outside. Since vinyl and other higher-quality plastic films are considerably more expensive, they must be applied carefully to your windows so that they can be reused through a number of winters.

Polyester glazing films will probably soon replace the use of polyethylene and clear vinyl for plastic storm windows. Polyester films are low in cost and are highly transparent and durable. When used inside protective layers of glass or acrylic, they should last 15 years. 3M Corporation has developed a composite film called Flexiguard 7410, which has an exterior acrylic layer tough enough to withstand the outside elements.

Some plastic films have a very high transmittance of solar radiation—93 to 95%. A

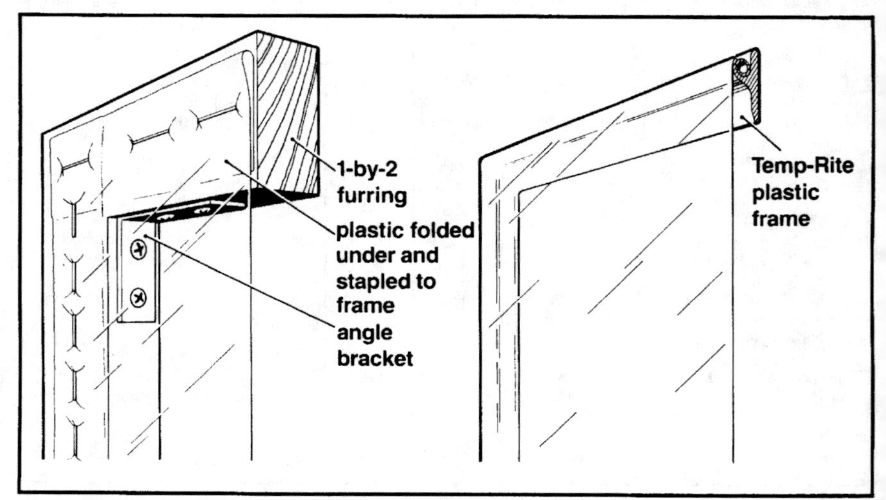

Figure 3-7: Frames for plastic glazing films.

film layer can be applied to both sides of a wooden frame, adding two layers of glazing and two extra air spaces, and still have a transmittance of close to 90%.

You can quite simply construct a window sash from wood and plastic film. Make a frame from wood furring strips of 1-by-2 stock as detailed in Figure 3-7. Then staple the plastic glazing film to the frame. The whole assembly can be installed each fall and removed each spring, either on the outside of your window as an exterior storm sash, or between two layers of glass if you already have a storm sash.

Some very convenient frames for stretching out plastic films and mounting them over your window are also available from several manufacturers. These frames have a plastic snap-in spline that holds the plastic film taut. Mounting a plastic film on one of these frames is much simpler than stapling it to wood. One frame called the Temp-Rite, is available through the *Sun Catalog*, Solar Usage Now, Inc., Box 306, Bascom, OH 44809. Another plastic window-film frame is in the developmental stage, soon to be released by Thermo Tech Corporation, 410 Pine Street, Burlington, VT 05401.

All plastic glazing materials are much softer than glass—even the rigid acrylics will scratch easily. Plastic films are more vulnerable, particularly to puncture by sharp objects. Films should not be placed where children, tree limbs, or small animals will damage them. Clear plastic films are most protected when placed between two panes of glass and most vulnerable when facing the outdoors or used on interior windows near the floor.

Rigid Plastic Glazings

A clear, rigid acrylic plastic has been available under the trade name, Plexiglas, for a number of years. This material is easily cut and special glues weld it together without even a visible joint, making it popular for display boxes, terrariums, dust covers, and other practical yet decorative items.

Acrylic plastic is being used more and more for storm doors and storm windows. Although it is soft and scratches easily, it does not shatter when broken. Acrylic sheets can be cut with a handsaw, jigsaw, or band saw and then simply drilled and mounted onto a window sash with wood screws, washers, and thin weather stripping or attached and sealed with heavy, double-stick tape.

A number of acrylic storm-window kits are available at hardware stores. These usually contain a plastic frame that snaps together to hold the acrylic sheet in place during the winter and comes apart to make removal easy for the summer. These are handy for casement and awning windows which open by swinging out because the acrylic glazing can be added to the interior side of the window and not interfere with these windows' performance. Plastic storm-window kits are available from:

Perkasie Industries Corp.
50 East Spruce Street
Perkasie, PA 18944

Plaskolite, Inc.
1770 Joyce Avenue
Box 1497
Columbus, OH 43216

Reynolds Metals Co.
E. T. Stevak
P.O. Box 27003
Richmond, VA 23261

The plastic frames for these storm windows are not particularly attractive. You may prefer to attach a sheet of acrylic to a wooden stop in your window opening with screw-mounted wood stops. Clamp the acrylic sheet tightly to this sash or moisture may accumulate behind it.

For better insulation on the interior of your window you can build a wooden frame, and stretch a clear film on the exterior side and attach clear acrylic on the interior side. Seal this double glazing over the window to prevent moisture problems.

A number of other types of rigid plastics are available today, including fiberglass-reinforced polyesters, double-layer acrylics, and polycarbonates. These plastics generally do not allow clear vision to the outside and therefore are not used over ordinary windows. They are popular in attached solar greenhouses, though, and could be used over bathroom windows or in spaces where privacy is desired.

Tinted and Reflective Plastic Films

A variety of plastic films are available that adhere to the inside surfaces of window glass. These films are commonly called "solar control films," somewhat of a misnomer since they merely screen out sunlight instead of providing an active control of sunlight in response to seasonal variations. They have limitations similar to those of high-iron glass, discussed earlier in this chapter. Consisting of two or three layers of metallic, clear, or tinted plastic material, these films can be applied easily to your window glass with a pressure-sensitive or water-activated adhesive. Solar control films can screen out as much as three-quarters of the sunlight striking the glass (depending on film type), and therefore should be used only on windows that do not provide useful cold-weather heat gains, but are a problem in hot weather. Thus, they can be used to reduce summer heat gains only on east- and west-facing windows. These films usually increase the R value of glass by only about 35%, which means that they are not useful for substantially reducing heat loss.

One film, made by 3M Company, has a substantially higher R value due to a highly reflective surface that reflects heat radiated from room surfaces back to them. Called P-19 film, it is an expensive material having an R value of 1.47 (U = 0.68) when applied to single glazing. Still, it does not perform as well as double glazing either in R value or in winter sunlight transmitted.

If you're interested in year-round sunshading you may be interested in these films. They are available from:

Dun-Ray
Dunmore Corporation
Newtown Industrial Commons
Pennsylvania Trail
Newtown, PA 18940

Kool Vue
Solar Screen
53-11 105th Street
Corona, NY 11368

Llumar
Martin Processing Inc.
P.O. Box 5068
Martinsville, VA 24112

Nunsun Sumsun
Standard Packaging Corporation
National Metallizing Division
Cranbury, NJ 08512

Plastic-View Film
Plastic-View Transparent Shades, Inc.
P.O. Box 25
Van Nuys, CA 91408

Reflecto-Shield
Madico
64 Industrial Parkway
Woburn, MA 01801

Sungard
Metallized Products
224 Terminal Drive S
Saint Petersburg, FL 33712

Heat Mirror Films

A major function of window glazing is to get heat into a room and keep it there, or at least prevent a significant portion of it from being lost to the outside. In Chapter 2 the transmissivity of window glass was discussed for three bands of radiation: visible light, near-visible infrared, and room-temperature infrared (see Figure 2-10). Fortunately, glass is opaque to room-temperature radiant energy. However, if window glass can be modified to reflect heat back into a room rather than absorb it, as shown in Figure 3-8, it is far more effective in reducing heat losses.

To create a film that is clear to solar radiation but reflects radiant transfer back out through glass is no easy task! Such a film, termed "heat mirror film" is under development. A heat mirror surface is formed by depositing a very thin film of metallic material, sometimes gold or silver, on a clear plastic film or a layer of glass. If the metallic film is kept to an optimum thickness, it reflects room-temperature radiation but still has a high transmittance of incoming light. Double-glazed windows for new construction may be manufactured in the near future with a heat mirror on inside surfaces. Clear plastic films coated with heat mirrors are also becoming available

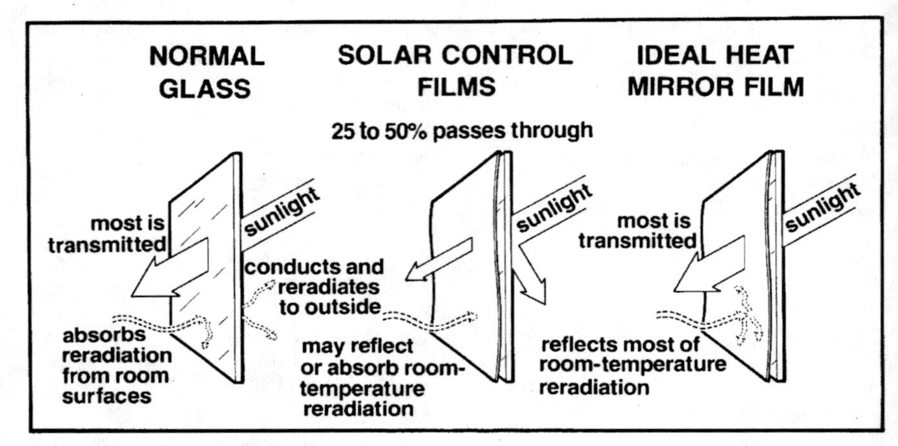

Figure 3-8: Heat mirror principles.

and can either be stretched over a frame or applied to glass with an adhesive backing, in the same manner as solar control films.

The use of a heat mirror film on single-glazed glass is inferior to adding another layer of glazing. Not only are R values and solar transmission poor, but glass covered with this film will become very cold and condensation will form on the interior. Surface moisture will cause the heat mirror to lose its reflective properties and subsequently its resistance to heat transfer until the surface is dry again.

Heat mirror films are not effective when facing outdoors on an exterior pane of glass. The most logical place to add heat mirror film on any window with a storm sash or with access to the space between glazings is on the back side of the inner sash, facing into the air space. Since the films developed so far have poor resistance to wear and abrasion, they will be protected from scratches in this air space. The highest thermal resistance is achieved by stretching a plastic heat-mirror film midway between two layers of glass. Schemes such as this have generated R values of 5 and greater!

Thermo Film Corporation, 385 Sherman Avenue #3, Palo Alto, CA 94306, is a new corporation formed to manufacture and market heat mirror films. Inquiries about availability and cost can be directed to them.

Heat mirror surfaces are still in an early stage of development and it will be at least several years before they are widely available. Most production-grade heat mirror films add about the same heat resistance to a window as an additional layer of glazing. Often these films also reduce the solar transmission as much or nearly as much as another layer of glazing, although some laboratory-grade films have performed better. Competing with glass on a dollar per dollar basis will be difficult.

Summing Up Your Options

There are many clear, plastic glazing materials that are now available for window applications. Low-cost plastic-film storm windows which can easily be fabricated and installed on site are the most promising. Some of these plastic films are clearer than ordinary glass and are lower in cost.

Glass, however, is the only glazing material that should not need to be replaced during the lifetime of a building. The durability and hardness of glass along with its transparent sparkle and clarity have made it one of the most popular and poetic materials in buildings for centuries. Because of its unique qualities, glass will probably remain the preferred glazing material in building for a long time to come.

Double glazing is thermally superior to single glazing on windows in just about any location. Triple glazing is thermally superior to bare double glazing on all windows except those that face south. However, the best performance is achieved by combining double glazing and movable insulation. The remaining chapters in this book describe how such movable insulation can be added to your windows.

Chapter 4

Choosing a Window Insulation Design for Your Home

Windows and the light they provide are two of the most important design elements in creating a quality interior space in your home. A window insulation design for your home should also be carefully planned to complement the other elements of your home's interior. Movable window insulation provides a control for the amount of sunlight entering your home and can enhance the quality of your home's interior, but it can also add clutter and detract from your home's appearance. A professional interior designer *may* be helpful to you, but unfortunately, most of them do not yet fully understand window energy control, and their advice may conflict with your window insulation goals. For names of interior designers who specialize in thermal window treatment, see the list of designers in Appendix V, Section 3.

A survey of all the windows in your home will reveal: which windows are important providers of light; which windows are used very little or are located in infrequently used spaces; which windows have the necessary clearances from furniture, plants, and adjacent walls to allow the addition of a shutter, shade, or curtain; and which windows should remain transparent to allow for early morning sunlight or a special nighttime view. You will probably want to insulate your larger windows, but you may want to leave some of the smaller ones transparent to allow for continual awareness of what is happening outside. As you survey the windows in your home or in the design of a home you are planning to build, assess the amount of time you are willing to spend each day operating the window insulation systems you are considering. Many indoor gardeners have stated that there is an optimum number of house plants an individual can care for. Beyond that number, the plants will suffer from neglect. The same can be said for movable window insulation, and there is no point in spending time and money on manual window insulation systems that you are not going to take the time to operate. Any windows having only single glazing undoubtedly deserve your initial attention, but, other things being equal, the largest windows are the ones you should focus on.

Orientation vs. Size

As was mentioned in Chapter 2, the orientation of a window is a very significant factor with regard to summer heat gain, but it is of minor importance with regard to winter nighttime heat losses. North-facing and other windows that are directly exposed to cold winter winds have a slightly greater heat loss per square foot of glass than windows that face south or are sheltered from the wind. However, these differences are small enough so that they do not normally enter into engineering methods of estimating heat losses.

The size of a window is one of the main factors that governs the rate of heat loss from a window—a glass area of 2 by 4 feet will lose twice as much heat at night as a window with a glass area half that size. Therefore, large south-facing windows have a greater need for night insulation than small north-facing windows, even though this appears contrary to the general principle that greater protection is necessary on a north wall. In a mild southern climate, a practical solution to window insulation in a home where the largest windows face south is to install movable window insulation on the south windows and triple glazing on the smaller north windows. In more severe northern climates, you may want to install window insulation on all your windows.

***Photo 4-1:** Decorative window insulation. These shutters are part of a modern solar home in Quechee, Vermont. The beautiful fabric which covers them adds a warm touch to the interior. Edge seals would make these shutters more effective.*

Windows facing southeast, southwest, or sloping to the south have tremendous summer heat gains that are difficult to block with exterior shading devices. Window insulation that includes features to prevent summer heat gain as well as to reduce winter heat losses will provide added comfort and savings year-round.

Another option that works well in winter, particularly with the pop-in type shutter discussed in Chapter 5, is to cover only the lower portion of a window with insulation, leaving the upper portion exposed so that natural light can enter. Figure 5-13 shows how this option can be modified to act as a hot-air collector. Windows facing north that are used very little during the winter can be covered with an insulation panel that is left in place throughout the entire heating season.

Each window in your home is unique in its wall location, view, orientation to the sun, and relationship to your daily routine. From the many options for movable insulation which follow in Parts II through VI of this book, you should find a design to meet your needs in any specific situation. Design features common to all types of window insulation include operability, R values, cost, edge seals, moisture protection, durability, safety, and appearance. The rest of this chapter discusses these features.

Operability

Since most of the window insulation systems installed in residences are manually operated, ease of operation is one of the most important considerations in choosing a system. Unless the window insulation is used in a location where it is left closed most of the time (a north-facing window in a guest or sewing room, for example), you must be comfortable opening and closing it every day. If you have to wrestle with the system to get it to close properly, you will probably leave it open most of the time, and this will result in only marginal energy savings.

The choice of an operating mechanism for a system depends to a large extent on the height of the window you are protecting, the access which you have to it, and the clearances around the window from furniture, plants, or adjacent walls. All forms of movable insulation must be moved out of the window proper when they are opened. The availability and location of space for daytime storage often determine the means of operation. Movable insulation for skylights or clerestories which are high and out of reach must be operated differently than a system for windows which you can reach. Greenhouses have their own special problems.

Automatic systems are considerably more expensive than manual ones, but they are easier to operate and can be set to open and close automatically at optimal times. South-facing glass areas with an automatic window-insulation system can bring the maximum solar heat gain into your home, even while you are away from it, without losing this heat back to the outside. Some automatic systems such as the Beadwall (manufactured by Zomeworks) are so dynamic as to be almost breathtaking. In this system Styrofoam beads are blown between two layers of glass when protection from heat gains or losses is needed, and the beads are drawn into concealed storage units when it is desirable to let sunlight enter the living space (see Chapter 9). There are also heat-sensitive systems on the market which open whenever the sun shines on them. In the Skylid (see Chapter 13), a heat-sensitive Freon canister tilts insulated louvers to an open or closed position without any motors, switches, or relays. Other automatic systems employ small electric motors activated by a light sensor or differential thermostat.

The convenience of automation can be an illusion, however, since the time spent maintaining labor-saving devices often nullifies the time or work they save. Try to keep your home's movable insulation design as simple as possible. We all know the pitfalls of a push-button world.

R Values

When adding insulation to your walls or attic the first inch is the most important, and the energy return yielded from each additional inch added becomes less and less. If an existing wall already contains 1 inch of insulation, you must add one more inch, doubling the R value, to cut the original heat loss in half. But if you wanted to cut that reduced heat loss in half again, you'd have to add two more inches of insulation, to again double the wall R value. You quickly reach a practical limit (in the case of the wall, about 6 inches in most climates) where the thermal design conflicts with other goals in building design. The same is true for window insulation.

Figure 4-1 is a graph of the daily heat losses through a square foot of glass on a 30°F. average temperature January day, using movable insulation of various thermal resistances for 14 hours per day. The R values at the bottom of the graph do not include the R value of the bare window. Each R value of 1 can be thought of as a ¼-inch layer of polystyrene beadboard or an additional ¾-inch air space. Note the following:

1. The graph rapidly levels off. The greatest reductions in heat loss occur in the low R value range of 1 to 3. An R-3 shutter on double glazing saves 38%.

2. The medium-range R value of 3.5 to 8 reduces heat losses somewhat further to a high of 50%. R-5 with 45% reduction, a point where the curve flattens out, appears to be an optimal value.

3. Above R-8, the graph shows very small further reductions in heat losses. As long as the movable insulation is left open for 10 hours during the daytime, the heat loss reduction can never exceed about 58%.

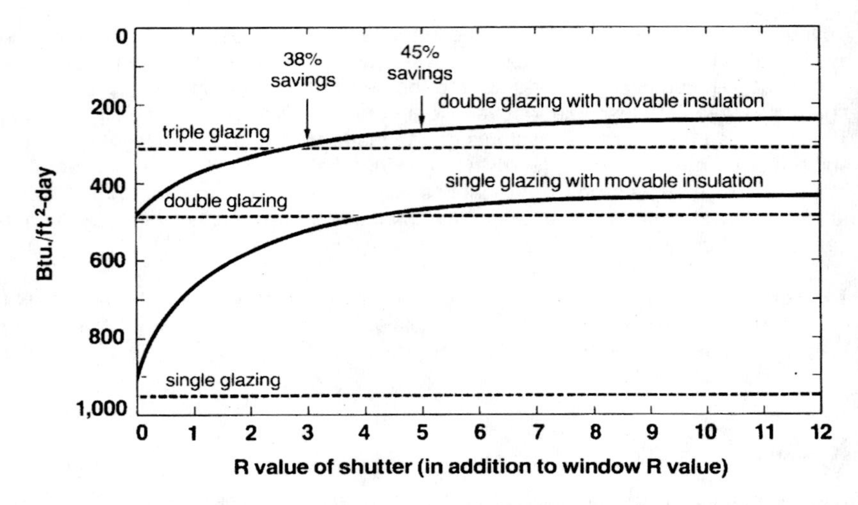

Figure 4-1: Heat losses through various window assemblies (ΔT = 35°F.).

4. The heat loss for an R-5 panel over single glazing is near that of bare double glazing. However, an R-5 panel over double glazing is definitely superior to triple glazing.

The high-R-value syndrome has caused some window insulation products to become thermally overdesigned to the extent that ease of operation and cost effectiveness begin to suffer. To summarize Figure 4-1, the additional R of 3 to 5 provided by movable insulation is quite adequate with double glazing. Although some improvement is observed up to about R-8, over that it makes little difference at all. Even though your neighbor buys an R-8 system, it may not be superior to your R-4 system, particularly if you close yours nightly while he leaves his open.

Cost

The optimal window insulation system provides you with the most benefit for each dollar spent. Benefits are not limited to heat loss reductions but also include aesthetic enhancement of your interior, the control of natural light, and privacy. The shading of unwanted heat gains in the summer, another benefit, provides a cooler summer interior and dollar savings for those using air conditioners. Some exterior-mount movable insulation shutters double as sun reflectors when open during the winter to reflect more heat in through the window and therefore increase the solar energy provided by south-facing glass.

The cost of a window insulation system ranges from 30¢ per square foot of window to well over $10 per square foot. The energy savings generated by some of the lower-cost systems will pay for the system in less than a year. But 4 to 5 years is an average payback period, and some of the more costly systems may take longer than 15 years to pay for themselves on strictly an energy-dollar basis (see Table 4-1).

Table 4-1: Heating Dollars*,† Lost Through Windows‡ per Winter for Oil, Gas, and Electricity§

(for 4,000 to 8,000 Degree-Day Climates)

Degree-Days (cities)	Oil (per gal.)			Natural Gas (per therm)			Electric Resistance (per kwh.)		
	$.70	$.90	$1.10	$.35	$.45	$.55	$.04	$.05	$.06
4,000 (Baltimore, Md., Asheville, N.C., Portland, Oreg.)	91	116	141	63	80	99	147	184	220
5,000 (Dodge City, Kans., Trenton, N.J., Pittsburgh, Pa.)	113	145	176	79	99	124	184	230	275
6,000 (Lincoln, Nebr., Akron, Ohio, Chicago, Ill.)	136	174	211	94	117	149	221	275	330
7,000 (Billings, Mont., Flagstaff, Ariz., Buffalo, N.Y.)	158	202	246	110	139	174	258	321	385
8,000 (Green Bay, Wis., Burlington, Vt., Missoula, Mont.)	181	231	282	126	161	199	294	367	440

*Figures are rounded off to nearest dollar.
†Double the dollars for single-glazed windows.
‡Figures pertain to a 1,300 $ft.^2$ insulated structure with storm windows amounting to 20% of floor area.
§As of 1979 prices.

Aesthetic enhancement is usually underrated when considering the economics of window insulation. Few decisions about home furnishings are made based strictly on cost-benefit comparisons; window insulation should not be evaluated solely this way, either. One to five dollars per square foot of window is often spent on standard curtains and shades, which provide practically no thermal protection. If the window insulation you select replaces the need for decorative window accessories, you can deduct the amount you would have spent on these items from the cost of your system. As usual, the hardest items to assess economically are those that you cannot assign a dollar value to or those related to your enjoyment and pleasure. A good movable insulation system is one that you will enjoy as you would a well-made chair or garment.

Edge Seals

Air flow behind movable insulation reduces the thermal effectiveness of the system if it is allowed to enter the home and mix with warm room air. This is particularly true of high R value systems which demand very tight air seals. The effects of poor edge seals are magnified on insulation panels for small windows where the perimeter-to-area ratio is low. The edge seals in very large window insulation panels can be somewhat looser because the length of the perimeter cracks through which air can enter is small compared to the area covered by the panel.

It is generally easier to seal the edges of rigid shutters than the edges of flexible membranes such as shades and curtains. Shutters can be made to fit snugly and weather stripping can be added to the edges to make them seal even more tightly. Shades and curtains rely on tracking devices, clamping strips, and cloth fasteners such as Velcro, a material which sticks together like a cocklebur on a wool sock, to insure a tight seal. Magnetic seals are a very promising development for all types of window insulation and are currently used with rigid and flexible thermal barriers.

In winter, when you want to keep heat inside, the edge seals on the bottom and sides of an interior window insulation system are most critical since leaks here create a cold air pocket that does not mix with room air. But a top seal is also important in preventing warm room air from being drawn into this air pocket and to prevent moisture from condensing on the window glass. A top seal is even more critical for keeping heat out in summer because the solar-heated air behind the window insulation will tend to rise into the room if there is no top seal.

If window insulation is applied right onto the window glass, edge seals are not as necessary as they are in systems where the insulation is more than 1 inch from the glass. Nightwall magnetic clips (produced and sold by Zomeworks), used on polystyrene foam, position a panel only 5/64 inch from the glass surface and, therefore, edge seals are not necessary to prevent convective losses.

Tests done on ordinary vinyl window shades* over single glazing showed a 25% reduction of heat losses when the shade was 1 inch from the glass but only a 12% reduction when the shade was 3½ inches from the glass. When tracks were added to the shade, heat loss reductions were 34% and 29% in the same respective positions. The additional R values of these shades were low—all less than 0.5.

As the R value and thermal effectiveness of a system rise, the temperature behind the panel becomes lower, causing more convective losses around the edges if the panel is not tightly sealed. More research is necessary to better determine the effect that the distance between insulating panels and window glass has on the panels' performance and to confirm edge seal requirements for various thicknesses of window insulation.

Moisture Protection

In homes that maintain a relative humidity of 40 to 50% you will often find moisture condensing on the cold window glass behind the insulation unless the window insulation provides a barrier to the flow of moisture. To prevent moisture migration, thermal curtains should be lined with a moistureproof material such as polyethylene or vinyl. Thermal shutters often have a foil layer to act as a vapor barrier.

Most homes can only maintain a humidity of about 30% throughout the winter. In moderate climates a humidity in the 30% range will not cause a significant moisture problem on double glazing covered with movable insulation unless air is readily convecting from the living space to the window glass.

In northern climates where subzero temperatures are common, the condensation of moisture on window glass can be a serious problem even at an inside relative humidity of 25 to 30%. Even without window insulation, the rotting of windowsills from moisture condensation is a common problem in northern climates. This problem is often aggravated by ordinary drapes which make the window glass cooler and at the same time allow room air to convect onto the glass. Figure 4-2 shows the temperatures at which condensation will form on window glass at various relative humidities. Temperature-humidity combinations near these values will produce moisture condensation when movable insulation is added, unless the insulation contains a foil or plastic vapor barrier and the edge seals are very tight.

*By Maureen M. Grasso of the University of Texas and David R. Buchanan of Cornell University.

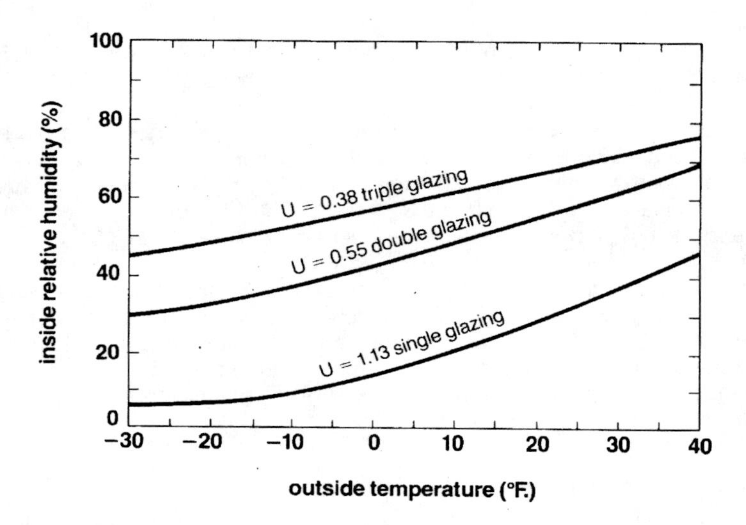

Figure 4-2: Relative humidity at which visible condensation will appear on the inside surface of bare window glass. (Adapted with permission from 1977 Fundamentals Volume, **ASHRAE Handbook & Product Directory.)**

People generate moisture in the home from perspiration, cooking, and bathing. The smaller a home, relative to the number of occupants, the more humidity will be generated. A building that is sealed tightly against air infiltration also traps moisture more readily than the leaky construction of the past, thus allowing a higher level of humidity to be maintained inside the dwelling. Both compact house plans and tight construction are desirable for economic and energy reasons. Spaces with humidities between 30 and 50% are also more comfortable. Movable insulation systems with proper moisture seals can assist compact, energy-efficient designs by protecting windows against moisture condensation.

The humidity will be highest in areas with water fixtures such as bathrooms and kitchens. Adequate ventilation here reduces moisture problems but caution should be taken not to use excessive ventilation that will lose precious heat to the outside. For example, an exhaust fan in a bathroom should be run only when showers or baths are taken.

Windows covered with insulation day and night for weeks on end deserve more attention to the problem of moisture buildup than windows where the insulation is removed daily. If a small amount of frost forms on a double-glazed window overnight, it usually melts and evaporates within an hour or two after the window insulation is removed the next day. But if the insulation is not removed, frost may build up and become a problem.

Moisture is a more serious concern when there is window insulation over the interior of single glazing, than when it is over double glazing. On cold winter nights, the single-glazed glass becomes very cold, and frost can build up rapidly on the interior side. If a large moisture buildup occurs daily, serious damage to the wood trim and wall beneath a window can result.

If moisture buildup is a problem in your system, a good look at the moisture barrier and edge seals of your window insulation is in order. A stopgap to this problem is to catch the moisture long enough for it to evaporate or to direct it into a small container. The 3M Company's Scotch Brand Weatherstrip described in Chapter 2 can be attached at the bottom of window glass to catch moisture as shown in Figure 4-3.

In at least one instance, a movable insulation product has solved a moisture problem on window glass instead of creating one. Insul Shutter, a company in Silt, Colorado, makes a very sophisticated interior shutter of hardwood veneer filled with urethane foam (see Chapter 8). This shutter was installed in a home in a severe winter climate where the single-glazed aluminum window sashes allowed condensation to drip off the glass. Since these airtight shutters have been installed, there has been no nighttime condensation on the windows whatsoever.

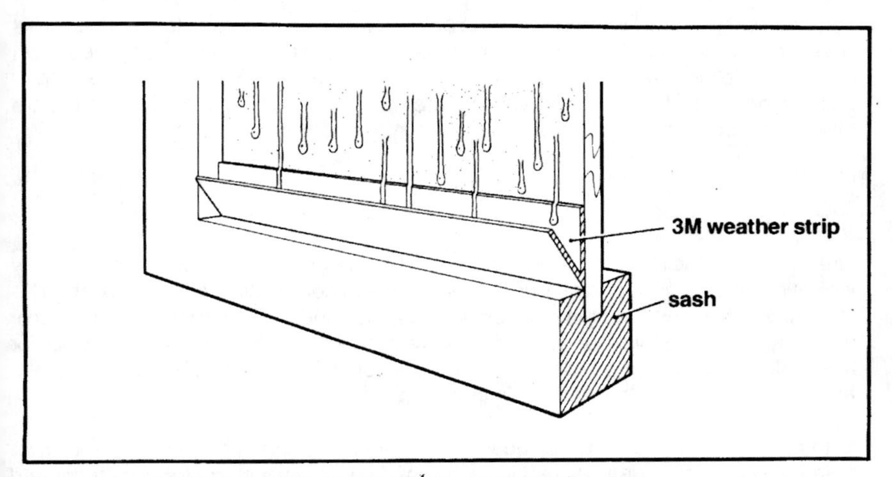

Figure 4-3: A stopgap to prevent moisture condensation damage.

Durability

Durability is a very important feature in any system which has a long-term payback, but even a low-cost system should be designed to provide as long a life as possible. If we are to become a society that takes pride in building durable items that are sustained by renewable energy resources, every product and building produced should have optimum wear resistance and long life.

By its very nature, movable insulation has parts that move and eventually wear out. The operation of a system causes wear on the hinges, pivot points, edge seals, and tracks. These areas deserve careful examination in any system you are considering for purchase. Fabrics can easily become frayed where they feed into side tracks or slide along the floor. Weather strippings on hinged panels are often the first components to be replaced. A well-designed system will allow the wear points to be replaced while the rest of the system stays intact.

Sun and degradation are also some things to be wary of, particularly with fabrics that become brittle and crack after a period of time. Table 6-1 describes the resistance of a number of fabrics to sunlight. Outdoor window insulation systems must withstand a much more vigorous attack by the elements. Not only must the parts subject to wear be highly durable, but the whole system must be able to withstand wind, rain, sun, insects, birds, etc.

It is likely, however, that abuse by people and pets is a more serious cause of wear and reduced window insulation life than are the abuses of the outside elements. This wear factor varies widely and should be weighed according to the habits of your own family. Insulation that extends to the floor is much more vulnerable to abuse from children and pets, but even systems that are well out of the reach require gentle care and occasional cleaning or maintenance to optimize their life span.

Safety

Fire safety is usually the most important safety concern with the addition of window insulation since most types of window insulation are combustible to some extent. The ideal situation would be to install noncombustible window insulation in a home built with only noncombustible materials. A more practical approach is to watch out for materials that flame up rapidly. Window insulation should not impose any more of a fire hazard than the other materials in the home.

With any material you consider questionable it is suggested that you do your own flame spread tests in a well-ventilated room. Take a small strip of the material and ignite it with a match or with the flame in your fireplace. A pair of pliers may be useful to avoid burning your fingers. Observe how the material burns, particularly when held vertically. Be sure that you do this in a fire-safe area and avoid breathing the fumes.

Heavy shutters which hinge on the top and swing upward are another potential danger as they could fall and injure someone. If this type of system is used, it should always contain some type of counterweight or braking mechanism. Toxic chemicals are encountered in some of the components for home-built systems. 3M Company's Plastiform magnetic strip, while it is so handy, contains poisonous lead compounds, and children should not be allowed to put it in their mouths. Pointed edges and hardware at eye or head level on movable insulation are other very important safety concerns. Any such projections should be padded, installed above head height, or avoided altogether.

Appearance

The appearance of a window insulation system will probably be the most important feature in determining the type of system you choose. By all means find a system that will be attractive in your home, but don't be blinded by appearance alone. Most systems described in this book can be modified with a coat of paint or a decorative fabric for that unique touch you are seeking. We all have distinct likes and dislikes in dress and decor. Make it your system!

As you probe the various ways to insulate your windows in the remainder of this book, you may discover a whole new approach to the problem. There is a great deal of unexplored territory in this field, and if you turn your imagination loose, you can give a real boost to the state-of-the-art. Be sure to read the details and specifications carefully, however, for many of the systems described will not work properly if important details are not heeded. With full knowledge of the details, perhaps you can devise your own way to provide yourself with a warmer and uniquely comfortable home.

Plastic Foams and Fire Safety

Some time ago, manufacturers claimed that plastic foams were fire-retardant and would not support combustion. This claim was supported by a standard test done at Underwriters Laboratories, Inc. As a result, plastic foams were left exposed in the interior walls and ceilings of many residential garages, basements, and other informal spaces. Fires occurred in several residences where the exposed foams burned explosively, resulting in the deaths of the occupants. The test methods were reviewed, and it was discovered that, although the foams only smoldered in the standard test, they burned violently if a continuous flame supported their combustion until they reached a critical temperature.

Equally serious to the problem of rapid combustion of household plastics are the noxious fumes they give off when burned. Because of new, stricter regulations on the acceptable rate of flame spread for plastic products, the chemical industry now adds fire-retardant chemicals to many plastics. These chemicals do reduce the rate of combustion of many plastics, but they also increase the toxicity of the gases when the plastics do burn. Ironically, the combustion products of smoldering plastics are deadly, but with faster rates of combustion, the gases are much less toxic. This is a hard trade-off to assess and is compounded further by the fact that many of these fire-retardant materials are highly carcinogenic (cancer causing).

In the case of a home fire today, you want to get out particularly fast to avoid breathing the toxic gases given off by burning plastics. It is of little avail to worry about a small amount of foam plastic in window shutters when your carpeting, furniture, upholstery, plumbing, appliances, and many finishes contain plastic. Of course every precaution should be taken to prevent a household fire from occurring in the first place. Additional precautions should be made to insure that a fire will spread slowly enough to allow all occupants to escape. Fires that begin slowly can sometimes be put out with an extinguisher.

Aluminum foil over plastic foam will reflect heat away from the foam and help to retard flame spread until the fire gets hot enough to destroy the foil. For this reason I recommend that plastic foam used for interior thermal shutters be protected by aluminum foil. The foil is very thin and can also reflect radiant heat, increasing the insulating value of a movable insulation system. An optional method of reducing flammability is to apply two coats of intumescent (fire-retarding) paint to the panel, particularly if it is not faced with foil.

The use of foil-faced plastic foams in your interior may not meet some fire codes, but we need to reassert the notion that personal safety comes from an understanding and respect for the technology which we are embracing—what some call "common sense." It does not come from mere government regulations, which among other things allow the licensing of faulty nuclear plants that not only endanger us, but many generations to come.

Part II

Movable Insulation Inside Your Home

Movable insulation *inside* the dwelling has many advantages. The system can be as simple as a thermal curtain which hangs behind your existing curtain and curtain tracks. Movable insulation on the inside—whether it be a shutter, shade, or curtain—is protected from the harsh wind and rain outside. Therefore, the hardware and seals in the system can be much lighter than those required for outdoor systems.

It is easy and pleasant to operate movable insulation indoors. Adjusting your insulating shields each morning and night from within a warm and dry shelter is certainly preferable to going outside to face the elements. The exterior of your home remains simple and uncluttered as the careful integration of exterior shutters is not required.

Interior systems have a few disadvantages. The home's inside appearance will be altered although this can be for the better. Hinged shutters may interfere with furniture. Fire safety requirements are more stringent indoors than on the exterior. The system must also withstand abuse from children, pets, and sometimes adults, and must not endanger their safety.

Chapter 5

Pop-In Shutters

Pop-in shutters are rigid, lightweight panels that can be pressed into window openings at night and removed during the day. They are easy to construct, low in cost, and very effective at reducing window heat losses. This approach to window insulation was developed and first marketed by Steve Baer of Zomeworks in Albuquerque, New Mexico, using his Nightwall system of attaching beadboard to window glass with magnetic clips. Holding a sheet of insulation board very close to smooth window glass, Nightwall clips take advantage of the resistance to air flow that is provided by an air film layer on the surface of ordinary window glass.

Pop-in shutters can be used over just about any accessible window. On windows facing north, these mobile shutters can be left in place both day and night, except at times when the use of a space requires daylighting or a view to the outside. This characteristic makes pop-in shutters ideal for occasionally used spaces such as laundry rooms, dens, or guest bedrooms. Because the panels are lightweight and easily moved, they are by no means limited to semifixed locations. On south-facing windows in living, dining, and other frequently used rooms, pop-in shutters can simply be removed each morning to their respective storage locations and pressed back into service at night.

To be effective, pop-in shutters must prevent cold air near window glass from convecting into a room. Two different strategies in panel design serve to stop this air flow. In the *edge-seal* method (see Figure 5-1a) the panel fits tightly into the window opening with a foam strip or some other type of weather stripping around its entire perimeter. In the other method, a *glass-hugging* panel (see Figure 5-1b) is spaced only 1/16 inch or so off the window glass, eliminating the need for edge seals. The narrow gap between the glass and panel creates so much friction against air flow that air can only move through the gap very slowly, like molasses in January. This greatly diminished air flow is not a significant factor in the thermal effectiveness.

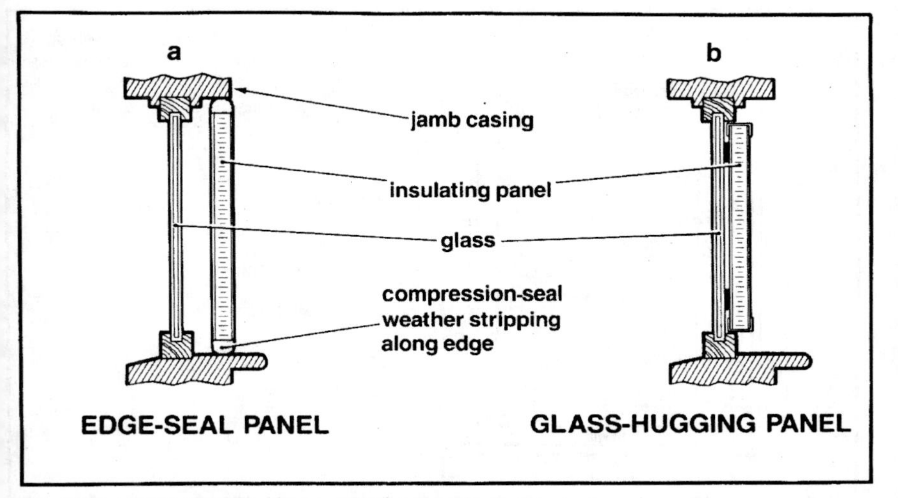

Figure 5-1: Reducing air convection behind pop-in panels.

An analogy might be drawn between these two kinds of barriers by considering air flow and ways of controlling water runoff on hilly land. The edge-seal type of panel is similar to cleared land on which water gushes down to a small flood-control dam that stops the flow at one point only. This point must be very tightly sealed or a pressure or head of water will build up and try to push the water through the gate. The glass-hugging panel has flow resistance over its entire surface, similar to a wooded landscape where flooding is rarely a problem since there is no buildup of water.

Measuring a Window for Shutters

Before you construct a rigid window shutter you must accurately measure your window. Figure 5-2 shows how to measure a window in preparation for building a shutter to fill the window opening. The glass opening in a window is usually squared off and requires only height and width measurements, but the window sash opening is often out of square, particularly in older homes. The glass-hugging shutters discussed in this chapter are usually cut to the glass size, but some of the edge-sealing shutters in this chapter and most of the shutters described in Chapter 9 require very careful window sash measurements.

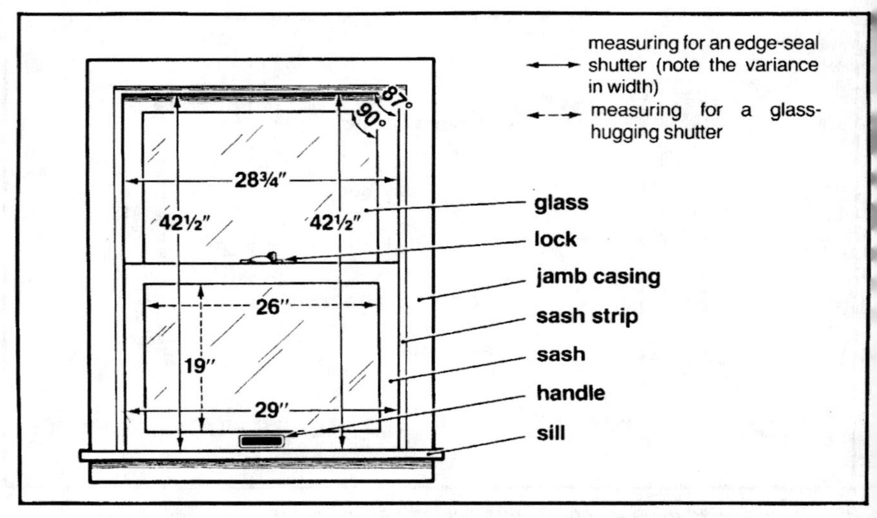

Figure 5-2: Measuring a window opening for a rigid window shutter.

A full-scale pattern of the shutter you plan to construct is very useful. It can be cut and mounted into your window opening to check the accuracy of your measurements and then used to draw an outline on a sheet of insulation board before you make the final cut. A cardboard pattern often saves you from making a bad cut and wasting an expensive sheet of insulation board.

Materials for Pop-In Shutters

Panels can be constructed from any of the foam sheathing materials available at building supply outlets or from corrugated cardboard which is usually available from furniture and appliance stores at no cost. Less-combustible materials can also be used, but they generally make a heavier, more cumbersome panel.

A good material for pop-in shutters is a foam isocyanurate sheathing sold under the trade names Thermax and R-Max (see Appendix III, Section 6). This material comes in 4-by-8-foot sheets with foil on both sides in ¾-inch and 1-inch thicknesses. Although a bit more expensive than polystyrene or polyurethane, isocyanurate foam has superior fire-resisting qualities and the foil layers add strength to this very lightweight material. Cardboard panels laminated from several layers of ¼-inch corru-

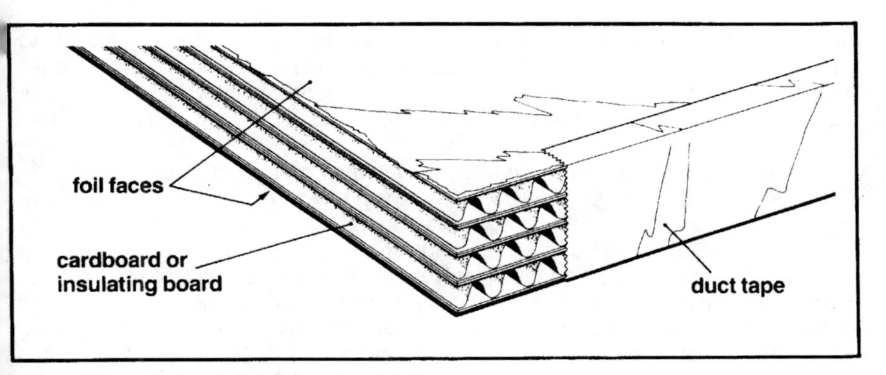

Figure 5-3: Taping the edges of a panel.

gated stock (as well as unprotected polystyrene or beadboard) should be faced on both sides with heavy-duty aluminum foil to protect against rapid flame spread. Be sure to put foil on both sides as foil on one side only is likely to cause warping. Wrapping the edges of the sheathing or cardboard with duct tape (see Figure 5-3) provides some additional fire protection and also protects them against wear.

Depending on the sizes of your windows, purchasing insulating boards in 4-by-8-foot sheets may present a problem of waste. For example, constructing a shutter for a 30-by-60-inch window requires one 4-by-8-foot sheet and wastes over half of the sheet (see Figure 5-4). However, with careful planning, the scraps can be used on

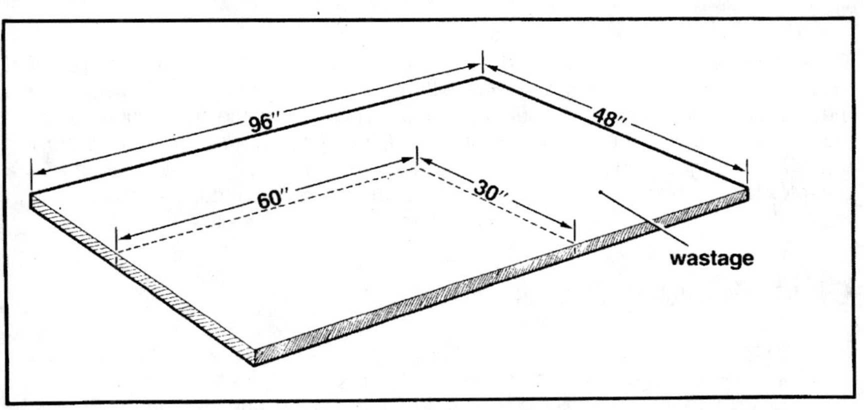

***Figure 5-4:* Wastage, which can be over half of a standard sheathing panel, can be used for another board.**

smaller windows or other insulating projects around the house. Another option is to glue the scraps onto a piece of corrugated cardboard to make a 1-inch-thick panel. Cardboard usually comes in pieces ¼ inch thick or thinner. Beadboard is sometimes available in ½-inch thicknesses that can be laminated into 1-inch panels.

When gluing layers of insulation board together, match the type of glue you use to the kinds of surfaces you are gluing. The uses of most glues are noted on their containers. Cardboard and other porous materials can be glued with white glues such as Elmer's Glue-All. This type of glue also works well with polystyrene (Styrofoam or beadboard). Some types of "panel adhesives," available in caulking tubes, are good adhesives for styrene, but others dissolve foam so be sure to read the descriptions on the containers before purchasing. Gluing aluminum foil to a foam panel or gluing two foil-faced boards together requires an adhesive such as 3M Company's Scotch Grip Plastic Adhesive 4693, which will adhere to smooth, metallic surfaces.

The best tool for cutting pop-in panels out of foam or cardboard is a utility knife with replaceable blades. Cut with the blade against a metal ruler or straightedge using a series of light strokes, each cut going a little deeper than the one before. Use a sharp blade and go slowly and carefully, and you will save time and money by not having to discard material and recut.

When the panel is cut it should fit freely into the window with just a little bit of play. Some of the foam sheathings such as Thermax board can be easily sanded on the edges for a close fit. If you have to sand a little, wear a mouth and nose filter and eye protection. Next, put duct tape on the edges and you will be ready to apply magnets or weather stripping.

The method you use to finish or decorate these panels is up to you. The greatest resistance to heat transfer is had with both foil surfaces left exposed. However, this may not be attractive in many interiors and to improve looks the side facing the room can be covered with a decorative fabric that is stretched across the panels, turned under at the edges, and pinned to the board with straight pins. On the other hand, the outer foil layer is often better left exposed, not only to increase the insulating value of the shutter in the winter, but also to reflect heat back to the outside in summer.

Glass-Hugging Pop-In Shutters

Tests show that a sheet of material placed very closely to a pane of window glass minimizes the flow of air from behind that material. One way to examine this principle is to blow a little bit of smoke onto a cold windowpane and then hold a piece of Styrofoam ⅛ inch away from this panel for 3 or 4 minutes. The smoke will become almost "frozen" by the cold, smooth glass. If you repeat this experiment with foam held ¾ inch off the glass, you will find that the smoke does not "freeze" but drifts slowly out the bottom.

How close does a panel have to be to the glass to take advantage of this "freezing" or slowing-down phenomenon? Steve Baer suggests the panel be placed as close to the glass as possible, and he designed his thin Nightwall clips accordingly. Tests done at Zomeworks show that even the 5/64-inch air space allowed by a Nightwall clip reduces the effective R value of a beadboard panel from what it would be if held tightly against the glass. Recent tests done by William Shurcliff, an honorary research associate at Harvard University and renowned inventor in the field of low-cost solar heating, indicate that some laminar resistance to air flow still holds true for spaces of ¼ inch or greater. In general it can be assumed that the closer a panel is to glass the less air flow there will be.

> Nightwall Clips—Available from Zomeworks Corporation, P.O. Box 712, Albuquerque, NM 87103, these magnetic clips do an excellent job of holding insulation board to window glass. A Nightwall clip is a thin, ½-by-6-inch metal strip connected to a flexible magnetic strip cut to the same size. Both strips have an adhesive backing (see Figure 5-5). The flexible magnetic strip is attached to the shutter and the thin steel strip adheres to the glass.
>
> To apply these clips, first mount the flexible strip onto the shutter with the mated steel strips facing out. Remove the paper backing from the steel strip, press the shutter onto the window, placing the steel strips against the glass. Then with the panel removed, firmly rub the steel strips with a spoon or hard object onto the window so they adhere firmly to the glass.

An alternative to Nightwall clips is to use a flexible magnetic strip on both the glass and pop-in shutter (see Appendix III, Section 11). The flexible strip is manufactured by 3M Company and is also available from Zomeworks. A pair of flexible magnetic strips holds the shutter at a slightly greater distance from the glass than Nightwall clips, allowing more air flow behind the panel. If you slide one layer of magnetic strip against another layer, you will find the magnets have poles every ¼ inch and that they will attract or repel each other, depending on how the poles are aligned. Therefore, if you use double magnetic strips, the tape will often end up ¼ inch or so from where you originally intended it to be. In William Shurcliff's fine book, *Thermal Shutters and Shades,** he suggests additional ways of holding panels onto glass.

Glass-hugging panels are the easiest type of pop-in shutter to cut and mount.

*See Appendix V, Section 1 for book description and complete bibliographical information.

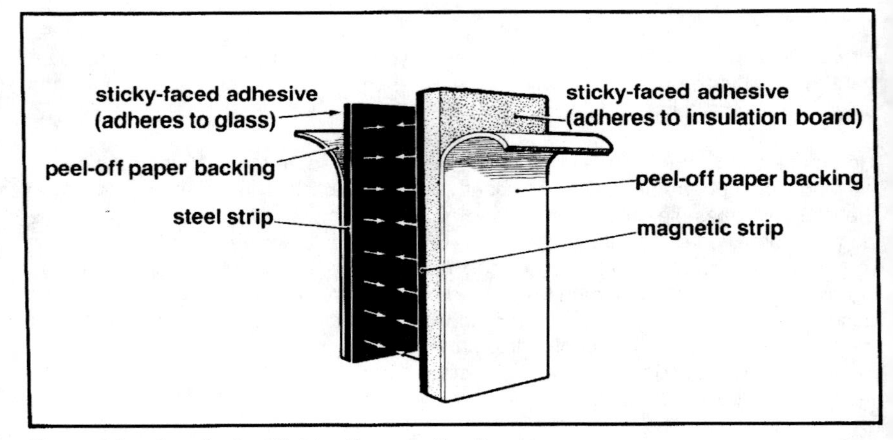

Figure 5-5: Detail of a Nightwall magnetic clip.

Because edge seals are not needed, there are greater tolerances when cutting the panels. Glass-hugging panels should be fit to the glass size rather than to the size of the entire window opening. This can be a big advantage with large sliding or double-hung windows where two glass-hugging panels can be used instead of one large, unwieldy panel that fits the entire opening.

A possible disadvantage of attaching shutters with magnetic clips is that the clips rest on the glass when the panel is removed, and some people may feel this blocks or clutters the view. A very careful and regular placement of these clips is necessary to avoid a cluttered appearance.

In double-hung windows facing north (and east- and west-facing windows that don't get significant direct sunlight for a good part of the day), two separate panels provide the option of opening the top one for daylight while leaving the bottom panel in place most of the time. However, it is important to remove window insulation daily on south-facing windows and others that receive a useful heat gain from direct winter sun.

Many older homes have windows with numerous small panes of glass, usually in the upper sashes (see Figure 5-6). This multipaned design makes it difficult to place a panel close to the glass. A simple solution, however, is to cut a ¾-inch beadboard section—thicker if necessary—to fit into each section of glass. Fasten these pieces to the glass with adhesive magnetic clips and then take another sheet of beadboard cut to the size of the entire window sash and glue it to these panels. When the glue dries, remove the panel and you will have a single pop-in shutter with pieces that fit into each section of glass.

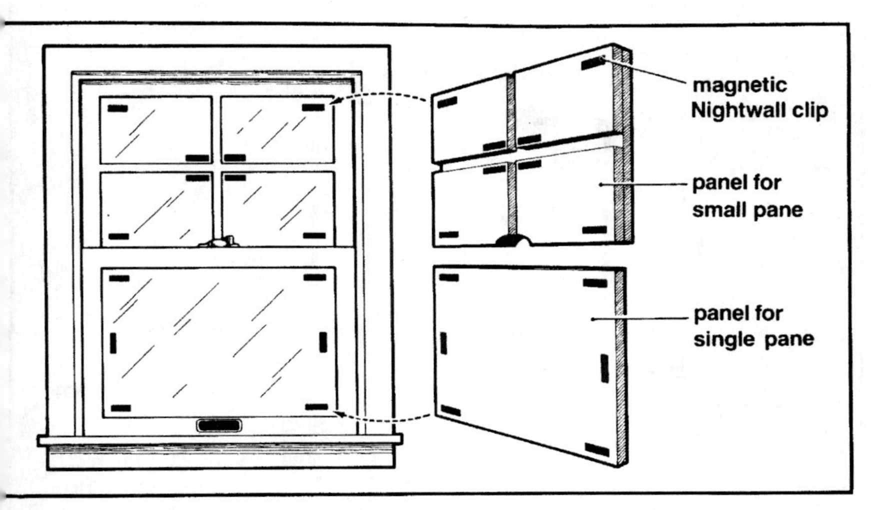

Figure 5-6: Nightwall panels on a double-hung window.

Edge-Sealed Shutters

Edge-sealed pop-in shutters have continuous seals around their perimeters that are formed by combining magnetic strip seals and compression-foam gasket seals. Figure 5-7 shows how a shutter which uses this system can be attached to your window trim. Cut several 3-inch-long strips of 3M magnetic tape and place them 9 to 12 inches apart on the edges of the insulation board where it is wrapped with duct tape. (Do not let small children put these magnetic strips in their mouths since they contain lead compounds.) Place mating magnetic strips on these strips, remove the paper backing, and press the panel against the window frame to deposit the strips on the wood trim. Leave gaps between the pairs of magnetic strips that have a combined thickness of ⅛ inch. A 3/16-inch-thick, self-adhesive foam gasket material sold for pickup truck camper mounts is ideal for filling in these gaps. It comes in a roll 1¼ inches wide and can be cut into ½-inch-wide pieces with a utility knife and straightedge. Apply this ½-inch-wide gasket to the insulation panel between the magnets. It compresses to ⅛ inch when the panel is in place. Seal the sides and top of the shutter with this magnetic strip-gasket combination, and seal the bottom with a vinyl-clad foam weather stripping. The Schlegel Corporation's products mentioned in Chapter 2 perform well here.

A disadvantage of this design is that it clutters the face of your window trim with magnetic strips. These magnetic strips are less visually pronounced if they are

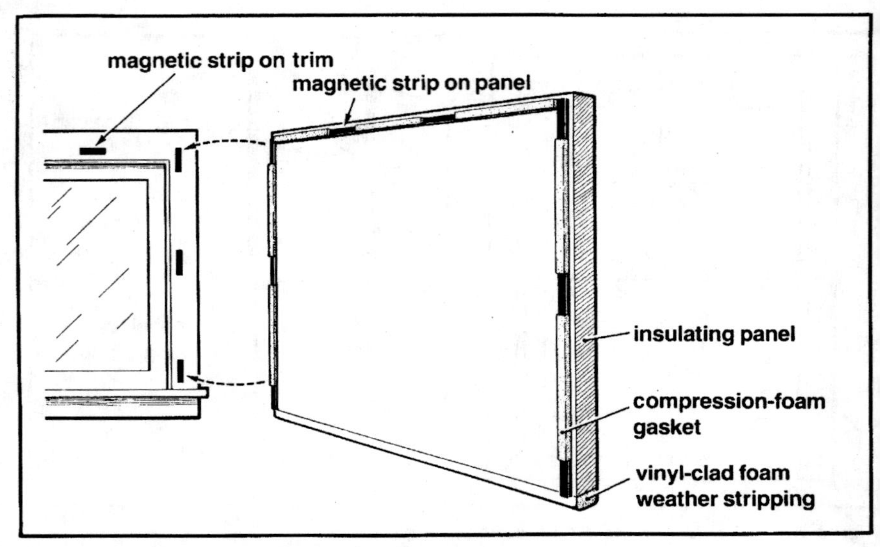

Figure 5-7: Edge-sealed pop-in shutter.

mounted on sash or shutter stops inside the window opening. This not only makes the appearance of the window cleaner during the daytime with the shutter removed, but also helps to protect the shutter when it is set back into the opening at night.

If your window has a sash stop facing the room with a square, flat edge, wide enough to accept a magnetic strip as shown in Figure 5-8a, an additional stop for the shutter is not necessary. If the interior-facing surface on this stop is too narrow to accept the magnetic strip, you can add a ½-by-½-inch stop as in Figure 5-8b. If the stop is canted or angled as in Figure 5-8c, it will need to be removed and replaced with rectangular moldings to make a section like those shown in Figure 5-8d.

When magnetic strips are applied to painted wooden stops, the adhesive often removes the paint after some time, allowing the magnetic strip to fall off. In such a case, a dab of epoxy or a couple of small carpet tacks will secure the strip.

Wooden buttons are handy in helping to remove flush-mounted panels from the window opening. They can be found in any hardware store and should be attached with large washers to keep them from pulling out of the foam.

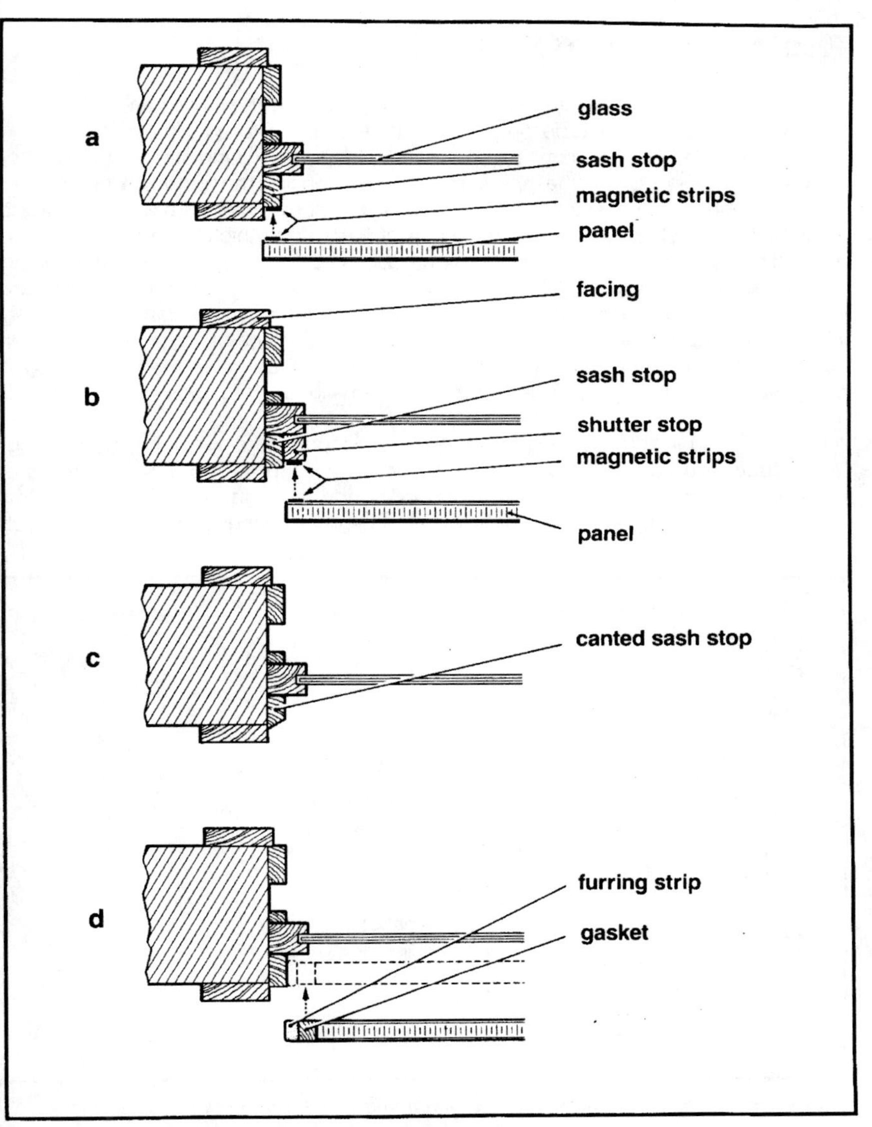

Figure 5-8: Window trim and stops for edge-sealed pop-in shutters.

Friction-Fit Pop-In Shutters

Friction-fit shutters differ from edge-seal shutters in that they fit so snugly into the window opening that they do not need magnetic strips to hold them in place. A tight-fitting compression-foam gasket around the entire perimeter of the shutter allows one to slide it into the opening with a friction fit. The window opening will require a sash stop similar to the stops shown in Figures 5-8a or 5-8b with flat, square surfaces for the edges of the shutter to slide into. A ¼-inch-thick furring strip must be applied to the outside edge of the foam board to keep the shutter from binding when it compresses the edge seal. These furring strips should be glued to the foam board with a panel adhesive compatible with the foam. The furring strips can be held firmly in place with duct tape until the glue dries. A 3/16-inch truck camper gasket is then applied to the edges. Although this type of gasket does not have a vinyl cladding, it has a wide band of strong adhesive and remains flat under stress.

A friction-fit pop-in shutter requires very exact measurements. When cutting the insulation board, be sure to allow ¼ inch for the wood strip and ⅛ inch for the compressed 3/16-inch gasket all the way around its edges—a total of ¾ inch. To allow these clearances, the insulation board should be cut to dimensions that are ¾ inch

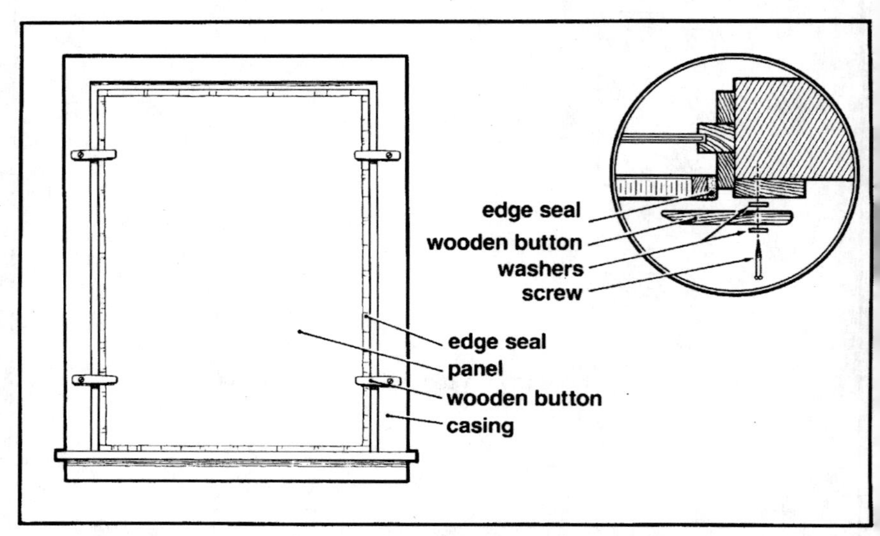

Figure 5-9: Wooden buttons added to keep shutters from popping out.

←*Photo 5-1: Jack Ruttle's friction-fit pop-in shutter.*

↓ *Photo 5-2: A single piece of bent cold steel makes it easy to pop in friction-fit panels.*

less than the window opening. This type of panel will seal very tightly, and if your windows are leaky they may pop out during windstorms from air pressure behind the panel. If the panel does pop out, you can add four wooden buttons or clips at the corners as shown in Figure 5-9. In any case you should be sure to caulk and weather-strip the window to reduce the infiltration problem at its source.

Jack Ruttle found that friction-fit pop-in shutters proved to be the simplest and least expensive way to insulate the double-hung windows in his Emmaus, Pennsylvania, house (see Photo 5-1). Jack stapled decorative fabric to the interior side of ¾-inch Styrofoam and trimmed it with 1-inch square wood molding to give the panel strength and protect the Styrofoam edges from wear. Compressible foam weather stripping glued to the outside edges of the molding allows these panels to fit snugly inside the window frames and also compensates for those windows that are slightly out of square. Photo 5-2 shows a device that Jack picked up from *Thermal Shutters and Shades* by Bill Shurcliff.* This simple piece of bent cold steel allows Jack to shoehorn the bottom of the panel in and out of the window.

*See Appendix V, Section 1 for book description and complete bibliographical information.

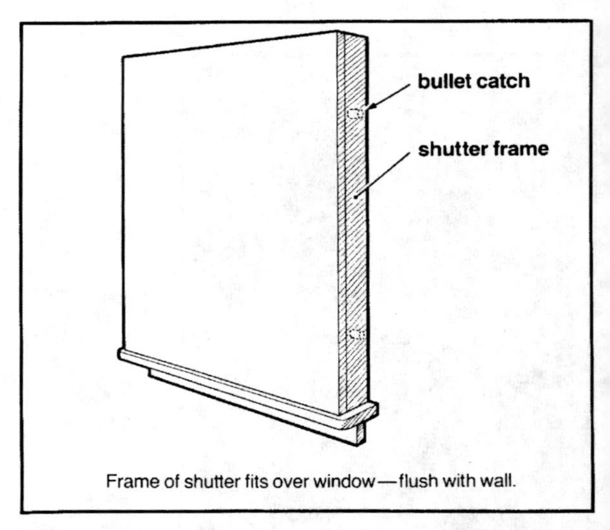

Frame of shutter fits over window—flush with wall.

Figure 5-10: Shutter design by The Center for Community Technology.

A Shutter for the Muppets

Another edge-sealed design is shown in Figure 5-10 for those who would rather avoid the use of foam plastics because they are a potential fire hazard. Developed by The Center for Community Technology in Madison, Wisconsin, this panel is constructed of a 1-by-2 wooden frame and ½-inch wood-fiber wall sheathing. This sheathing can double as a bulletin board to accommodate characters like the Muppets as shown in Photo 5-4. Details of the bullet catch described below are shown in Figure 5-11.

Behind the sheathing are 2 inches of fiberglass insulation or polyester fiberfill covered with tightly woven cloth to contain these fibrous materials. The frame mounts over the outside edge of the window trim and is held in place with bullet catches—a type of latch that has a spring-loaded ball bearing. When mounted over a window, the insulation is compressed slightly against the windowsill, forming a good seal.

A Heat-Collecting Pop-In Shutter

Pop-in panels on north-facing windows can be left in place except when light or a view is desired, but south-facing windows must be allowed to take full advantage of

←Photo 5-4: The Muppet shutter, ready to pop in the window.

↓ Photo 5-3: The Muppet shutter under construction.

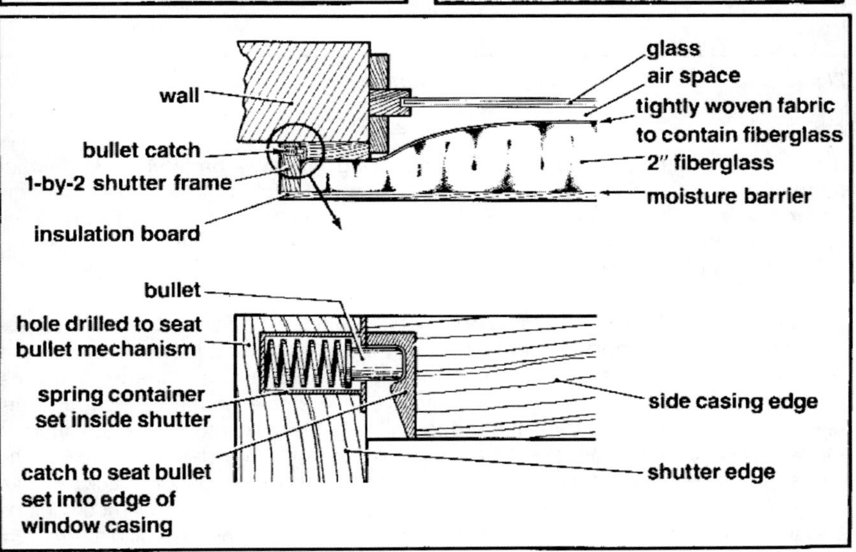

Figure 5-11: Details of the bullet catch.

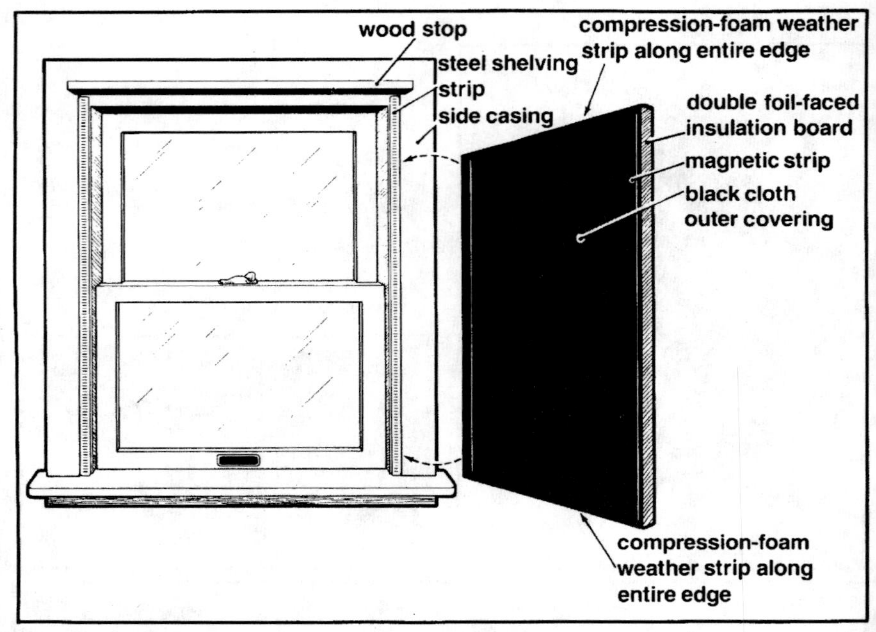

Figure 5-12: The heat-collecting pop-in shutter.

the winter sun. A pop-in shutter for a south-facing window—preferably the double-hung type—is shown in Figure 5-12. This shutter is designed for windows in rooms where one is "up and gone" in the morning, such as in the bedroom. It can provide: (1) a convective solar heat gain in the winter, (2) nighttime insulation, (3) morning sunlight, and (4) summer ventilation, in the various modes discussed a bit later.

This pop-in panel is cut 5 inches shorter than the window opening. A foam compression-type weather stripping is applied to the top and bottom edges of the panel, and a ¾-by-¾-inch wood stop is added to the top or head casing of the window. The back side of the panel has a flat black acrylic or polyester fabric surface, and the interior-facing side is covered with a decorative fabric. Steel strips are attached to the window face trim with glue or countersunk screws. The panel adheres to the strips by means of the 3M magnetic tape that is attached to the panel's back side. An off-the-shelf option is to use steel shelving-support strips (the kind that mount inside bookcases and support the end of a shelf with small clips). They are also made from aluminum and brass, but the steel kind is easily identified with a magnet.

During the winter, there are distinct nighttime and daytime modes for using the shutter, shown in Figure 5-13. At night the panel can be placed tightly against the windowsill and provides thermal protection for about 90% of the window area. The gap at the top provides a hint of any outside nighttime activity and allows early morning daylight to enter the room. In the morning, the panel is raised 3 inches to draw air in the bottom and out the top, allowing it to act as a solar collector during the day.

During the spring and fall when window protection is not needed due to mild temperatures, the panel is removed and stored. As outside daytime temperatures increase and protection from window heat gain is necessary, the panel is added once again. This time it is located toward the top of the window opening and pressed tightly against the furring strip mounted on the head casing. The gap at the bottom permits light to enter the room. On a double-hung window the upper sash is dropped to induce ventilation. If the sash is fixed or the use of air conditioning does not allow ventilation, the black cloth should be removed to expose the foil face of the panel, which will reflect incoming sunlight back out the window.

This shutter, while certainly a good idea, is only an experimental design at this point and needs to be further tested and improved. In cold climates leaving the top open can generate condensation problems. An optional flap, which is hinged with a cloth tape, allows the shutter to completely seal over the window on winter nights instead of using the open top shutter. Double foil-faced, rigid insulation board is recommend-

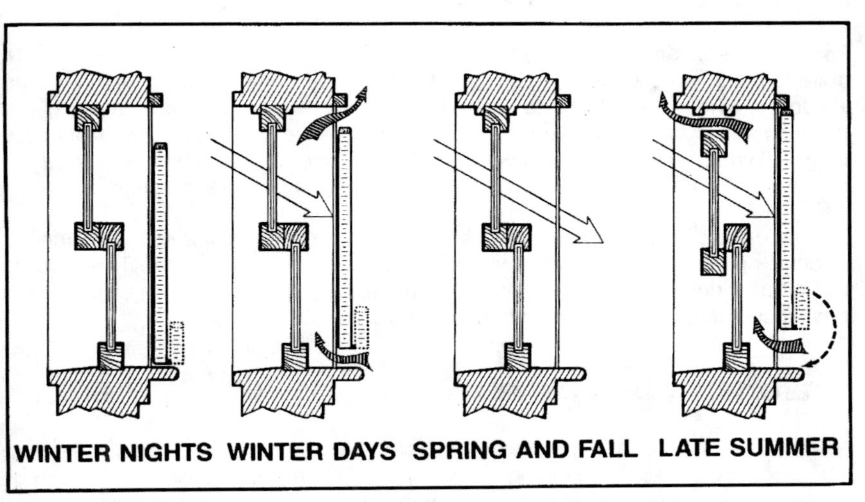

Figure 5-13: Modes of operation—the heat-collecting shutter.

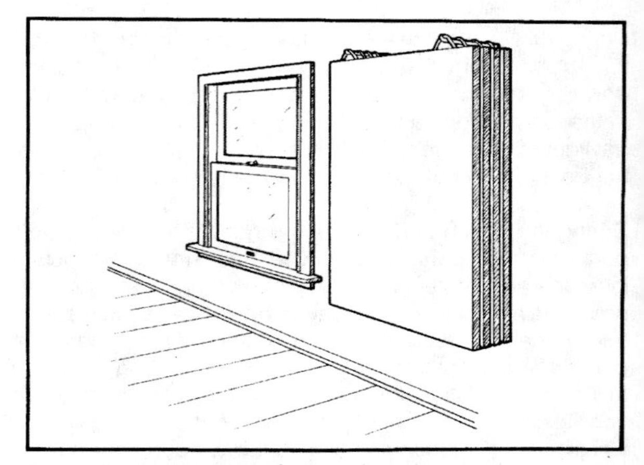

Figure 5-14: Hanging shutters from loops.

ed for this type of panel to resist warpage from heat and sun exposure. Although extensive testing of its warp resistance in this application is lacking, it is probably the most suitable material for this shutter design.

Daytime Storage for Pop-In Shutters

Pop-in shutters can be stored behind doors, furniture, or just about anyplace where there is blank wall space. A good location for these panels is on a wall near the window at window height. Loops used to hang these shutters on hooks can be glued to the top of the shutters with epoxy or a high-grade glue. By installing pairs of hooks on a wall you can store one or several panels at a time. Shelf brackets fixed high up and out of the way on a wall provide a convenient place to hang panels.

Nearby closets or spaces behind doors can also be used to store pop-in shutters. Another option is to construct a storage pocket in a vacant corner. This storage pocket provides a handy hideaway for the panels and the top of it can be used as a shelf, counterspace, or, if supported well, as a bench.

Seasonal Pop-In Panels

Many homes today have large north-facing windows that serve no purpose at all during the winter except providing an expensive, frozen view. These windows can be protected with pop-in panels different from the type that must be moved daily.

Photo 5-5: Robert Gray's pop-in panels to insulate glass patio doors.

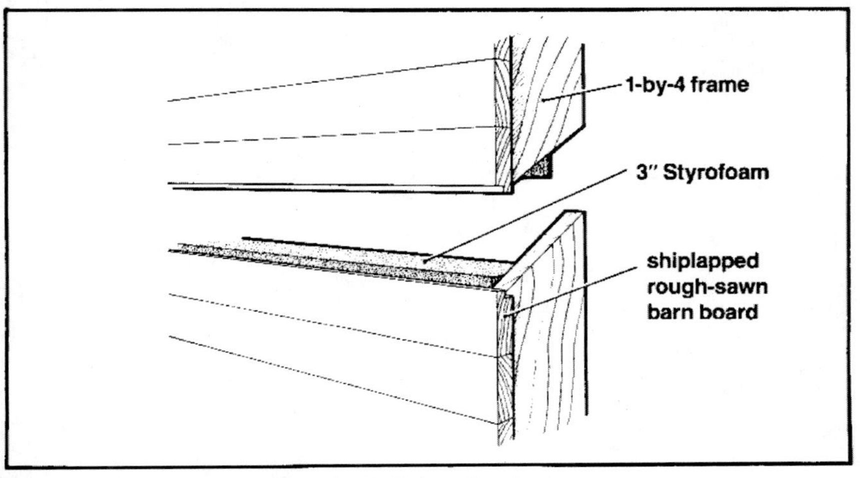

Figure 5-15: Detail of shiplap joint in Robert Gray's shutter.

Because they are moved only two or so times a year, seasonal pop-in panels can be larger and heavier than ordinary ones and can have thicker insulation, often 3 or 4 inches of foam instead of the standard 2 inches.

Robert Gray of Mason, New Hampshire, has a home with 24 linear feet of glass patio door in the dining room, most of which faces east and north. To make a warmer winter interior and reduce the tremendous drain on his heating bills, Bob constructed a series of pop-in panels shown in Photo 5-5. With the large window openings in this room, Bob developed a clever detail, as shown in Figure 5-15, to form a tight panel, which appears continuous even though it is assembled from sections no wider than 3 feet each. The panels are made with 3-inch-thick Styrofoam and are faced with attractive rough-sawn barn boards. They are simply mounted behind an overhead wooden strip or cornice board that holds them in place.

Chapter 6

Thermal Curtains—Blankets That Fold

Thermal curtains* are a controversial item among window insulation advocates. For over two decades, "thermal" curtain liners have been sold that do virtually nothing to stop heat flow. Those who advocate serious thermal window treatment have encountered a public-at-large who think they have the problem solved with their drapes and "thermal" liners. As a result, the term "thermal curtain" has become meaningless and many window insulation buffs treat the subject of thermal curtains with disdain. The thermal curtain, however, may be the late sleeper in movable window insulation development.

Store-bought curtains cost from $2 to $5 per square foot, and custom draperies sometimes cost considerably more. This expenditure is primarily for decorating purposes, to some extent for privacy, and marginally for light control. The choice of a lining fabric is usually tainted by claims of thermal performance which have little truth or merit. As a result, the common curtain or drape is an expensive, interior design component that often has little practical value. However, by redesigning the window curtain so that it performs effectively as window insulation, the cost of a curtain for decorating purposes can be returned to the owner in fuel savings.

The richness of texture and color which curtains provide makes them a primary element of interior design. Although decorative fabrics are also used on shades and shutters, the graceful way in which a fabric freely hangs with natural flowing waves from an overhead rod or track is unique to the curtain. Herein lies the challenge of redesigning the curtain into an effective thermal shield or membrane for windows. It

*The terms thermal curtains and thermal draperies are used interchangeably in this chapter to mean a hanging thermal blanket that draws open by folding into accordion-type pleats. Roman shades are included in this chapter because they fold vertically, similar to the way curtains fold horizontally. Systems with similar flexible blankets, but which roll out of the way instead of folding, are included in Chapter 7, which covers thermal shades.

must retain its identity as a soft and graceful fabric over the window, yet incorporate the following new energy features:

1. The curtain must seal tightly to the window frame on the bottom and sides and, to some degree, to the top to prevent convection with room-temperature air. If the curtain is well sealed around its lower perimeter, the air behind it is cooled by the window and settles into this pouchlike space. Seals on the bottom and sides only prevent room heat from convecting into the space behind the curtain; moisture can still enter from the top. Without a top seal, overnight frosting is likely to form on the window glass in cold climates or in high humidity rooms.

2. The curtain must contain a layer of fabric or material that is impervious to air flow and moisture. Fabrics that prevent air flow include any tightly woven material, such as nylon and polyester. To prevent the movement of moisture through the curtain to the window glass, many thermal curtain designs include a plastic layer of polyethylene or vinyl. A plastic vapor barrier is recommended for curtains in high humidity spaces and in cold northern climates. In moderate climates, a tight layer of nylon is often adequate as a vapor barrier for curtains used in rooms that do not have high humidity.

3. The curtain material(s) should provide resistance to heat transfer either with multiple air spaces, aluminized foil layers, a fibrous filler material, or minimally with a thick, tightly woven material like the wool in heavy blankets.

The above requirements clearly put some constraints on a curtain's design. The thermal curtains covered in this chapter are mostly for those who have simple tastes, although a wide variety of fabrics and patterns can be displayed in the material facing the room. If these simple curtain designs conflict with the interior decor of your home, plan to use a neutral-colored thermal curtain behind your decorative curtains. Consulting an interior designer may be helpful, but be sure that the designer understands the thermal requirements of your curtains.

In addition, the fabric selected must be able to withstand exposure to sunlight, heat, and occasional moisture. This is especially a concern on the window-facing fabric, which may get a large amount of direct sunshine in the summer. The interior-facing fabric must maintain an attractive appearance without fading. A wide range of fabric types are available under dozens of trade names. Table 6-1 shows some characteristics of commonly used drapery fabrics when exposed to sunlight and heat.

Resistance to shrinkage from moisture absorption is one of the most important aspects to consider when selecting a fabric for a thermal curtain or drapery. Fabrics

TABLE 6-1: Stability of Fibers to Environmental Conditions*

Generic Fiber Type Trade Name & Company	Effect of Sunlight	Effect of Heat	Effect of Moisture
Cotton	Gradual loss of strength, gradual yellowing	Excellent resistance to degradation by heat	Medium absorbency
Linen	Gradual loss of strength	Discolors at high temperatures	High absorbency
Silk	Moderate loss of strength, affected more than cotton, depends on dye and additives	Less affected than wool	Medium absorbency
Wool	Loss of strength, gradual fading	Loses softness from prolonged exposure	High absorbency
Acetate Acele (Du Pont) Celanese (Celanese) Estron (Eastman)	Slight loss of strength, little color loss	Little degradation	Low absorbency
Acrylic Acrilan (Monsanto) Creslan (American Cyanamid) Dolan (Hoechst) Dralon (Bayer) Leacril, Orlon (Du Pont) Zefkrome, Zefran (Dow Badische)	Very little loss of strength, no discoloration	Little degradation	Low absorbency
Glass Beta (Owens-Corning) Fiberglas (Owens-Corning) PPG (Pittsburgh Plate Glass)	None	None	None

*Prepared by Denise A. Guerin, interior designer.

TABLE 6-1 (cont.)

Generic Fiber Type/ Trade Name & Company	Effect of Sunlight	Effect of Heat	Effect of Moisture
Leather	No loss of strength, slight discoloration	Embrittlement, stabilized by care	Low absorbency
Modacrylic Dynel (Union Carbide) Elura (Monsanto) Kanekalon (Kanekafuchi) Leavil, SEF, Verel (Eastman)	Very little loss of strength	Little degradation	Low absorbency
Nylon Antron (Du Pont) Antron II (Du Pont) Cadon (Monsanto) Caprolan (Allied Chemical) Cumuloft (Monsanto) Nomex (Du Pont) Perlon (Bayer) Qiana (Du Pont)	Gradual loss of strength, little color loss	Little degradation	Low absorbency
Olefin Durel (Celanese) Herculon (Hercules) Marvess (Phillips Fibers) Meraklon, Polycrest, Polypropylene (Thiokol)	Moderate loss of strength, gradual embrittlement, can be stabilized	Embrittlement, moderate decomposition	None
Polyester Dacron (Du Pont) Fortrel (Celanese) Kodel (Eastman) Tergal (Rhodiaceta) Terital, Terylene (ICI) Textura (Rohm & Haas) Trevira (Hystron) Vyron (Beaunit)	Very gradual loss of strength, no discoloration	Little degradation	Low absorbency
Rayon—Viscose Avril (FMC) Coloray (Courtaulds) Enkrome (American Enka) Fibro (Courtaulds) Jetspon (American Enka)	Gradual loss of strength, affected more than cotton	Little degradation	High absorbency

TABLE 6-1 (cont.)

Generic Fiber Type/ Trade Name & Company	Effect of Sunlight	Effect of Heat	Effect of Moisture
Rayon—High Wet Modulous Zantrel (American Enka)	Gradual loss of strength, affected more than cotton	Little degradation	High absorbency
Rayon—Cupramonium Bemberg (Beaunit)	Gradual loss of strength, affected more than cotton	Little degradation	High absorbency
Vinyl Naugahyde	No loss of strength, slight discoloration	Gradual embrittlement	None

that absorb moisture tend to shrink or stretch when they dry out. Even a preshrunk cotton fabric guaranteed to shrink less than 3% may create serious problems in an 80-inch-long, floor-length drapery. The drapery extends all the way to the floor when installed, but a year later may be 2½ inches above the floor. The fabrics in Table 6-1 listed as having low absorbency shrink or stretch very little. Acrylic, polyester, glass, nylon, and modacrylic are the most durable and dimensionally stable fabrics available for thermal curtains and draperies.

Working with Existing Curtains

Some curtains can be modified into more thermally effective shields for your windows. With others it is pointless to try. Tier or cafe curtains and sheer curtains which have a thin, open weave are a lost cause in controlling window heat losses. They are best left as is with additional thermal drapery placed behind them.

Heavy draw draperies on traveling rods usually provide some thermal protection over windows, particularly if they are floor length and are made of a tightly woven fabric that is impervious to air flow. If the bottom edge of the drapery is in contact with the floor, a barrier is formed against convective air flow. Adding a weighted chain or a sand-filled pouch to the bottom of the drapery as shown in Figure 6-1 helps to insure an effective air seal.

Sealing the bottom of a drapery against air convection by dragging it along the floor works fine except when a floor register or baseboard radiator is located under a window. The heat from registers or radiators must convect directly into the living space. If this heat is allowed to enter the space behind the drapery, it will be quickly lost to the outside through the window. Metal and plastic divertors are sold in building

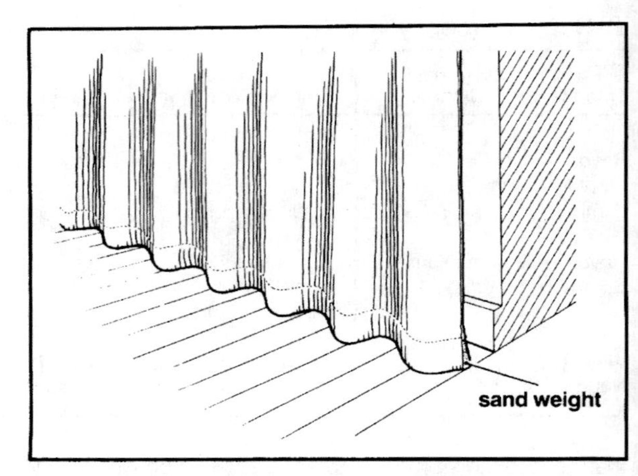

Figure 6-1: Weighted hem on floor-length drapery.

and home supply outlets to direct heat from floor grills away from windows and into the room. Figure 6-2 shows how a wooden divertor can be made to fit over either floor registers or baseboard heaters. This attractive divertor not only serves to direct heat into the room, but also acts as a hopper to collect the weighted bottom of the drape. It

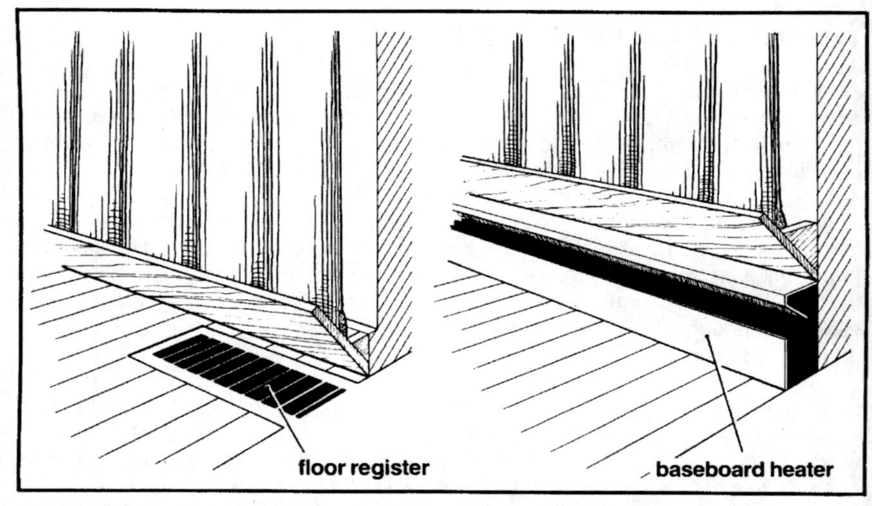

Figure 6-2: Homemade divertors for floor and baseboard heat sources.

also makes your home safer by meeting the requirements of many fire codes which state that draperies must clear electric heating units by several inches.

Adding a foil-faced, reflective liner fabric is often the simplest way to increase the thermal effectiveness of an existing curtain or drapery. The radiant heat transfer in the air space behind the drapery is decreased by an added layer of Foylon, Astrolon III, or Mylar on the back side of the curtain. Another air space is created since this lining material usually hangs freely, apart from the folds of the drapery. It can either be pinned to the drapery beneath each hook, or, if a hem is sewn in at the top it can be hung directly from the drapery hooks. The reflective liner should lie flat when closed whether or not the drapery has folds from header pleats (discussed below).

Bottom Seals

Because air, cooled by a window, tends to fall to the floor, the bottom seal on a curtain is the most important one. A versatile bottom seal for draperies is very difficult to design. No single method suits all window situations. A variety of bottom seal options are described below:

1. Permanent magnetic strip—This system provides an excellent seal that is easy to operate in most cases. A magnetic strip adheres to the window side of the liner hem and mates with a magnetic strip on the wall or on trim below the window. Care in installation is required so that the magnetic strips mate properly (see Figure 6-3a).

2. Elastic cords or metal spring rod (suggested by Clare Moorhead of Conservation Concepts, Ltd.)—Dress elastic or elastic cord can be used to make a bottom seal for narrow windows if the windowsill projects at least 1 inch from the wall. The elastic cord attaches to a cup hook (a 1-inch screw or nail can also be used), which is mounted just below the windowsill on each side of the window. The curtain is firmly sealed by pulling this cord out, tucking the bottom of the curtain under it, and letting it snap back in place (see Figure 6-3b). "Sash rods," which are long metal springs, can be used in the same manner as the elastic, although adjustments are sometimes required. They can be obtained in drapery hardware departments.

3. Bar with brackets (suggested by Bill Shurcliff)—Brackets can be installed on each side of the windowsill to receive a bar or rod. The brackets should be carefully located so that the bar presses the curtain snugly against the windowsill when it is lifted into these brackets. This bar can be a decorative curtain rod to match the curtain rods above. When the curtain is open this rod can hang from the brackets by a chain or rope (see Figure 6-3c).

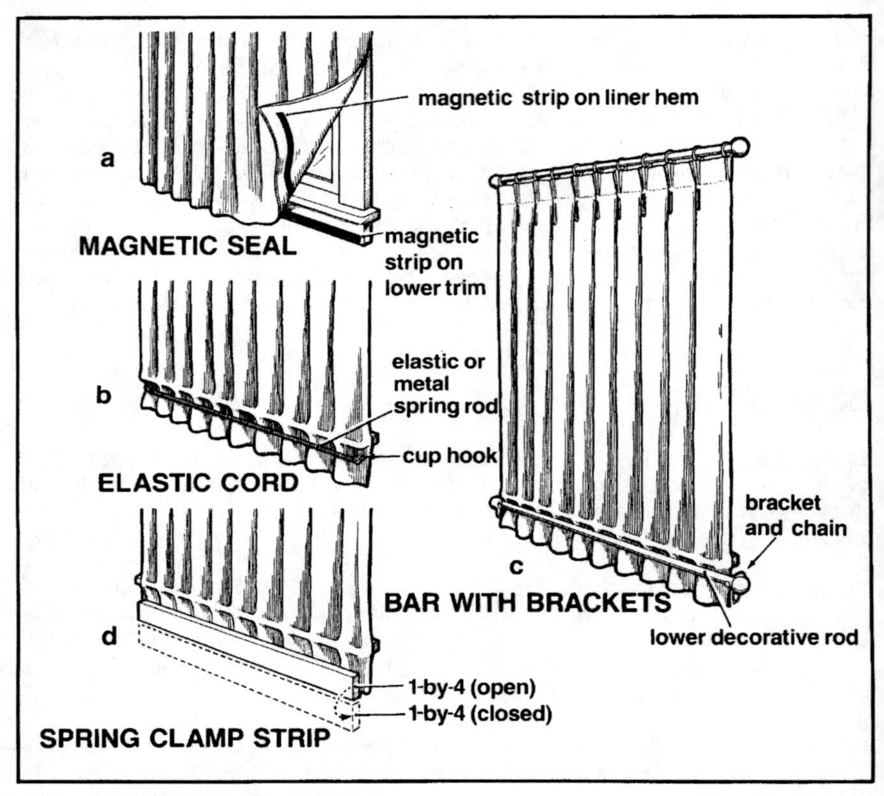

Figure 6-3: Bottom seal options.

4. Spring clamp strip—A 1-by-4 wooden strip can be mounted horizontally on the wall below the sill with spring-loaded (self-closing) hinges. The hinges for this wooden clamp strip should attach to the wall just below the bottom hem of the curtain. When the curtain is drawn, the wooden strip is opened to catch the curtain hem and then closed to clamp the curtain tightly to the wall with the spring action of the hinges (see Figure 6-3d).

5. Weighted hem—A drapery which extends to the floor can be weighted so that it forms a fairly good seal along the bottom as shown in Figure 6-1.

Most large draperies have folded header pleats along their top edge as shown in Figure 6-4. These pleats help the drape to fold regularly and also create graceful

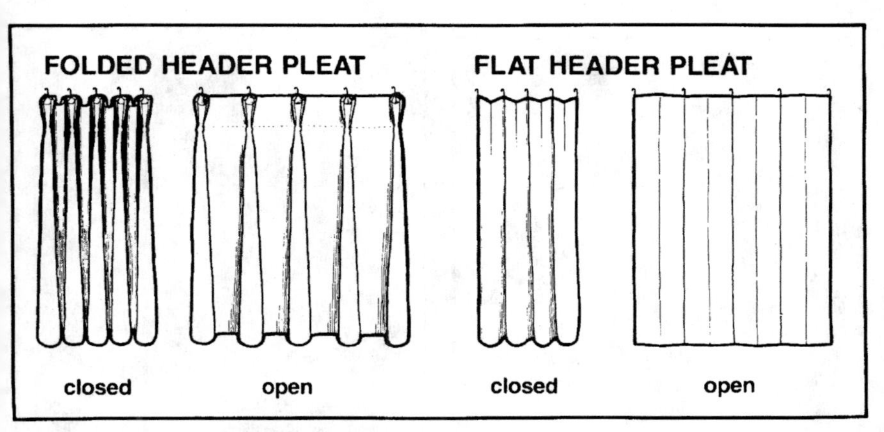

Figure 6-4: Flat vs. folded header pleats.

folds even when it is closed. With these folds, the bottom of the drape has an extra length of material and is difficult to seal. On drapes which drag along the floor to seal, this extra material is not a problem. A hopper like the one in Figure 6-2 can also handle folded draperies if care is taken to insure that the bottom edge is tucked into this slot. However, when a drape seals to a wall, windowsill, or any flat vertical surface behind it, its bottom must be flat.

Drapes or thermal liners can also be hung with flat header pleats (what some describe as "without any pleats") so that they draw into a flat position when closed. However, a drape with flat header pleats doesn't always fold regularly when opened.

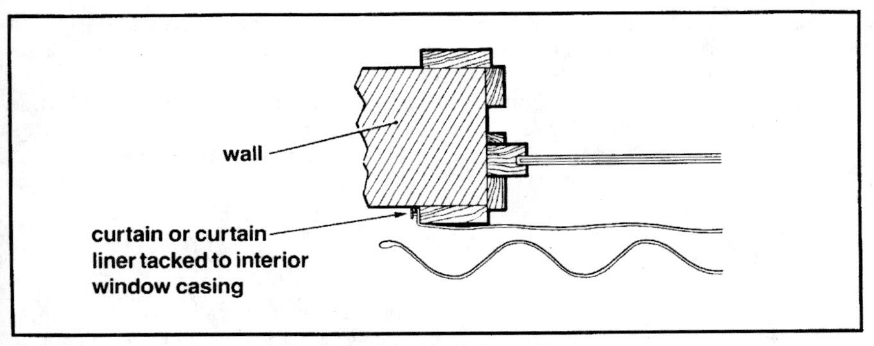

Figure 6-5: Tacking the sides of a curtain to the wall.

Photo 6-1: A clamp strip along the sides of a curtain.

Photo 6-2: A reflective drapery liner made from Foylon 7001.

To remedy this you can iron creases into the top edge of the drape to help it fold evenly as it is closed. Curtain hooks should then be placed at every other crease along the rear edge of the drape so that it can draw properly.

Side Seals

To reduce air drafts along the sides of a curtain, the fabric should be fastened to the wall or wooden trim along both sides of your window. The curtain or separate insulated liner can be fastened to the wall with small carpet tacks every 8 inches, as shown in Figure 6-5, or clamped with a wooden molding strip and small wood screws. If the curtain must be released periodically, a 1-by-2 clamp strip with spring-loaded cabinet hinges can be used. Photo 6-1 shows how this clamp strip can be applied to the sides of a curtain.

Center Seals

Air gaps are common where two curtains meet. Because there is nothing to apply pressure to behind this joint, a convenient center seal for a curtain is difficult to design. To make a curtain energy effective, it must be tightly drawn together here by some means. Snaps, zippers, and tie strings are sometimes used, but they are definitely for those who have patience since they require some time to manage. They are also limited by the arm reach of the operator.

Velcro is a marvelous fastener for knapsacks, purses, and a wide range of other uses. However, this hook and loop, cockleburlike material is not as successful on curtain seals as one might expect. It is quite expensive and often grabs two curtain halves in a crooked fashion. Like the tale of the tar baby, one portion of the curtain often gets stuck while you're freeing another portion from its grasp.

On a thermal curtain I installed in my home (see Photo 6-2), I attached strips of 3M's Plastiform magnetic tape along the center hem where the two curtain sections join. I cut the strips in 1½-inch lengths, punched small holes through them, and then stitched them to the curtain at about 8-inch intervals. The adhesive alone held the strips to the curtain for awhile, but due to the continual flexing of the curtain, the strips eventually fell off and needed to be stitched to the center hem.

This magnetic seal was quite easy to engage and disengage. However, due to the gaps between the magnets, the curtain allowed a small amount of air to seep through the center seal. At the suggestion of Clare Moorhead, who developed the Warm-In Drapery Liner, I tried running Plastiform tape along both hems for a continuous magnetic seal. Even though this flexible strip is a bit stiff, it worked quite well as a center seal. You can also form a center seal by sewing a steel washer every couple of inches along one seam and attaching the Plastiform tape to the other seam.

New, highly flexible magnetic strips are now under development for curtain seals by firms such as the 3M Company. The new strips are more flexible and workable on a curtain than the present 3M strip. Magnetic seals show much promise in reducing heat losses at the edges of curtains.

Clare Moorhead, Conservation Concepts, Ltd., Box 376, Stratton Mountain, VT 05155, has worked with magnetic edge seals in numerous configurations. You may wish to consult with her for up-to-date and effective methods of using them.

Top Seals

A top seal is strongly recommended for any high-performance thermal curtain. In high humidity spaces, a top seal helps to prevent condensation problems on the window glass during the winter. A top seal can also make the bottom and side seals more effective since cool air escaping around the lower edges of the curtain must enter from above. A top seal blocks its entry.

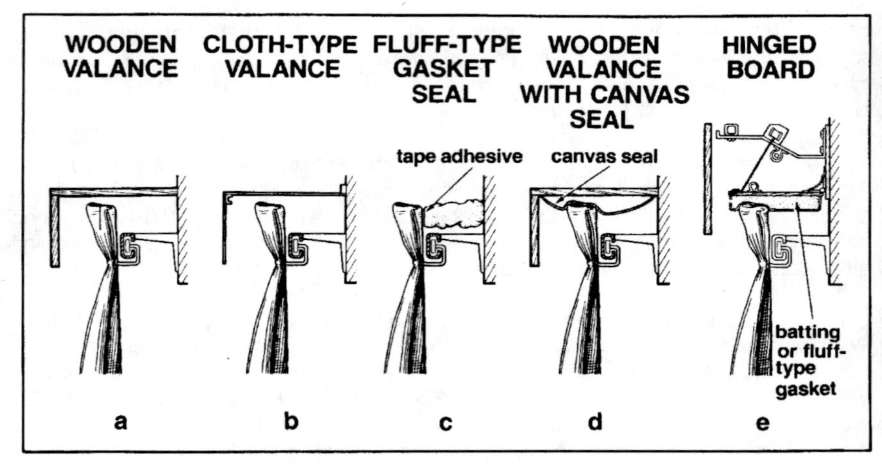

Figure 6-6: Top seal options.

A top seal also helps prevent window heat from entering a room on a hot sunny day when draperies are closed. When the sun shines on a closed drapery with a white liner, about 50% of its heat is reflected back to the outside. The other 50% is trapped between the window and the drapery and readily convects into the room if there is no top seal. Venting the space behind the curtain or drapery is also very helpful in reducing window solar-heat gains. A drapery with a top seal over a double-hung window that has its upper sash lowered for ventilation is an effective way to combat summer heat gains.

The most common top seal for window drapes is the wooden valance shown in Figure 6-6a. Although the valance blocks most of the convection at the top of a window, it has definite limitations. The top of the valance can be fitted more tightly to the curtain than the front where you must allow space for the curtain to fold when it is drawn open. Wooden valances should be mounted as close to the top of a curtain or drapery as possible without causing it to bind when in operation.

Cloth valances (see Figure 6-6b) similar to the wooden kind can also be used. They should be of a nonporous fabric to stop the movement of both air and moisture. Because cloth valances are flexible, they can often be mounted to fit more tightly to the top of the drapery, without binding, than can wooden ones.

A third option (see Figure 6-6c) for a top seal is under development by David E. Russell, Consulting Engineer, 110 Riverside Avenue, Jacksonville, FL 32202. This patented design employs a very low-density, closed-cell, foam fluff gasket that fastens to the top of the drapery with an adhesive tape to fill the space between it and the wall. This gasket, called a Gossamer, must be highly compressible to allow it to fold with the drapery when it is opened. Drapery tracks are usually 2 to 3 inches from the wall. It will probably take some time to develop this special product since the type of foam needed must compress properly when the drape is opened and completely fill the space between the drape and the wall when it is closed.

The fluff-type gasket would probably perform better on the top edge of a curtain or drapery under a valance. Any filler material can be used to seal the top of a curtain under the valance if it doesn't generate much friction. An excellent way to reduce air convection through this gap is shown in Figure 6-6d. A heavy strip of canvas is stapled to the underside of the valance to fill this space.

Figure 6-6e shows another design to fill this space, by Dale B. Gerdeman, 1403 Fifth Street, Las Vegas, NM 87701 (patent pending). This invention has a hinged board with polyester batting or a fluff-type gasket on its bottom edge so that it rests on top of the curtain. To allow the curtain to open, the board is raised by a pull cord that swings the board up to where it catches on a releasable latch. The latch is tripped by another pull cord when the curtain is closed, letting the board fall into place by gravity.

Reflective Fabric Curtains

Reflective fabrics are effective at resisting heat transfer when they face into an air space. One or several layers of reflective materials such as Foylon, Astrolon III, or Mylar make an efficient thermal curtain that has very little bulk.

Figure 6-7, as well as photos 6-3 and 6-4, shows a beautiful curtain installed in the home of Barbara Putnam, an architect and home builder, in Harrisville, New Hampshire. This curtain has a layer of Astrolon facing the outdoors and an attractive cloth fabric facing the room. The curtain is two stories tall and the top is attached to a curtain rod with sliding rings (no top seal on this curtain). It is permanently clamped to one side of the window with a wooden strip. The other side is fastened with a hinged clamp strip. The bottom of the curtain hem is weighted with a chain that slides along a trough constructed of 1-by-6s.

Up to now home energy experts have been mostly concerned with the loss of Btu.'s (British thermal units) from homes in the winter. In a humorous note, Barb describes a new way of examining these heat losses: "When it's closed, it gets a potbelly. It really needs the weight of the chain to keep it in the trough. Bob and I invented a new term for use in winter heat-flow situations. Instead of Btu.'s escaping, we speak of Itu.'s—Irish thermal units—coming in. It's Itu.'s which make the curtain belly out."

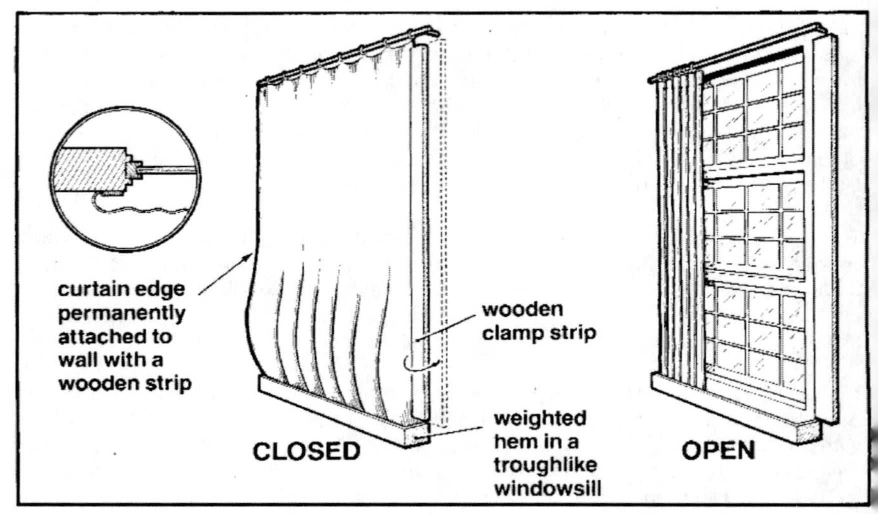

Figure 6-7: Barbara Putnam's curtain.

Photo 6-3: Barbara Putnam's three-tier insulated curtain.

Photo 6-4: Detail of the bottom of the Putnam curtain. The bottom edge of the curtain, weighted by a chain sewn into the hem, sits in a wooden trough.

A Plastic Bubble Curtain Liner

Warm-In Drapery Liner—Conservation Concepts, Ltd., Box 376, Stratton Mountain, VT 05155, offers a kit of partially assembled components for the construction of effective thermal curtain liners. They also make custom liners upon request. The Warm-In liners are made with a layer of bubble polyethylene—the kind used for packing material in shipping boxes—sandwiched between two layers of cotton material. This system creates several layers of dead or near-dead air space as shown in Figure 6-8b. The liners are light, flexible, translucent, and they can be machine washed in cold water on a gentle cycle without bleach. These liners are hung from the standard, traverse drapery rod either on an independent rod or behind draw draperies. When closed, the liners lie flat with a 2-inch overlap between panels.

Clare Moorhead, a graduate of Boston Architectural Center, has spent several years developing this system. Continuous magnetic strips in the leading edge of each liner mate when the liners are closed, creating an effective seal. Many subtleties in the use of magnetic curtain seals have been explored at Conservation Concepts, including the ability of various magnets to adhere through a layer or two of cloth, and the care magnetic seals require to maintain a long life. A curtain with a magnetically sealed Warm-In liner is shown in Photo 6-5.

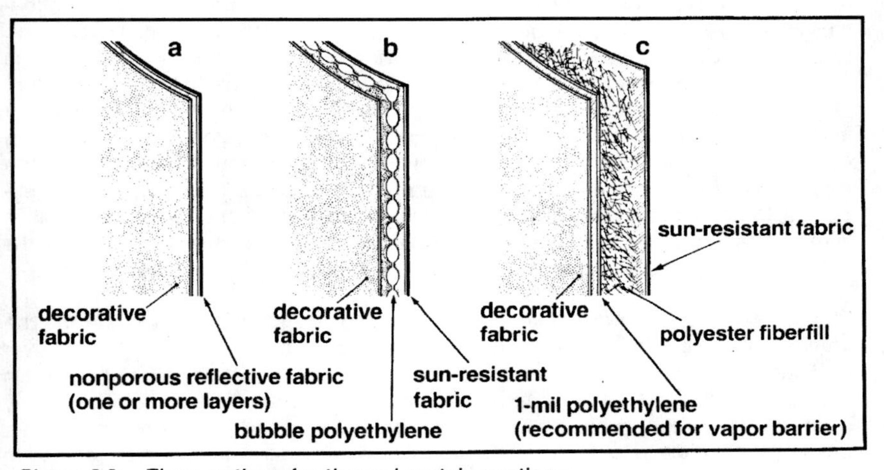

Figure 6-8: Three options for thermal curtain sections.

Photo 6-5: The Warm-In Drapery Liner.

Photo 6-6: Kathi Hannaford's PolarGuard curtain.

Fiberfill Curtains

Photo 6-6 shows a three-layer thermal curtain constructed by Kathi Hannaford of Albuquerque, New Mexico, for her home. This curtain combines a plain cotton-muslin window liner with a decorative cotton print facing into the room, and a 1-inch layer of PolarGuard sandwiched in between these two fabrics (see Figure 6-8c). The curtain drags along the floor to seal at the bottom but has no seal at the sides or top. The curtain provides added comfort for the dining room, protecting occupants from the large glass door's cold surfaces. During cold weather, some frosting and icing occurs on this door where the glass meets its aluminum frame; nevertheless, the door provides most of the daytime heat needed in this small, 1,000-square-foot residence. Perimeter seals and a 1-mil, polyethylene vapor barrier under the decorative layers of fabric would eliminate most of the moisture problem and also make the curtain more thermally effective.

Photo 6-7 shows a curtain which I constructed for my study three years ago. The vertical seams shown in the photograph were recently added. With a 1½-inch layer of PolarGuard between an inner cotton layer and outer polyester layer, this curtain was quite bulky and folded unevenly at the bottom. To get the curtain to draw evenly on a traverse drapery rod, the vertical seams were added at 4 inches on center, causing the curtain to fold like an accordion. However, this reduces the resistance to heat loss at the seams where the cloth is stitched together.

Photo 6-7: A PolarGuard curtain constructed by the author.

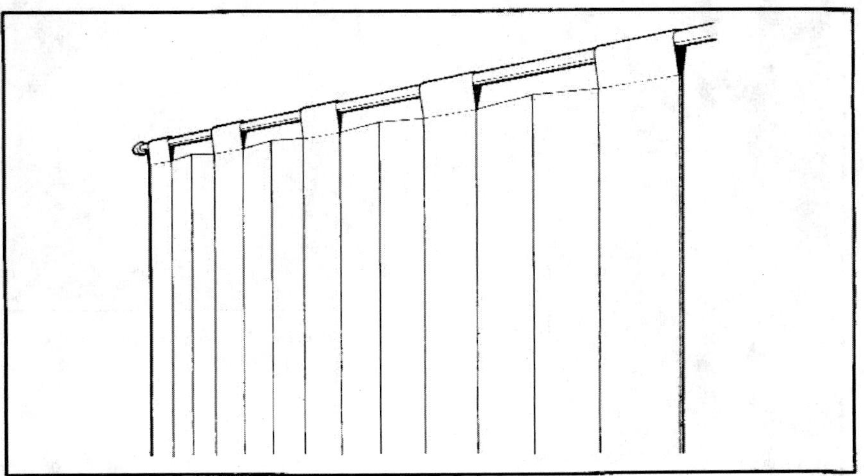

Figure 6-9: The Window Blanket.

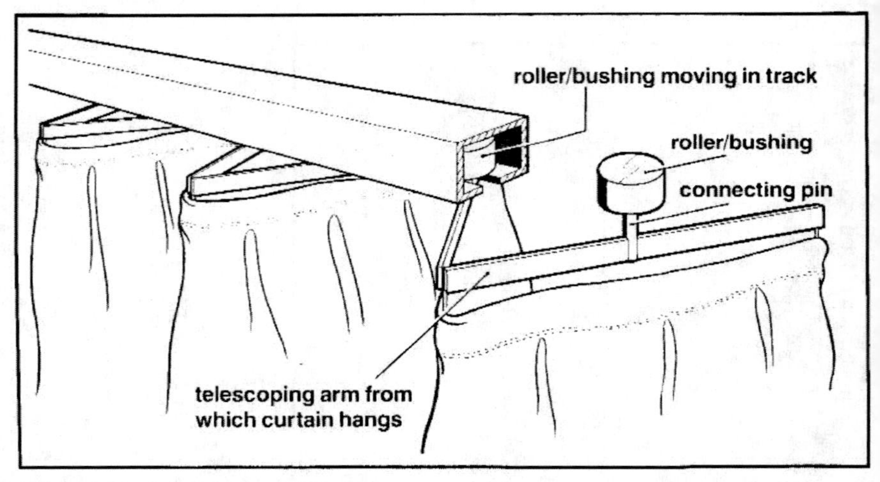

Figure 6-10: Modified curtain track.

Much of the problem with getting the curtain to move from a flat position to a folded one is caused by the 1½-inch thickness of PolarGuard. A ¾-inch layer of PolarGuard works better but has less resistance to heat flow. Two other fiberfills that can be used

↑ *Photo 6-9: Roman shade in David Wright's residence.*

←*Photo 6-8: Roman shade by The Center for Community Technology.*

are Hollofil by Du Pont and Thinsulate by 3M Company. See Appendix III, Section 7 for the address of the outlet for Hollofill.

> Window Blanket—Window Blanket Company, Inc., Route 1, Box 83, Lenoir City, TN 37771, makes a curtain with a room-facing, decorative cotton fabric, which is machine-quilted to an insulated lining. The fiberfill used is Dacron Hollofil 808, which has an R value of about 2. Figure 6-9 illustrates how this curtain attaches to a drapery rod. The panels are 45 inches wide and 84 inches long.

Many of the problems that arise from folding thick curtains could be solved by using a modified curtain track (see Figure 6-10). The amount of volume that even a 2-inch-thick curtain fills when folded is surprisingly small. This curtain rod has telescoping arms to fold the curtain as it opens and to pull it flat when the curtain is closed. With this track, large south-facing glass walls could be covered with a single drape two stories high, and if necessary, would draw neatly into 12-inch-wide folds at one end of a room.

Roman Shades

Thermal roman shades are constructed of the same materials as thermal draperies but fold vertically rather than horizontally. The shade folds at regular intervals from the bottom up as it is gently raised by means of pull cords. Photographs 6-8 and 6-9 show two roman shades—one designed by Nancy Korda and Susan Kummer of The Center for Community Technology, in Madison, Wisconsin, and the other by solar architect David Wright for his home in California.

The Korda-Kummer design (see Figure 6-11) is constructed of PolarGuard polyester fiberfill, lined on the exterior-facing side with a sun-resistant fabric and covered on the interior-facing side with a vapor barrier and decorative cloth. The vapor barrier can be either 1-mil polyethylene or 2- to 3-mil vinyl. Vinyl is more durable but 1-mil polyethylene is cheaper and easier to work with. A cotton or polyester preshrunk fabric makes a good covering for the side facing the room.

This design is sewn together along the edges and then quilted together at approximately 9 inches on center each way. Half-inch plastic rings are attached to the back of each quilting tie. Nylon twine, tied to the bottom ring, runs up the shade through a row of rings and then through a ½-inch screw eye at the top. The cords then run horizontally to a hook eye at the side of the shade and down to a cleat that holds the shade up when opened.

The bottom of the shade rests on the windowsill when closed so that it seals. The sides are clamped to the window side casing with spring-loaded hinges on 1-by-2

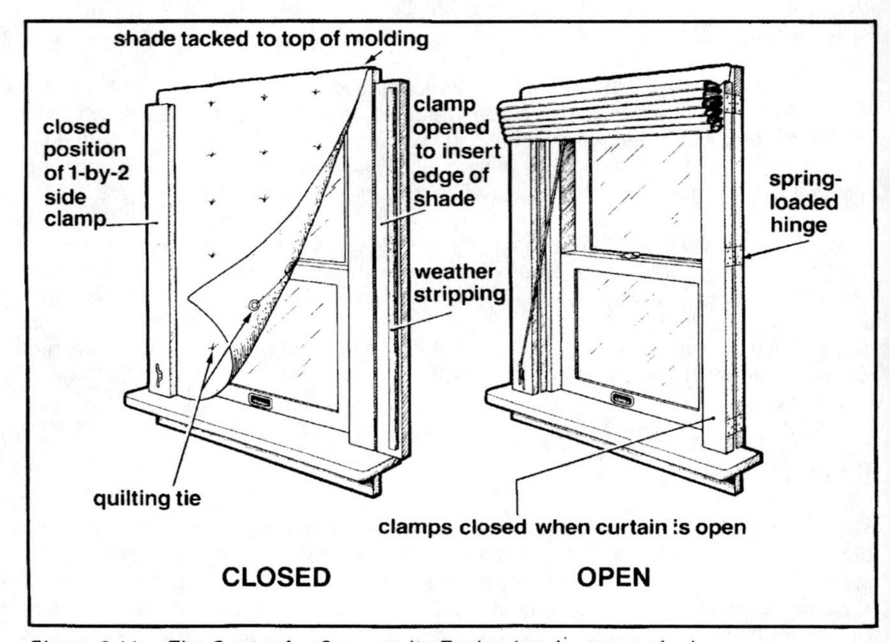

Figure 6-11: The Center for Community Technology's roman shade.

wooden strips. A step-by-step process on how to construct this shade design by The Center for Community Technology (CCT) is explained in Appendix I, Section 1.

David Wright's roman shade design (see Figure 6-12) is constructed of three layers of Foylon covered inside and out with a heavy cotton fabric. The shade has "sail tracks" along its edges to hold the edges of the shade as it draws up and down (there are no rings as in the CCT design). The bottom of the shade has a wooden batten strip to which the drawcords attach. A ⅛-inch, nylon cord loop attaches the batten strip behind the curtain along each side. This cord runs through screw eyes at the top and down one side where it attaches to a cleat when the shade is raised.

The windows in this home lean about 20 degrees from vertical. This causes the shade to hang down slightly on the 1-by-6 and helps to reduce air convection along the sides. The sloped windows here also increase the problem of excessive sun during the summer, but the shade has been designed to vent this solar gain. In the

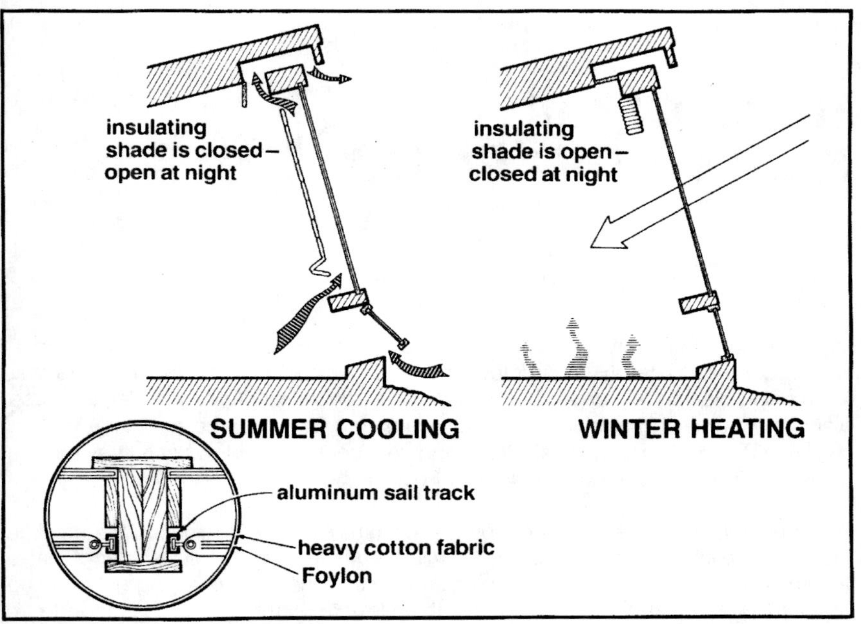

Figure 6-12: David Wright's roman shade.

winter the shade is lowered all the way to the floor to prevent air leakage at the bottom, and during summer the shade is lowered to just above the bottom vent window. This vent window is opened, along with a vent in the ceiling above the shade. The summer heat that is generated behind this shade rises up through the ceiling and out through a soffit vent, while cooler air is pulled in down below.

Chapter 7

Thermal Shades—Blankets That Roll

Most roll shades are very convenient to operate by simply pulling or releasing a cord. They require a space only a couple inches wide, unlike shutters which require that large swing clearances remain unobstructed by people, furniture, and plants. House plants are often hung or placed on shelves near windows to receive natural light as shown in Photo 7-1. Because shades hang very near the window sash, they can be drawn at night to protect plants against frost without disturbing the foliage.

When raised, window shades occupy a minimum of space, eliminating the daytime storage problems characteristic of other window insulations. The storage and spatial advantages of window shades, however, carry with them the challenge of creating an effective thermal barrier from only a thin, flexible membrane. If it is made to be wrapped onto a roller of reasonable dimensions, the shade cannot be too thick or bulky. Some shades expand when rolled down to provide a maximum thermal resistance, and deflate or compress when rolled up to minimize their bulk. Another challenge in shade design is to reduce air flow and heat losses around the edges.

The thermal shades in this chapter use many of the same flexible materials as the curtains in Chapter 6, but instead of folding the fabric, they are rolled up onto a roller. A somewhat arbitrary distinction is made in this book between curtains and shades, based on whether a fabric is folded out of or rolled up from the window opening.

Single-Layer Shades

The single-layer cloth or vinyl shade, common in most homes today, provides only about a 25% energy savings when placed 1 inch from single glazing, and only a 12% savings when 3½ inches from the single-glazed window. The heat loss reductions from common shades on double glazing are even lower than this. The resistance of

***Photo 7-1:** A roll shade can be pulled over a window without interfering with plants.*

the window shade can be as much as doubled by replacing the thin, white material of common shades with a foil-faced material.

Foylon, made by Duracote Corporation, is an excellent reflective material for improving window shades. Designed specifically for greenhouse shades, Foylon is soft and pliable, similar to the cloth fabric that may be on your existing shades. Foylon can be cut to the size of your existing shade, seamed with a wooden strip on the bottom, and used to replace the fabric of your shade as shown in Figure 7-1a, or added to your existing shade as in Figure 7-1b. In either case the Foylon material should be cut square on the end (an 18-by-24-inch carpenter's square will help here), and evenly taped or stapled to the shade roller. If Foylon is added to an existing shade, it should only be attached to the roller and should hang freely behind the shade material below. Two layers of shade material won't roll up at quite the same rate. If they are fastened together at the bottom they will wrinkle and cause it to jam.

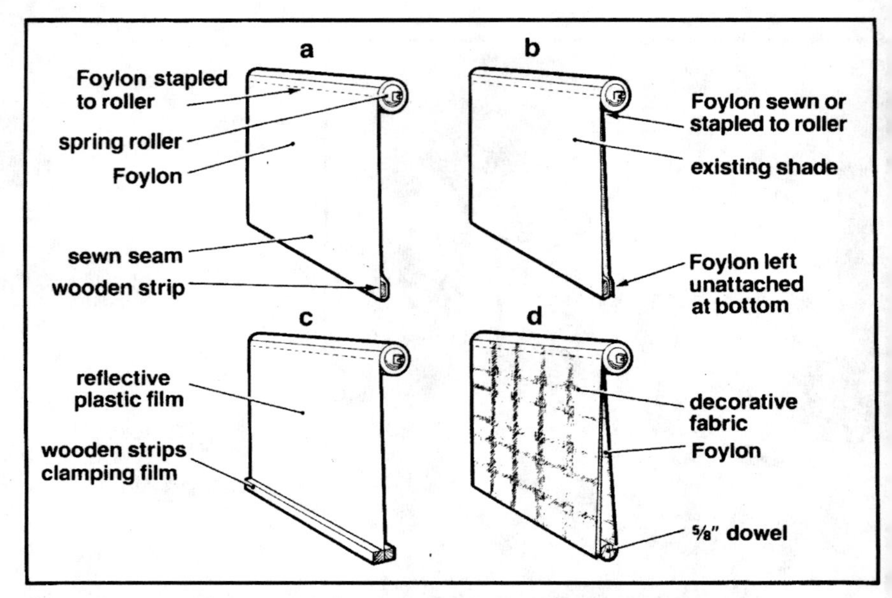

Figure 7-1: Modifications of the common window shade.

Another window shade material to cover windows which overheat during the summer is the reflective plastic film. One-way vision allows "optical clarity"—a tinted but undistorted view to the outside (see Tinted and Reflective Plastic Films in Chapter 3). Shades made from this film are transparent on the interior side but are mirrorlike from the exterior. They protect against the hot summer sun but can be raised to allow winter sunshine to enter. Although low in R value, these shades provide an R value superior to ordinary ones because their outer reflective surface reduces radiant heat transfer from the shade to the window glass.

NRG Shade—Sun Control Products, Inc., 431 Fourth Avenue SE, Rochester, MN 55901, makes a see-through reflective shade which includes a flexible plastic channel to seal the sides and a magnetic seal to close the bottom.

Plastic-View Shade—Plastic-View Transparent Shades, Inc., P.O. Box 25, Van Nuys, CA 91408, makes a high-quality, expensive, and reflective shade without edge seals.

See-through reflective shades which are lower in cost can be constructed at home as shown in Figure 7-1c. A piece of solar control film is simply attached to a shade roller, and a wooden strip is clamped to the bottom.

Multiple-Layer Shades

A series of thin layers of material insulates against heat flow due to the air spaces and air films formed by each layer. Highly reflective foil layers further reduce heat flow by stopping the radiant transfer between layers. Multiple-foil layers have been used for many years in the walls of commercial food lockers. More recently, thin, reflective layers of plastic film have been worn by astronauts for protection from extremes in temperatures in outer space. The modern space suit has 14 layers of thin, metallized fabric with nylon scrims between each layer. The moon walk would not have been possible without the lightweight, insulating capabilities of Mylar-type films.

Using multiple foil layers is an excellent way to provide high R values in window shades. Several layers of very thin, reflective material can be drawn onto a roller with

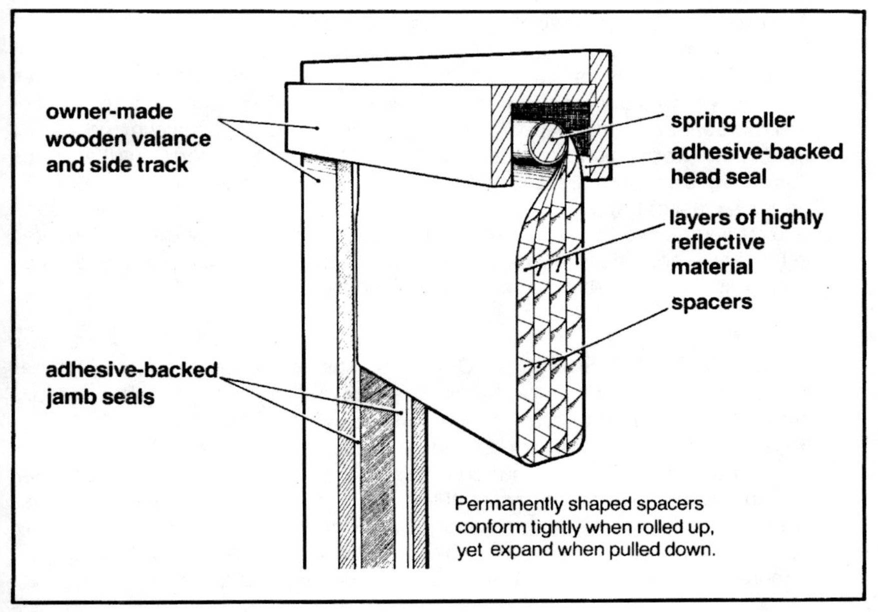

Figure 7-2: The High R Insulating Shade.

a minimum of bulk. A problem with foil layers is getting them to separate when the shade is down. If the material layers are tight against each other, heat quickly conducts through them as if they were one layer. If two layers of fabric are hung loosely together with some air between them, the shade is more thermally effective. However, for reflective layers of fabric to be most effective in a shade, they should be separated by a ½- to 1-inch air space. Several ingenious ways to separate these layers have been developed.

Figure 7-1d shows a design to modify your existing window shades with two additional layers of Foylon-type material. First, cut a layer of reflective material to twice the length of the shade. Then fold it in half and attach both ends of the material to the roller. Next, cut a ⅝-inch dowel rod to ¼ inch shorter than the width of the shade, and slide this dowel into the pocket in the bottom of the Foylon. The dowel holds the layers of reflective material apart, and its weight helps the shade to hang straight. As the shade is raised, the dowel rolls slightly, keeping the layers of shade even and free from wrinkles and creases. You can cover the shade on the interior side with a white or decorative fabric. This design works best when the shade is mounted into the window opening, which keeps the dowel from sliding out one end.

IS High R Shade—The Insulating Shade Company, Inc., P.O. Box 282, Branford, CT 06405, manufactures a shade with five layers of reflective plastic film (see Figure 7-2). The layers are separated with curved plastic strips which flatten when drawn onto a roller. These strips also help to reduce and localize convection currents between shade layers. The High R Shade is usually mounted in a homemade wooden frame. Details on the construction of the top (head) and side (jamb) wooden track are shown in Figure 7-2. If you prefer, a PVC (polyvinyl chloride plastic) track and frame are available from The Insulating Shade Company. An adhesive-backed plastic strip with a very soft, pliable edge is mounted in this frame so that it presses against the edges of the shade and forms an air seal. This five-layer shade itself has an R value of well over 10.

TTC Self-Inflating Curtain Wall—Ron Shore of The Thermal Technology Corporation, P.O. Box 130, Snowmass, CO 81654, has developed a shade which has four layers of highly reflective fabric (see Figure 7-3). When this shade is lowered, the air temperature differences on the two opposite sides of the shade generate air currents inside the shade that cause it to "inflate" and yield an R value of about 10. The TTC shade is raised and lowered by an electric motor. Due to the expense of the motor, this system is practical only for large glass areas. This shade was originally designed for the narrow and inaccessible space between south-facing glass and a masonry heat-storage or Trombe wall. A more complete description of this shade follows in Chapter 17.

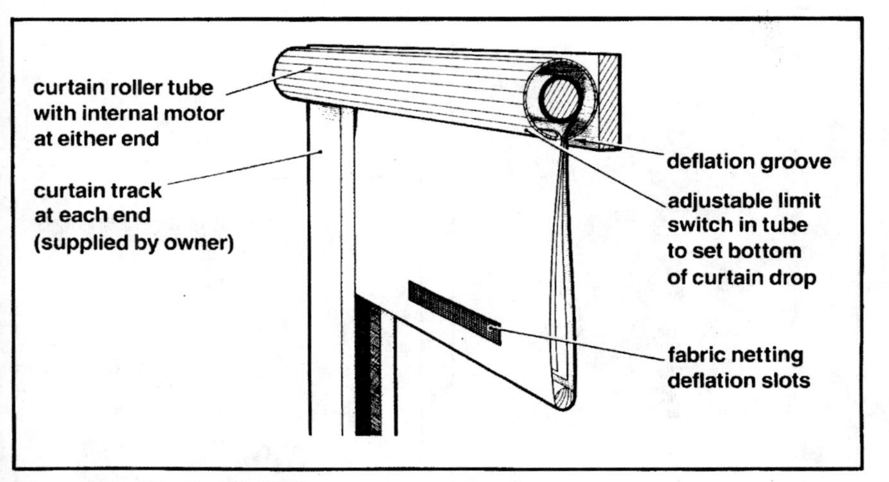

Figure 7-3: The Curtain Wall.

Fiberfill Shades

Goosedown is one of nature's best insulators. Its use in clothing and sleeping bags has enabled high-altitude campers to survive extremely cold temperatures. The tiny fibers in goosedown are a supreme insulating material which is very lightweight and highly compactible but also very costly. The high cost and limited availability has spurred the synthetic fiber industry to try to match its qualities with several new polyester fiber materials. PolarGuard by Celanese, Hollofil by Du Pont, and Thinsulate by 3M Company are among the best. The fiberfill used in window shades is usually only about ½ inch thick but thicker batts can be used for a higher insulating value. A. B. LaVigne of Seattle, Washington, has constructed shades from 1½-inch-thick fiberfill without experiencing any mechanical difficulties.

Photos 7-2 and 7-3 show homemade fiberfill shades—one made in Seattle by LaVigne, and the other made in Sandstone, West Virginia, by Peggy Rossi. Both shades were produced as a result of community-oriented self-help programs.

The shades in the first photograph were part of a project in Seattle aimed at introducing local residents to a variety of window coverings. Details on the construction of

Photo 7-2: Homemade quilted shade.

Photo 7-3: Patchwork quilt shade.

window insulation were presented at slide shows and tours were conducted through an energy-retrofit demonstration house. The project was performed by the Ravenna-Bryant/Wallingford Community Councils (LaVigne, principal) and was funded by the Seattle Light Department through the Neighborhood Conservation Demonstration Program.

The hand-stitched patchwork quilt in Photo 7-3 was made as part of a summer program at the Sandstone Senior Citizens Center in Sandstone, West Virginia. Figure 7-4 shows how the Sandstone patchwork quilt shade is constructed. Instead of using a spring-loaded roller, these shades are raised by pulling down on a cord which is wrapped around a free-turning ¾-inch dowel rod. A weight on the bottom edge of the shade drops the shade down while it rewinds the cord.

The quilted portion of this shade contains a layer of Astrolon, a layer of polyester fiberfill on each side of it, and a cotton-polyester-blend fabric facing the room and window—five layers in all. The Astrolon layer in the middle does not face an air space

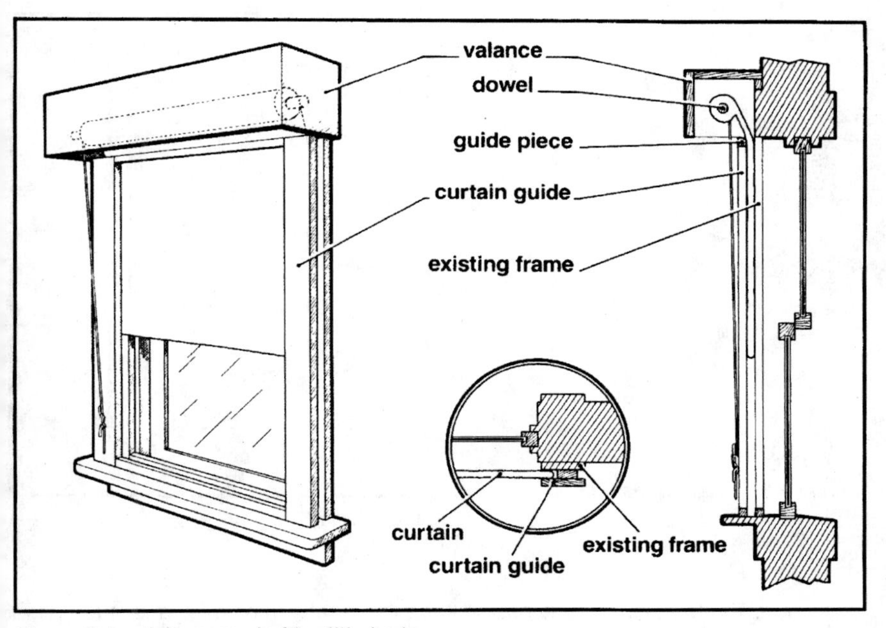

Figure 7-4: A homemade fiberfill shade.

and therefore does not enhance the shade's R value very much. However, it does serve as a vapor barrier. Complete, step-by-step instructions on how to construct this shade can be found in Appendix I, Section 2.

ATC Window Quilt—A fiberfill shade called the Window Quilt (see Figure 7-5) is made by the Appropriate Technology Corporation, P.O. Box 975, Brattleboro, VT 05301. This shade contains a ¼-inch layer of fiberfill on each side of a 4-mil layer of reflective plastic which acts as a vapor barrier, and two outer layers of cloth. All five layers are heat-welded together into a quilted pattern. The shade itself provides an R value of about 3.5.

The operation of this shade is very clever. The sides of the shade are bonded around a ¼-inch cord which runs down side tracks, making a very effective seal. Instead of a spring-loaded roller at the top, this shade has a rope and pulley which raises the shade (see Figure 7-6). The lower edge of the shade contains lead weights and a compression-foam strip to provide a good seal on the bottom

Figure 7-5: ATC Window Quilt.

edge. The top of the shade has an extra roller behind the main one to keep the shade firmly against the window head casing. A small jamb roller has diagonal notches to catch the cord and hold the shade at any desired position. This shade is a pleasure to operate.

Edge Seals and Tracks

The prevention of room air flow behind thermal shades is a critical factor in their effectiveness. Locating the shade so that it hangs very close to the window helps to minimize this flow. A simple way to reduce air convection on many common window shades is to reverse the roller as shown in Figure 7-7 so that when the shade is drawn the fabric is closer to the window glass.

A more effective way to stop edge drafts is with wooden clamp strips, constructed from 1-by-2s. These clamp strips are mounted with spring-loaded cabinet hinges. Be sure to use the kind that stay open on their own by way of a cam action in the hinge pins. Clamping the shade is much simpler if the clamp strips remain in an open position while the shade is raised or lowered (see Figure 7-8a).

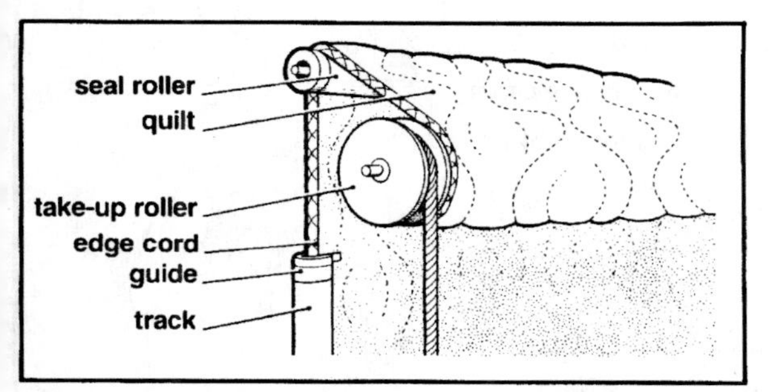

Figure 7-6: Roller detail of the ATC Window Quilt.

A third way to stop edge drafts is with a magnetic seal along the edges of the shade. A simple but crude way to magnetically seal these edges is shown in Figure 7-8b. Plastiform-type magnetic strips are attached vertically to the wood facing on the sides of the window, and a steel shelf bracket is placed so that it magnetically clamps over those strips to the edges of the shade. Plastiform magnetic tape can also be attached to the edges of the fabric on large greenhouse shades as described in Chapter 15. Magnetic strips which are lighter, thinner, and more flexible than Plastiform will soon be available for common window shades. Velcro also seals edges well.

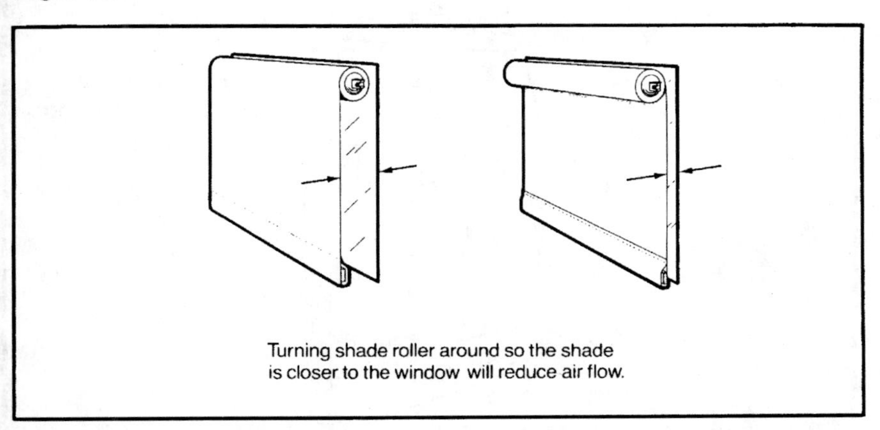

Figure 7-7: Reducing room air flow behind shades by decreasing space between shade and window.

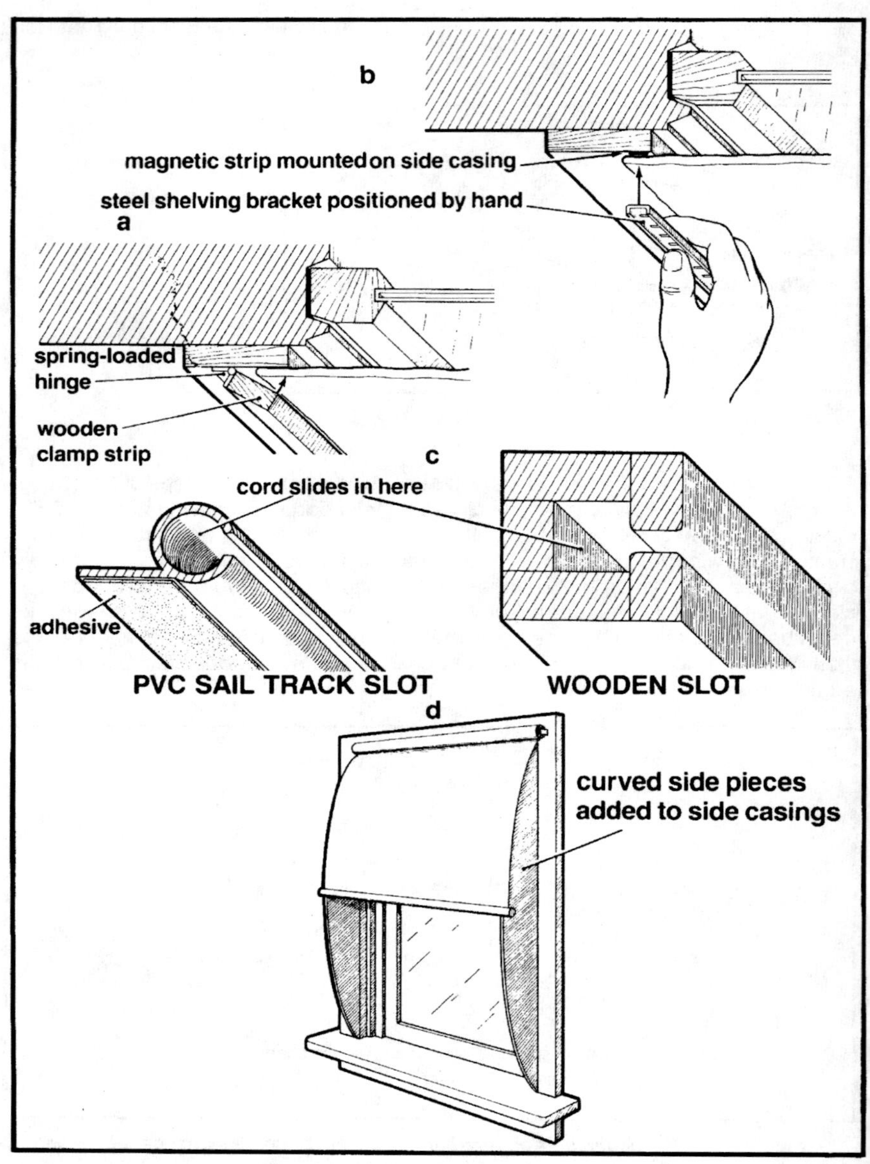

Figure 7-8: Options for edge seals for roll shades.

Sail-type tracks, which accommodate a continuous cord along the edge of the shade fabric, provide the best edge seals. The Appropriate Technology Corporation manufactures a beautiful, PVC extrusion for this purpose. Crude sail-type tracks can be approximated out of wood as shown in Figure 7-8c. However, wood is rougher than PVC, and care should be taken to make it as smooth as possible so it won't fray the fabric at the edge of the shade. Care should also be taken to make the wooden slots small enough to operate with a small cord. If the rope or cord drawn through this slot is larger than ¼ inch in diameter, it tends to bind and creates problems at the top of the shade when it is wound onto a roller.

One other approach is to stretch the fabric over curved pieces of wood cut from plywood or other rigid material on the sides of the window as shown in Figure 7-8d. This type of seal works best with shades having a floating roller on the bottom that moves up and down with the shade.

Finally, a soft-edge spring strip can be used to seal the edges of a shade. Soft-plastic spring strips are used for both the IS High R shade and the NRG shade.

Roller Mechanisms

The most common shade roller device is a spring-loaded roller with a pawl catch released by centrifugal force when the shade turns rapidly (see Figure 7-9). This device works quite well on lightweight, single-layer vinyl window shades but can cause a lot of problems on heavier shade designs. Many thermal shades must roll up slowly to deflate or compress properly on the roller. If these shades use a coil spring roller with a pawl-type catch, this pawl often grabs when the shade is partway up because the shade cannot be rolled up quickly enough, requiring it to be jerked down once again to disengage the pawl. To counterbalance the weight of a heavy shade, the roller spring must be set with considerable tension, making the quick downward motion to release the pawl more difficult. For this reason, the IS High R

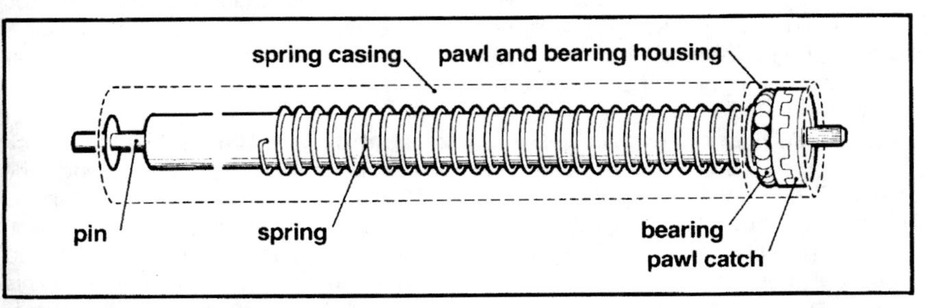

Figure 7-9: A spring-loaded shade roller mechanism.

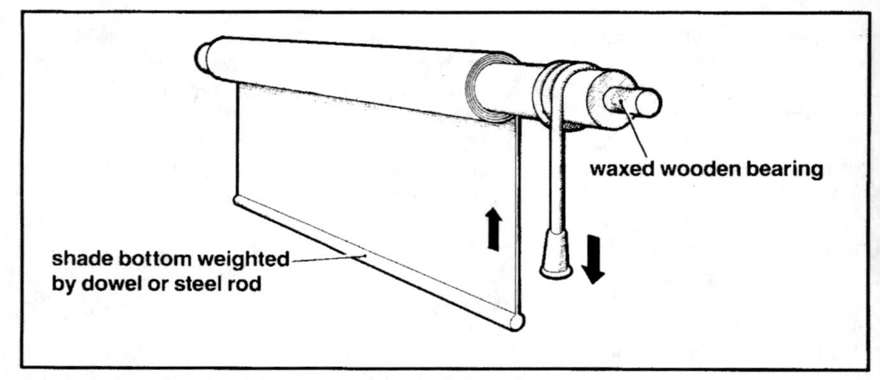

Figure 7-10: Shade with a top roller and pull cord.

shade design, originally a heavy, spring-loaded roller, has been changed to a chain-and-sprocket mechanism similar to the Mecho Shade System by Joel Berman Associates, Inc. (description follows).

The simplest type of roller mechanism for a home-built shade is a free-turning roller bar. The shade material wraps around in one direction and a pull cord winds around in the other direction as shown in Figure 7-10. This type of shade must have enough weight on the bottom edge so that it will lower itself by gravity. The shade is raised by pulling on the cord, and it is held up by securing the cord to a small cleat or hook. Bearings of some sort are required at each end of the roller bar. They can be very simple. A wooden dowel rod, waxed on each end and mounted into holes bored into wooden end plates, works quite well as a roller.

A wooden dowel is usually enough weight to cause a thin shade to fall, but heavier thermal shades often require weight added to their bottom seam. A steel or lead rod can be slid into a pocket here to provide this weight. A pouch of sand or buckshot can also be sewn into the bottom edge of the shade, not only helping the shade to fall but providing a good air seal at the bottom as well.

Shades can also be fixed to the top of a window and operated from a roller bar, which moves up and down with the bottom of the shade similar to a venetian blind (see Figure 7-11). A wooden dowel or steel bar is attached to the bottom edge of the shade. Pull cords start near the top of the shade, run down around this bar and back to small pulleys or screw eyes at the top of the window, across the head casing and back down one side. As the cord is pulled up, the bar and shade fabric roll up. This design sometimes works better in theory than in practice, though. The cords some-

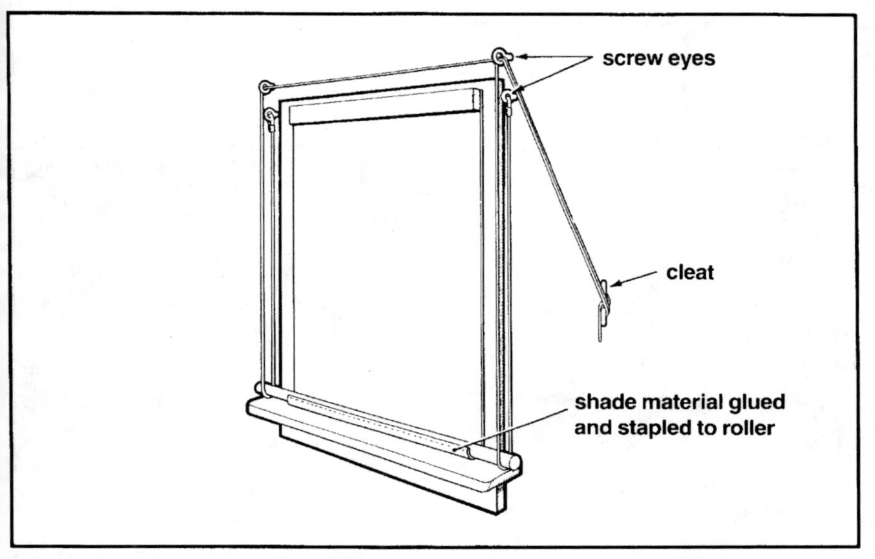

Figure 7-11: ***Shade with a floating roller at the bottom.***

times tend to slip on the bar, causing the shade to raise up faster than the roller turns. Such a shade works best with a rough or rubbery cord and a heavy bar to prevent slipping.

Sometimes the bar is eliminated altogether from this type of shade and it is simply rolled up by hand to a Velcro tab which holds the shade up at the top. Photos 7-4a and 7-4b show two shades that are raised in this manner.

Mecho Shade System—Joel Berman Associates, Inc., 102 Prince Street, New York, NY 10012, markets a very handy roller mechanism for window shades which allows you to adjust or change shade fabrics seasonally. The bar roller has a snap-in flexible plastic spline shown in Photo 7-5, allowing you to change layers of fabric without ever removing the bar! A chain-driven sprocket at one end of the shade turns the roller bar and raises or lowers the shade. Photo 7-6 shows this shade roller mechanism.

Shades made from solar control films or woven fabrics are available to fit this roller mechanism. A winter fabric with a high R value can be easily removed in the springtime and replaced with a reflective summer fabric. The Electro Shade, a more expensive line of motorized roller systems, is also available from Joel Berman Associates, Inc.

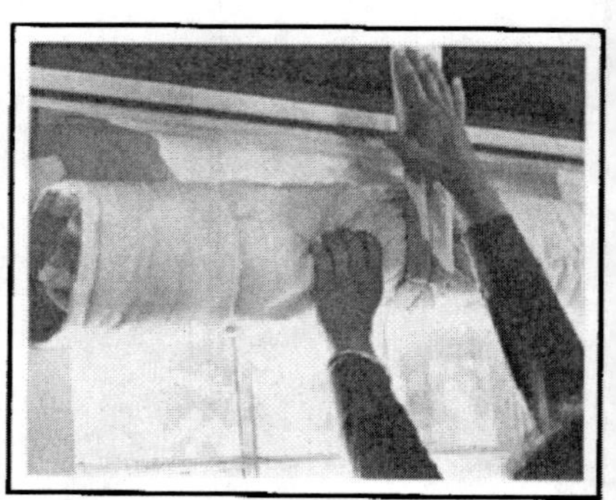

Photos 7-4a and 7-4b: Two shades that are hand rolled from the bottom. One is secured at the top with simple cords; the other with a tab that adheres to the casing.

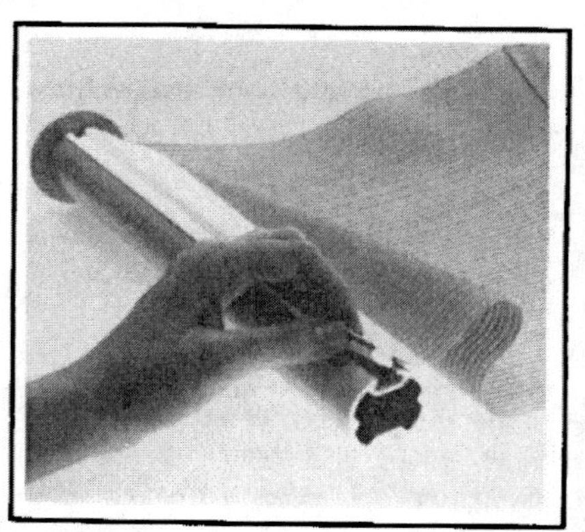

Photo 7-5: Detail of the spline on the Mecho Shade.

Photo 7-6: The shade roller mechanism on the Mecho Shade.

Dual-Shade and Triple-Shade Combinations

The shade options presented so far have been for various types of fabric combinations that work together on one roller. However, different seasons require that differ-

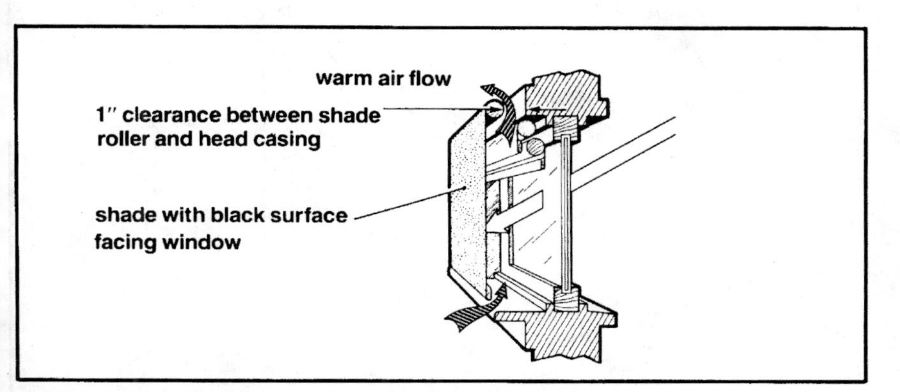

Figure 7-12: A combination of standard window shades.

ent types of material be used over the window. During the summer when the sunshine should be reflected to the outside, a shade with an outside aluminized surface works best. However, during the winter in spaces where too much glare from direct sunlight is a problem, a shade with a black outside surface can be used to absorb the incoming sunlight and convect heat into the room. This window-shade heat collector is about as effective at bringing in solar heat as is direct sunlight striking nonmasonry materials such as furniture and carpeting. However, it is not as effective for storing heat as is direct-gain sunlight striking masonry or water. The "convector" shade should be placed so there is a 1-inch clearance at the top of the window, allowing the heat to convect into the room.

Three independent, standard window shades can be mounted to most window openings as shown in Figure 7-12. These can be any combination of reflective, clear, sunlight-absorbing, or insulating materials. The way this combination of window shades can work together to meet a wide range of window energy needs during changing seasons and times of day is illustrated by the versatility of the Insealshaid.

> Insealshaid—This is a very clever shade unit with three separate shades for windows and glass patio doors. It is made by Arc-Tic-Seal Systems, Inc., P.O. Box 428, Butler, WI 53007. The unit comes complete with a reflective film shade on the window side, a clear film shade in the middle, and a black, heat-absorbing film shade on the room side. The four modes of seasonal day-night operation are shown in Figure 7-13. The shade housing has vents into the room that open automatically when bimetallic strips respond to temperatures behind the black shade that reach 82°F. or more.

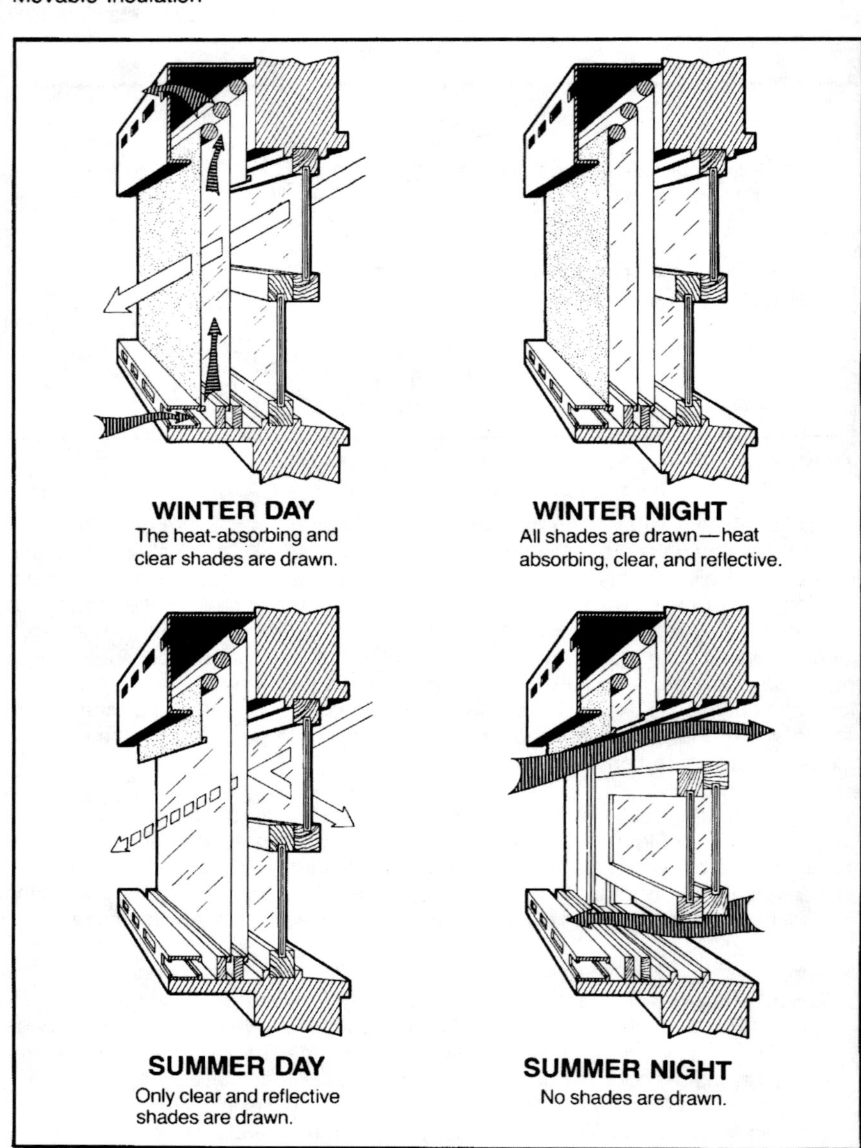

Figure 7-13: Modes of operation of the Insealshaid.

Slatted Shades

Another type of thermal window shade is composed of rigid slats which move up and down side tracks and onto an overhead roller. This type of shade is difficult to make at home, but there are commercial, interior slatted shades worth mentioning.

Thermo-Shade—Solar Energy Components, Inc., 212 Welsh Pool Road, Lionville, PA 19353, sells a white, PVC slatted-shade system which operates very smoothly and offers a clean appearance to modern interiors. The slats interlock with precision, allowing very little air flow between them. An exploded view of this system is shown in Figure 7-14. A valance can be added at the top. The Thermo-Shade can be hand-operated by a knob at the bottom, for small, easy-to-reach windows; a pulley sash cord is better for overhead windows. A 12-volt, motor drive system is also available. The Thermo-Shade can be used on a window of any angle or slope.

The Brattleboro Design Group Shade—Alan Ross of the Brattleboro Design Group, P.O. Box 235, Brattleboro, VT 05301, has designed a unique system of

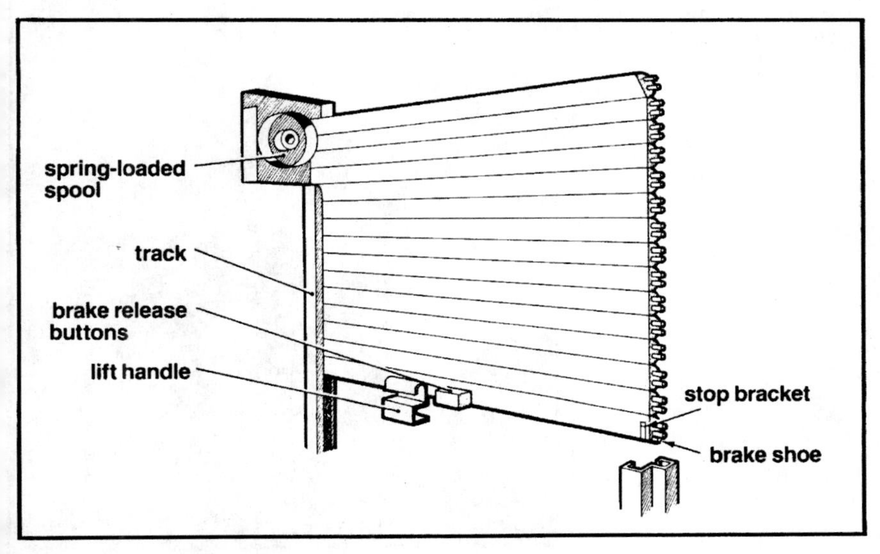

Figure 7-14: The Thermo-Shade.

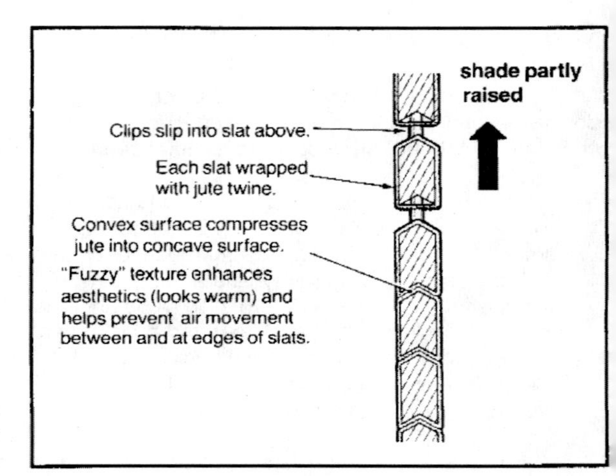

***Figure 7-15:** Thermal insulating blinds by Brattleboro Design Group.*

***Photo 7-7:** The Fusaro residence complete with thermal insulating blinds.*

thermal insulating blinds, which is now under development. This slatted, roll shade system has V-shaped slats that sit lightly on top of each other when the shade is closed (see Figure 7-15). These slats are joined by clips that allow them to separate as they are raised from above. The effect is very dramatic, as shown in Photo 7-7, of the Fusaro residence in Snowmass, Colorado.

Chapter 8

Thermal Shutters and Folding Screens

Chapter 5 discussed the various materials and methods used in the construction of pop-in shutters. The shutters in this chapter are similar to pop-in shutters except that they hinge or slide to "parking" locations on the edges of windows, rather than being removed from the windows completely. This chapter presents a wide variety of designs for a number of window and interior situations. Many require very exact measurements of your window openings, so follow the measuring instructions in Chapter 5 very carefully.

Both hinged and sliding shutters must open to a daytime position that does not block people's movements, other windows, or useful daylight, or get in the way of furniture, plants, and other room objects. If you made a cardboard pattern of your window opening, you may want to save it and make a duplicate to experiment with and help you visualize the needed clearances for each type of sliding and hinged shutter you are considering. By placing this cardboard panel in your window and referring to Figure 8-1, you can see the following options:

> The whole panel can be slid sideways if there is adequate space beside the window (8-1a). The panel can be hinged at the top to swing up to the ceiling (8-1b). The entire panel can also hinge on one side to swing to an adjacent wall (8-1c), but the arc of this swing is usually too large. Fold the cardboard panel vertically and try again (8-1d). This simple bifold panel works very well if there is a wall adjacent to the window. Next cut the panel in two vertically, along the line of the previous fold. The panels can slide to both sides of the window (8-1e) or they can be hinged to each side (8-1f). If the hinges are mounted properly, each panel will swing a full 180 degrees to the wall. Finally, if you fold each section again, the panels can bifold to each side (8-1g). This works well in deeply recessed windows, such as those set into old stone or brick walls.

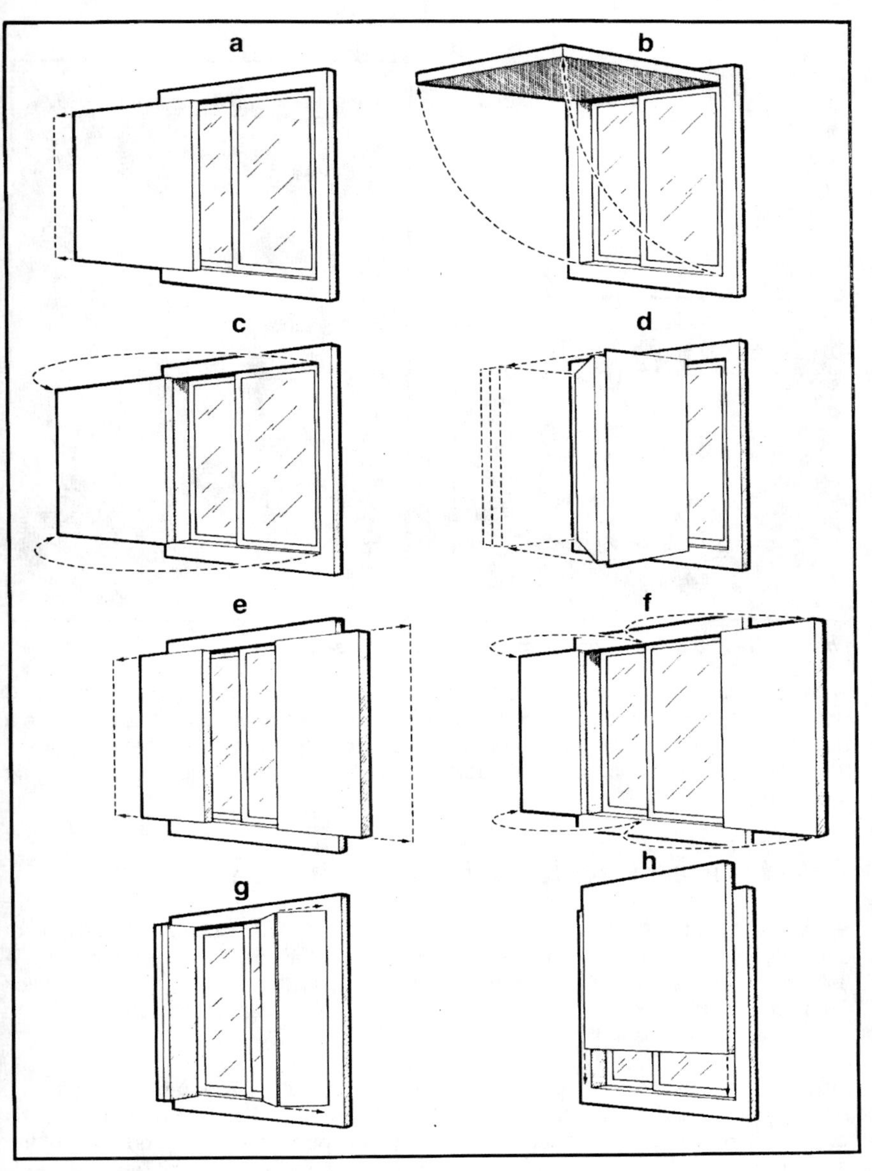

Figure 8-1: Exploring shutter options.

Photos 8-1a and 8-1b: Two examples of bifolding shutters.

These are the basic options you have with interior shutters. A few more can be added but are rarely used. For small windows you can use a panel that slides vertically like a guillotine (8-1h). Also, bifolds can be trifolds or quadrifolds if you want to experiment with more hinges; but as you add more hinges, the problem of sagging increases.

Side-Hinged Interior Window Shutters

Photos 8-1a and 8-1b show two bifold hinged shutters. The thickness of this type of shutter can vary from 1 to 4 inches according to the thickness of the frame used. A 1¾-inch-thick shutter panel is the most common, constructed with a 1½-inch-thick frame and ⅛-inch facings. If you plan to build a thicker shutter, be sure there are adequate clearances in the window opening for it.

Figure 8-2 shows the assembly of a typical hinged shutter. The frame is cut from a nominal 1-by-2 (¾ by 1½ inches, actual dimensions) butted, nailed, and glued at the corners. Sometimes a 2-by-2 strip is substituted on the side to be hinged to make the shutter stronger here and to allow for the attachment of larger hinges, but a ¾-

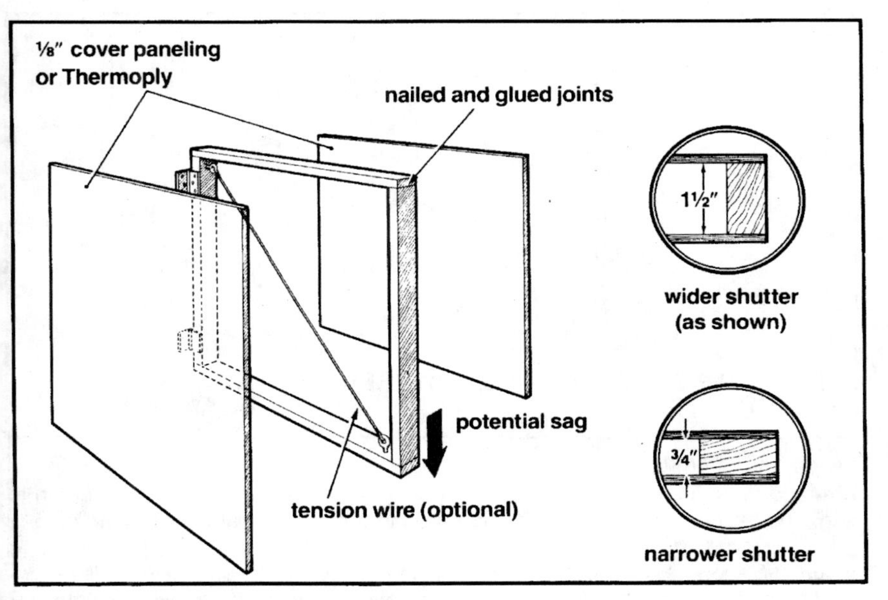

Figure 8-2: Shutter panel construction.

inch-wide frame is adequate to hold small cabinet hinges. Thinner shutters are sometimes made by turning frame members flat to make a 1-inch-thick shutter.

The most popular materials for the facings are ⅛-inch mahogany "luan" door paneling and ⅛-inch Thermoply board. Luan paneling makes a very attractive facing for shutter panels. Thermoply is a lower-cost, dense cardboard material, which is aluminized on one side and white on the other. The white side is very attractive when faced with a fabric that has an open weave. Thermoply is available at most large building supply outlets.

Rough-sawn plywood can also be used to face window shutters. Other facings include pressed hardboards such as Masonite, and ordinary fiberboard sheathing, which accepts thumbtacks. Whatever type of facing is used, it should have good tear resistance and shear strength, since those facings alone keep the shutter from sagging. Sagging problems are most common on wide shutters or on bifold shutters where weight on the free-hanging edge has the most leverage. If the shutter facing is likely to weaken and stretch, you may want to install a tension wire diagonally across the frame under the facing. If this wire is tight, it will minimize sagging, similar to the

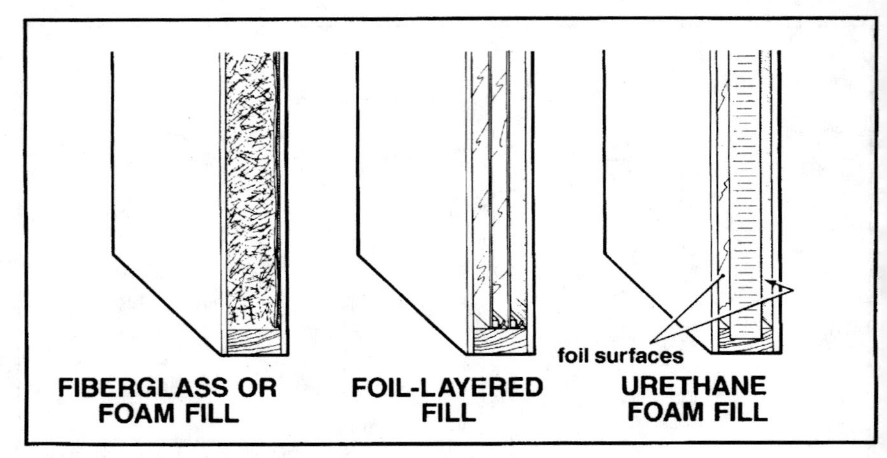

Figure 8-3: Shutter insulation — three options.

wire-and-turnbuckle rig often added diagonally to fence gates when they begin to sag. The facing should be nailed and glued to the shutter frame with white glue.

Figure 8-3 shows several options for insulating the core of these shutters. Use whatever type of insulation happens to be your favorite. Foam plastics have the highest R value but are somewhat of a fire hazard. Fiberglass insulation, which is fireproof, works very well. Another option is to secure a layer or two of aluminum foil (the heavy-duty kind) in the center of this panel, creating several foil-faced air spaces. If the facing has an aluminized surface, like Thermoply, you can face it into an air space in this core. A good rule to keep in mind is that one foil surface facing a nonfoil surface across an air space is practically as effective as two foil surfaces facing each other in the air space. The Insul Shutter (Figure 8-5) has a urethane foam core with a foil surface on each side of it, facing an air space. The combination of these insulating techniques yields a high R value.

The shutter shown in Photo 8-2 that was designed by The Center for Community Technology is translucent, allowing some sunlight to enter a room even when closed. The frame for this shutter is constructed with 1-by-2s joined at the corners with small, steel angle braces. PolarGuard is placed in this frame, and a translucent fabric is attached to each side. The minimum weight of the shutter, together with the steel angle braces, help it to resist sagging.

Hinged shutter panels should be cut to the same size whether they are to be hinged separately on each side on the window opening or joined in the center with hinges to

Photo 8-2: A translucent shutter filled with polyester fiberfill and covered with cloth.

make a bifold shutter. To measure these panels, your cardboard pattern will be helpful. The height of each shutter should be the same as the window opening minus 3/16 inch, top and bottom (a total of 3/8 inch). The width of each panel should allow 3/16 inch for weather strippings at the sides and 1/8 inch for the centerfold (1/2 inch total). (Complete step-by-step instructions on how to make this kind of shutter are given in Appendix I, Section 3.)

Foam weather strippings are ideal for the sides and centerfolds of these shutters. Schlegel Corporation, P.O. Box 23113, Rochester, NY 14692, makes a felt, brush-

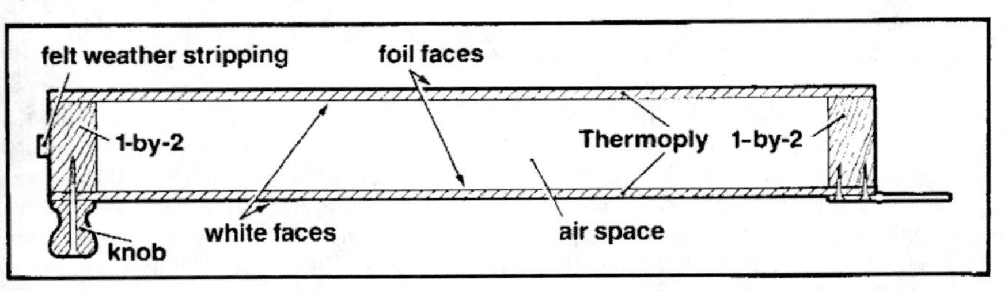

Figure 8-4: The Sun Saver.

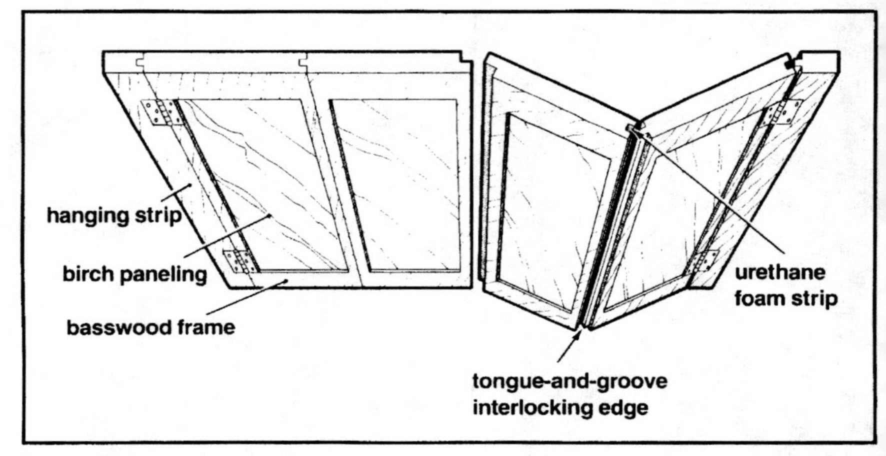

Figure 8-5: The Insul Shutter.

type weather stripping that works well on the top and bottom. The best air seal can be obtained by adding wooden shutter stops to the sides, bottom, and top of the window opening. The compression-type weather stripping should then be applied to the faces of these stops to form a very tight seal when the shutter is closed. Latches may not be necessary if your window is free from air infiltration leaks to the outside. However, if your shutter blows open during windstorms, any number of cabinet latches work well.

The Sun Saver—An interior shutter kit developed by Cornerstones is available from Homesworth Corporation, P.O. Box 565, Department MT-1, Brunswick, ME 04011. The materials to construct a pair of window shutters in this kit include four sheets of Thermoply board (two per shutter) 1-by-2 frame pieces, knobs, and hinges. The basic kit sizes of 32 by 48, 36 by 56, and 36 by 62 inches can be cut down to any window size. A section through a typical shutter is shown in Figure 8-4. The core of the shutter is left hollow with one aluminum side facing into an air space and the other into the window.

The Insul Shutter—Insul Shutter, Inc., P.O. Box 338, Silt, CO 81652, makes a highly effective shutter with tongue-and-groove interlocking edges (see Figure 8-5). This shutter combines a urethane foam core with a beautiful basswood frame and birch plywood panels. A high-density, resilient urethane foam strip enhances the seal provided by the interlocking edges. The shutter panels come in widths from 8 to 16 inches, in 1-inch increments, and are available in various lengths up to 80 inches. By combining a series of panels with a custom-cut "hang strip" you can fit a beautiful wood shutter to any size window.

Photo 8-3: Top-hinged shutters in the Chapman residence.

Top-Hinged Interior Shutters

Photo 8-3 shows hinged shutters that swing up to hook at the ceiling. These panels are constructed of 1-by-2s, foam insulation, and Masonite. They are so effective that the woodstove that serves as the backup heating system for this passive solar home in Snowmass, Colorado, rarely needs to be fired, even during extended periods of subzero weather.

In spite of its high thermal performance, the shutter design is far from ideal. The panels swing down with considerable weight and speed when released, and without

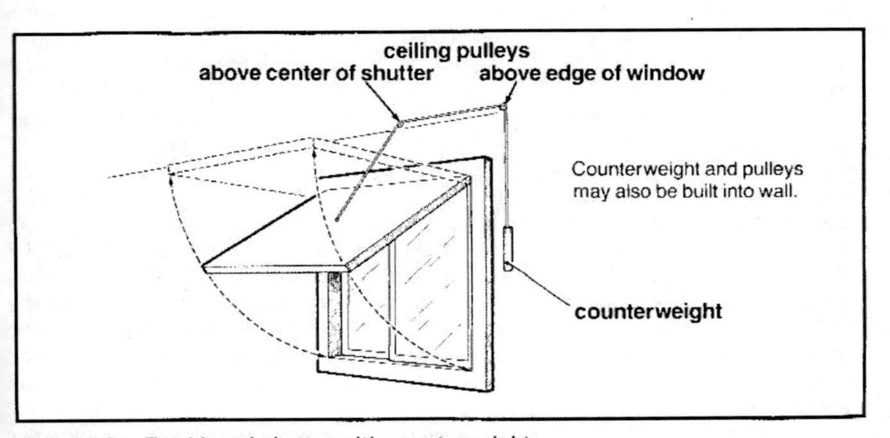

Figure 8-6: Top-hinged shutter with counterweight.

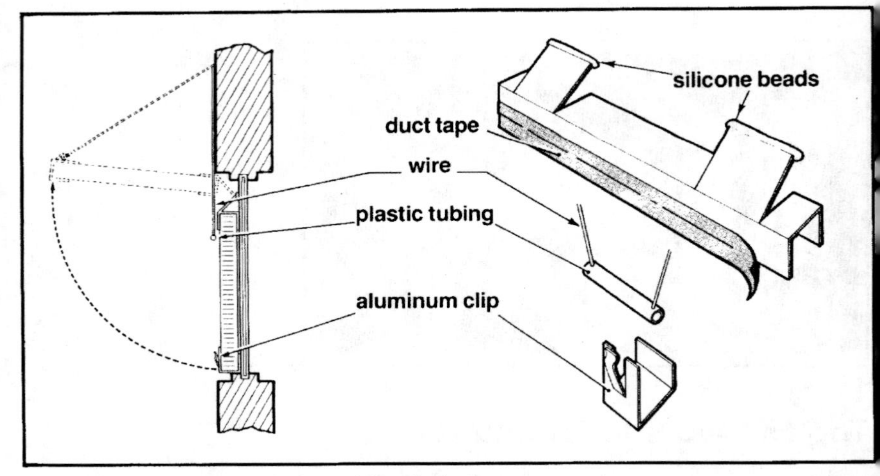

Figure 8-7: A featherweight top-hinged shutter.

a guiding hand, they could seriously injure someone. They also require the occupants to stand on a chair twice a day to latch them to the ceiling and release them—an awkward and potentially hazardous task.

Top-hinged shutters of this kind should have a counterweight to prevent them from accidentally falling and injuring someone (see Figure 8-6). The cord should attach to the center of the shutter, then run up through a pulley directly above the shutter at the ceiling, across to another pulley at the ceiling by the side of the window, and down to a counterweight. The cord should attach at the center of the shutter and not the bottom edge where it will get in the way of room traffic when the shutter is closed. With the cord attached at the center of the shutter, the counterweight should be slightly heavier than the entire shutter panel to balance its weight.

Counterweights are not needed if the shutter is so light that it is harmless should it strike someone. Bill Shurcliff has designed a top-hinged shutter constructed from Thermax board, which is so light that air resistance dampens its rate of fall. Figure 8-7 shows this design modified slightly to include a silicone hinge at the top. (Shurcliff's design uses another type of hinge.) An aluminum strip is cut and bent to the shape shown in the illustration and then is taped to the Thermax board with duct tape. The upper edge of both flange ears must be even and in a straight line so that the silicone hinge can be applied properly. The silicone hinge, developed by Mark Sherson of Zomeworks, is simply a thin bead of silicone caulking applied to the upper edge of the aluminum flange while it is held against the glass. A paper further describing this hinge is available from Zomeworks. The bead of caulking should be applied sparingly and flattened with your fingertip or a tool. The thinner the bead of caulking, the less likely it is to tear when the aluminum rotates.

In its closed position, gravity brings the shutter tight against the window glass. When it is swung open, it can be made to catch and be held from falling by a wire loop which is hung from the wall above the shutter.

A 3/32-inch steel rod (welding rod works well) is bent into the shape of a loop and a small piece of plastic tubing is fastened to the small end. This wire loop is then hinged to the wall so that when the shutter is raised, the plastic tube slides across the shutter until it catches on an aluminum clip which is taped to the bottom edge of the shutter. A gentle tug releases the shutter and allows it to close.

Interior Sliding Shutters

Sliding window shutters can be very convenient and attractive in many situations. Photo 8-4 shows a series of sliding shutters in the home of Dr. Luis H. Summers, an architecture professor at Penn State University. These white panels slide in front of a glass window wall which adjoins a patio, creating a feeling very similar to the "shoji" or sliding paper screens that cover openings in the walls of traditional Japanese dwellings.

Dr. Summers has tried these panels with two types of sliding tracks. The first panels were simply slid on wooden runners nailed to the floor as shown in Figure 8-8. They

***Photo 8-4:* A sliding shutter in the residence of Dr. Luis H. Summers.**

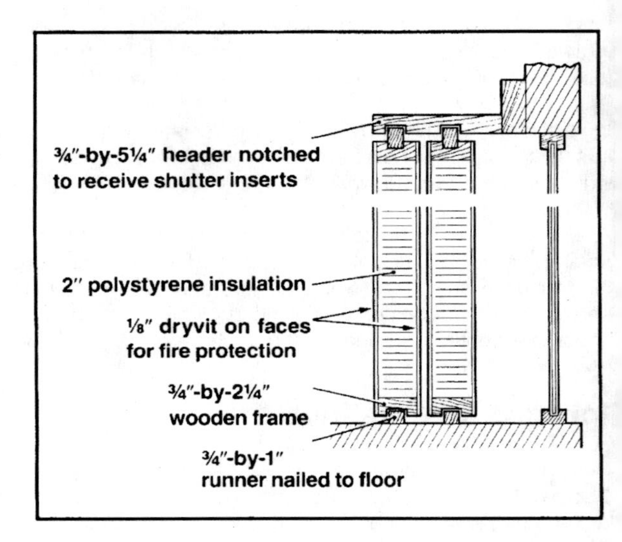

***Figure 8-8:** Dr. Summers's sliding shutters to cover a glass window wall.*

sealed the air flow quite well but had to be cleaned and waxed periodically because dirt would accumulate and cause them to jam. The second panels he installed were hung from Stanley overhead sliding-door tracks, and while they slide more readily, it is necessary to wedge a piece of foam under them to prevent drafts.

Panels that slide on hardwood rails can also be made for ordinary windows. Because they are smaller, and lighter, and are raised off the floor, the panels are farther away from dirt sources which can jam them. Figure 8-9 (modified for a window) shows how such a sliding shutter can be added to an existing window. Note how the edge seals are made to butt with a foam gasket. For wider windows the track can be extended to both sides of the window with a panel moving to each side. On a very large window a double track can be used so that the edges of the outer pair of shutters can interlock with the inner pair.

Nancy Korda and Susan Kummer of The Center for Community Technology have developed a sliding panel which moves along rings on an overhead dowel rod (see Photos 8-5a and 8-5b). This shutter is constructed with a 1-by-2 frame and insulated in the middle with cellulosic insulation in a plastic bag. (They avoid using plastic foams in their designs because they consider them a fire hazard.) The panels are faced with cardboard and an attractive fabric (see Figure 8-10). Roller cabinet catches hold the panel to compression-foam seals when closed.

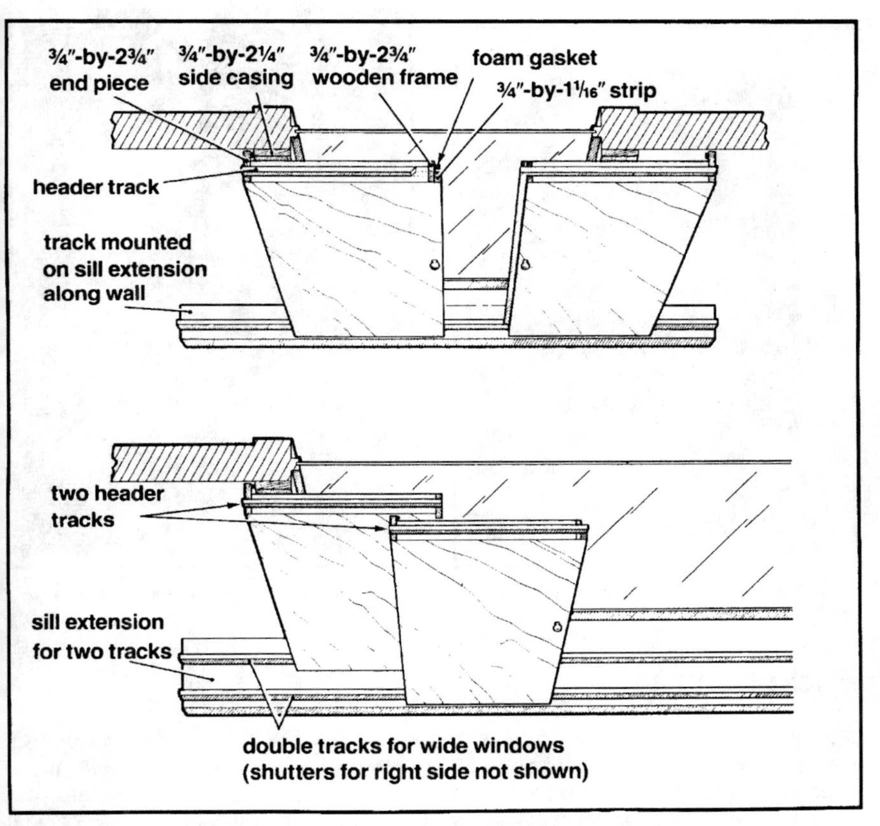

Figure 8-9: Sliding shutter designs for ordinary windows.

This type of sliding panel is similar to the pop-in designs in Chapter 5. It works well on windows that have relatively flat trim facings and even better on windows without any interior trim or projections. Foam panels and magnetic seals can be substituted for the roller-type cabinet catches.

The Shutter Shield—James Davidson of C. D. Davidson & Associates, Inc., P.O. Box 1293, Pontiac, MI 48057, has invented a sliding shutter system for windows and patio doors. Insulating panels slide to each side of a window on aluminum tracks. In new construction this system can be installed with pockets on each side of the window.

Photos 8-5a and 8-5b: Two versions of a shutter which slides on an overhead dowel, designed by The Center for Community Technology.

Pocket Shutters

Pocket shutters slide inside the wall and conveniently out of view when opened, leaving a clean and uncluttered appearance. However, if not planned properly, these shutters can rob the walls of needed insulation. It does little good to insulate your window at night and leave a pocket in the wall exposed to higher rates of heat loss.

Walls with pocket shutters can be constructed using 2-by-8 studs as shown in Figure 8-11. This allows a regular, insulated 2-by-4 stud wall to remain behind the shutter pocket, which can still support the headers for the window. With 2-by-4s flat to hold the wall surface facing the room, 2 inches are left for the shutter. By using ¾-inch steel studs, the total wall thickness can be reduced to 2-by-6 studs, which support the interior wallboard as well as the flat 2-by-4s.

Figure 8-12 shows a cross section through a large pocket shutter designed by Steve Merdler for a home in Sante Fe, New Mexico. These panels are extremely light even

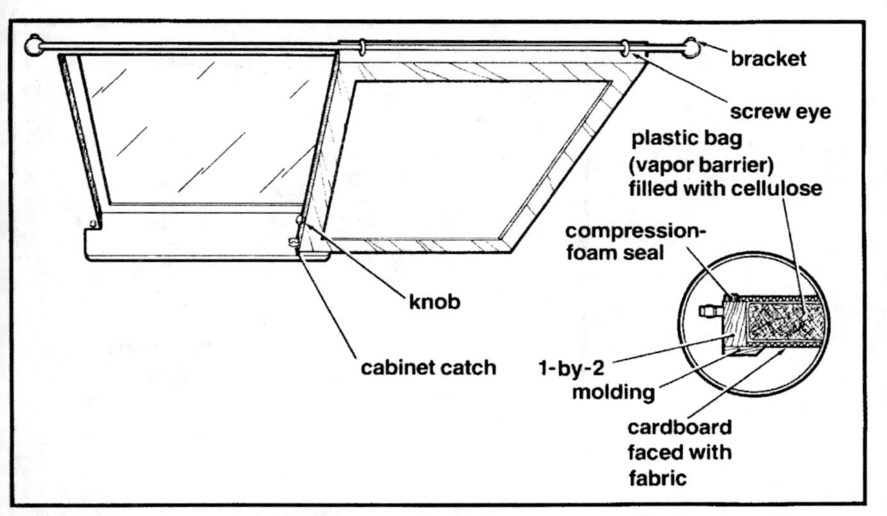

Figure 8-10: A sliding shutter design by The Center for Community Technology.

though they are 10 feet wide, 8 feet high, and 2 inches thick. Constructed of double-layer Thermax board, they are set into a dadoed slot in a frame which is ripped from a finish grade 2-by-4. To keep the bottom of the frame from sagging, a hole is drilled through each side of the frame every 4 feet or so, top and bottom, and a thin wire is wrapped around the panel to compress the frame onto the Thermax board.

The original design for these shutters specified an aluminized surface facing out. The owner changed his mind after these shutters were installed and painted them a dark

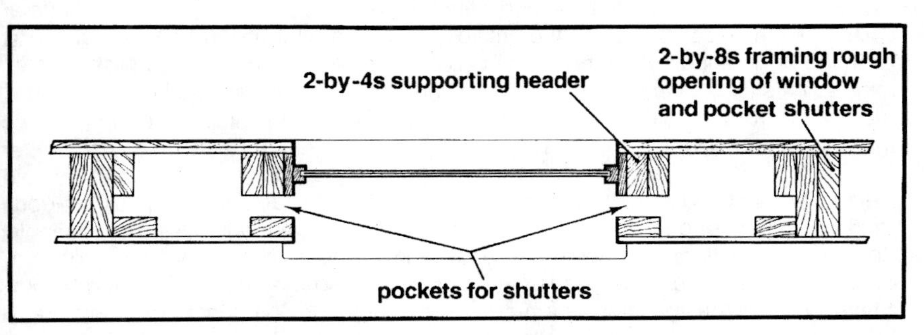

Figure 8-11: Eight-inch-thick wall for pocket shutters.

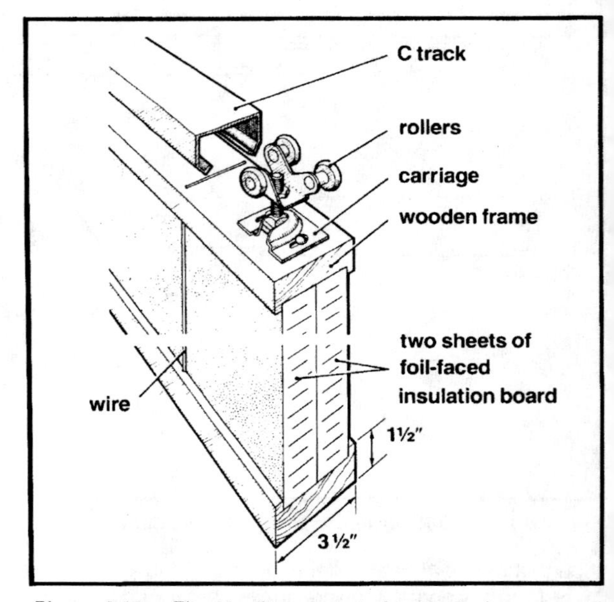

Figure 8-12: The Merdler pocket shutter design.

brown. Though facing due east, they were often left closed on summer mornings to reduce the morning heat gain. Sunlight was absorbed by the dark surfaces and heated the shutters, which caused the insulation board to bow out so that they could not be opened. To correct this problem, the shutters have been reconstructed to include a gap inside the frame so that the Thermax board can expand without bowing. Shutters that are placed behind glass to block direct sunlight should have a white or aluminized surface on the outside to avoid a heat buildup and subsequent jamming or deterioration of the shutter. (Be aware of the effect that a polished, aluminized exterior shutter face might have. Glare could be cast into the faces of pedestrians and, worse, drivers.) Venting the space between the shutter and the glass to the outside may even be necessary to prevent a buildup of heat.

If you plan to build pocket shutters, do not confuse ordinary sliding closet-door tracks with pocket door tracks. The closet-door tracks generally have only a single channel from which the wheels can easily jump. The pocket door tracks for dual cnannels have two or three wheels at each mount, so they rarely leave the tracks. Steve Merdler's design uses a high-quality, overhead pocket door track to avoid any

Photo 8-6: Muraled Panel Drape.

shutter jamming. Lawrence Brothers makes a roller that can easily be rotated to remove the shutter. Stanley Hardware makes several good pocket door tracks.

The Panel Drape—Made by Shelter, Inc., Box 108, Oakland, MD 21550, this pocket shutter is presently available only as a part of a prebuilt home package. Shelter, Inc., not only includes heavy insulation and energy conservation features in the homes they sell, they also suggest a design which will have south-facing window glass on a given site equal to at least 12% of the house floor area. Panel drapes with murals as shown in Photo 8-6, are an integral part of these homes.

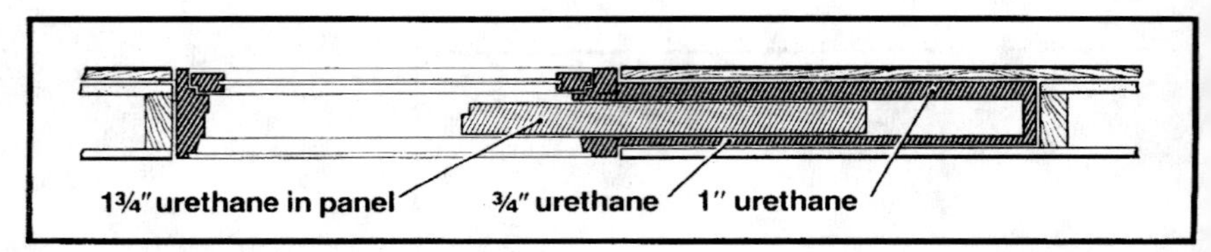

Figure 8-13: The Sunflake Window.

Sunflake Window—Available from Sunflake, 625 Goddard Avenue, P.O. Box 676, Ignacio, CO 81137, this integral window unit comes in both casement and picture window models that measure roughly 4 by 4, 4 by 6, and 4 by 8 feet. One-half the unit is a standard window and the other is a pocket and shutter. The unit is 4¼ inches thick, mounted in a standard wall having a 3½-inch stud and ¾-inch sheathing. The shutter pocket comes with 1 inch of urethane on the outside shutter and ¾ inch of urethane facing the room. When this unit is installed in a home, the empty shutter pocket has a surprisingly high R value of about 16. The sliding shutter panel is composed of 1¾ inches of urethane foam faced with a hard ABS plastic. The shutter also rates an R value of about 16 when closed over the window. Figure 8-13 shows a detailed cross section of this window unit.

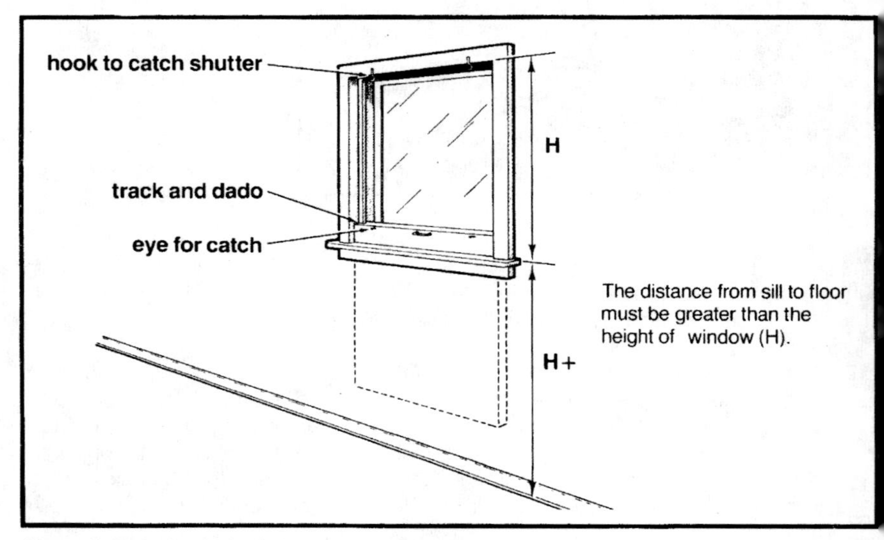

Figure 8-14: Vertical sliding pocket shutter.

Pocket panels can also be constructed in the wall beneath a window if there is room. To build a pocket shutter here, the distance from the windowsill to the floor must be greater than the height of the window. If the shutter panel is lightweight, it can easily be lifted vertically into a closed position where it is held by hooks or a catch as shown in Figure 8-14.

Pocket shutters can sometimes be added to existing walls that need insulation and repair. A new wall is created 4 inches out from the existing wall to create pockets for shutters. The old wall can be freely penetrated with holes to add wall insulation because it will be covered with the new wall. This reduces the size of the room 4 or 5 inches, however, and should only be done in large, old homes where interior space is not at a premium.

Sliding-Folding Panels

Hardware is available that allows shutter panels to both slide and pivot. With this hardware, panels can be joined with hinges to make an accordionlike shutter of just about any length. Photo 8-7 shows sliding-folding shutters designed by Steve Merdler that protect an entire south-facing window wall. The panels shown are 2¼ inches thick and can be constructed of 2-inch polystyrene foam faced with ⅛-inch mahogany veneer. Frames are of 2-inch-thick pine. Standard hollow-core doors can be substituted in this design if you can tolerate some reduction in insulating value. However, unless the edge seals are tight, much of the insulating value of a 2-inch-thick foam is lost to air convection anyway.

The sliding track available from Lawrence Brothers or Stanley Hardware is mounted from a valance that somewhat reduces summer heat gain. However, the valance

***Photo 8-7:** Sliding-folding shutter panel.*

here still allows air convection and needs a positive gasket or seal that does not hamper the operation of the panels. At the bottom of the panel is a wooden runner which the panels lean lightly against. Gravity helps form an air seal at the bottom, which can be improved by adding a strip of Schlegel Corporation's brush-type weather stripping (the kind used on sliding glass doors). The edges can be sealed by adding foam weather stripping between each panel.

This system is very easy to operate and allows for a very convenient displacement of numerous panels to a small area at one end of a room. A gentle shove sends the panels sliding the entire length of the room.

Chapter 9

Insulation Between Glazing and Interior Louvers

Movable insulation on the interior side of a window may interfere with plants, people, or furniture. It also can disturb or clutter an existing decorative window treatment. And movable insulation on the outside of a window encounters the external elements in their fullest intensity—direct sunlight, wind, rain, and ice. In the space between layers of window glass, however, movable insulation is both protected from the outside elements and out of the way of room furnishings and traffic.

Only in certain types of window assemblies, though, is it practical to insulate between the glazings. Most sealed, double-glazed panels have only ¼ to ½ inch of space between the panes of glass, not adequate for insulation. Also, the panels must remain totally sealed against moisture. These window units will not accommodate the devices that are required to fill or remove insulation from between the glazings like the Beadwall system, a clever invention discussed in this chapter. Windows with independently mounted layers of glass, or glazings that are separated by two or more inches, have the most potential for employing between-glazing insulation. The ordinary window with a separate storm window fits these requirements very well.

Several potential problems should be recognized when exploring between-glazing insulation designs. The moisture condensation problem, a frequent occurrence in the space between glazings, is aggravated by operating mechanisms for window insulation. The principles outlined in Chapter 3 for maintaining the tightest seal at the interior layer of glazing to prevent window fogging must be stressed here once again. Cords, pulleys, and other devices used should have close tolerances and proper seals to prevent the entry of moisture. Operating mechanisms that are moistureproof are difficult to construct with simple tools.

Another challenge in a between-glazing insulation design is to provide a way to remove it during the daytime so that it does not block entering daylight or the view. Roll shades occupy a minimum of space when rolled up and are often less obstructing than other types of insulation. Just a single-layer shade with highly reflective film, facing into the living space, substantially increases the thermal protection of a window assembly.

Wall pockets, discussed in Chapter 8, help to alleviate the problem of a shutter obstruction. However, they are very hard to repair since they are both behind the window glazing and inside the wall.

The Beadwall is a system that very effectively skirts both moisture and displacement problems. Moisture is not a problem because the space between the glazings is completely isolated from the room air. The insulating beads used in between the windows are completely removed to remote storage bins by vacuum motors that are activated by an electric switch or a light sensor.

Some windows do not require the daily movement of insulation; they can be insulated each fall and uncovered in the spring. Such are the most likely candidates for between-glazing insulation in an existing home.

Seasonal Insulation Between Glazing

North-facing windows particularly in winter, have little value in rooms where there are also south-facing windows or in rooms that are rarely occupied. If these north-facing windows have storm windows over them they have a very handy space for the addition of fixed window insulation during the winter season.

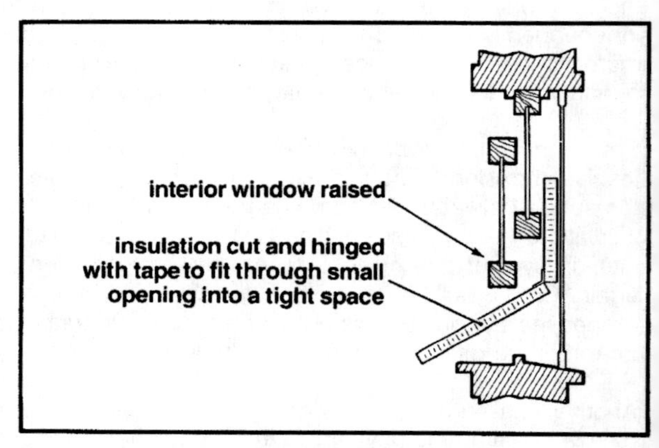

Figure 9-1: Hinged seasonal foam panel.

Foam insulation board can be cut and slid into this space for winter insulation (see Figure 9-1). If there is not already ample daylight from other windows, the board can be cut about 6 inches shorter than the window height so that light can still enter the room with only a small increase in heat loss. Daylight can also be allowed to enter by cutting a hole or "window" in the foam board. A piece of cellophane should be taped over this hole to prevent air convection between the front and back of the board.

Fiberglass, the type used for wall insulation, is another material that can be used to insulate the window-storm window space. However, fiberglass is awkward to install, objectionable to the touch, and unsightly.

Another way to insulate the space between a double-hung window sash and storm window is to add a loose, insulating filler. Styrofoam insulating beads or pellets, like those used in the shipping of fragile items, can be used to fill this air space (see Photo 9-1). To add loose insulation, lower the upper sash of the window (see Figure 9-2) and pour the beads or pellets into the space between the window sashes until it is full. Each spring open the lower sash, remove the pellets, and store them in a box until the next fall. If natural light or a view is needed through the filled window, a small space inside the frame from filling with pellets, creating a porthole in the pellet-filled window.

Photo 9-1: Styrofoam packing beads—poured into a window by John Kubricky—reduce heat losses but still allow some light to enter.

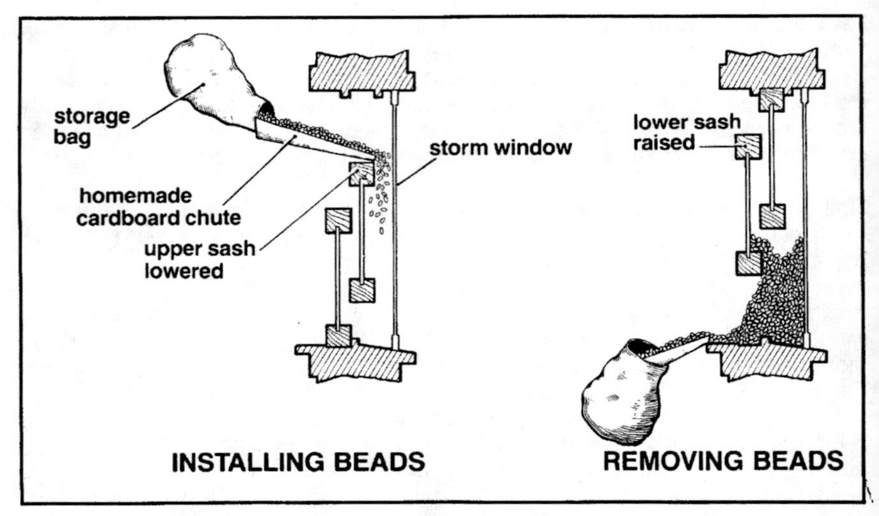

Figure 9-2: Filling the space between sashes with Styrofoam beads or pellets.

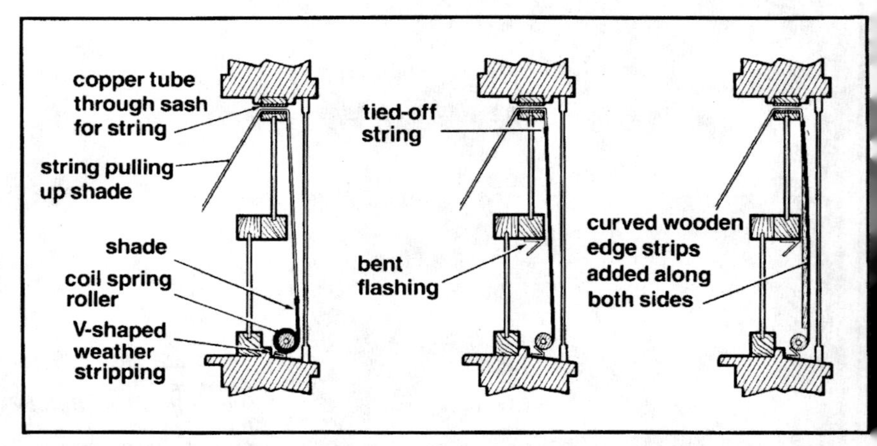

Figure 9-3: A reflective shade between window and storm sashes.

Shades Between Glazing

A window shade can be mounted at the bottom of the space between a window and the storm sash as shown in Figure 9-3. (This design is based partly on a patented invention of Norman B. Saunders.*) The shade roller should have a coil spring to keep constant tension on the cord as the shade is rolled down. Mount a small piece of copper or plastic tubing in a hole drilled through the top sash. Then run a cord from the shade through this tube to a small catch or cleat on the window jamb in the room. Copper is preferred for this tube since it can be splayed on the ends.

This shade provides two air spaces between the layers of glazing instead of just one. It is most effective when made from an aluminized or reflective material such as Foylon or Mylar. A fabric that is reflective on both sides is best in this case since it prevents the transfer of radiant heat across the air space on both sides of this shade. This reflective shade is also a very effective screen in summer against incoming solar radiation (insolation).

With some double-hung windows, this shade will tend to catch on the bottom edge of the upper sash. A strip of roof flashing can be cut, bent into a V shape, and attached to the sash to prevent the shade from catching on the sash.

The width of the shade should be equal to the width of the window opening so that air flow around the edges is kept to a minimum. Curved strips can also be added to the sides of this space to further reduce air flow. The tension of the coil spring in the shade roller will keep the edges of the shade material tight against the curved strips. However, these strips may impede window operation when ventilation is needed.

Other Between-Glazing Systems

The Beadwall—From what at first glance appears to be a sterile and uninteresting material—polystyrene beads—emerges the Beadwall, a very beautiful and visually dynamic window treatment. Photos 9-2 and 9-3 show two Beadwall installations—an interior view of the Beadwall at the Pitkin County Airport near Aspen, Colorado, and an exterior view at the Benedictine Monastery in Pecos, New Mexico. Styrofoam beads occupy the 3-inch cavity at night, and during the daytime they are removed to a remote bin. These beads fill and empty the space in seconds, with a churning motion almost as dramatic as an ocean surf.

The Beadwall, a patented invention of David Harrison, is marketed by Zome-

*1976 U.S. Patent No. 3,952,947.

Photo 9-2: Interior—Beadwall installed at the Pitkin County Airport, near Aspen, Colorado.

Photo 9-3: Exterior—Beadwall installed at the Benedictine Monastery, in Pecos, New Mexico.

works. It has an R value of 10 and can be operated with the convenience of a switch. Figure 9-4 graphically explains the system. Tanks, usually constructed from double, 55-gallon drums, store the beads on sunny winter days. Near sunset, a sensor switches a vacuum cleaner motor on, which creates pressure in the tank and drives the beads through a PVC pipe. They fill the window space through a supply nozzle at the top of the window. When the space is filled, the motor shuts off. The blower simply reverses each morning to pull the beads out the bottom of the window through graduated openings in a "bead drainage" duct and then back through the PVC pipe into the bin. By using sock check valves and a reversible motor, only one PVC pipe is needed for transporting the beads both ways.

Plans to construct the Beadwall are available from Zomeworks, but before you order yours, be aware that there are many specification subtleties to this system, including header hole sizes, antistatic agents, glass strengths required for blower pressures, etc.

The beads are very sensitive to static electricity, and, in a number of Beadwall installations jamming has been a serious problem. Too much antistatic agent in the beads can be worse than too little. Humidity affects the amount of static electricity generated by the beads, and their treatment seems to vary in different climatic regions. Consulting a person who has installed a Beadwall in your own region will be helpful in fine-tuning the system. However, the Zomeworks plans and specifications contain much essential information, so do not try to build a Beadwall system without them. Components for the Beadwall are available from Zomeworks Corporation, P.O. Box 712, Albuquerque, NM 87103. EGGE Research, Box 394B, RFD 1, Kingston, NY 12401, also plans to market a kit for constructing the Beadwall system.

Louvers or Blinds

Venetian blinds are commonly used on the interior side of a window for light control but also can be employed between layers of window glass. For the most part, blinds are quite effective at controlling the amount of solar radiation entering a space, but the thin, metal strips offer little thermal resistance. A closed venetian blind reduces the heat gained through a window by 50%, but reduces heat losses through double glass only about 20%. A shiny, metallic surface on one side of the blind increases insulating capacity by reducing the radiant heat transfer.

> Rolscreen Pella Slimshade—This window unit, made by Rolscreen Company, Pella, IA 50219, comes with a thin, flat venetian blind mounted between two layers of glazing within an operable casement sash. A small, knurled dial in the lower corner of the window sash lets you rotate these blinds from a horizontal position where they are barely visible, to a closed vertical position where they block all light and vision to the outside. These blinds cannot be raised or lowered and have a low R value.

Thermal louvers which are wider than blinds can be effective inside a room, particularly in high clerestory windows where clearances are no problem. The Reflective

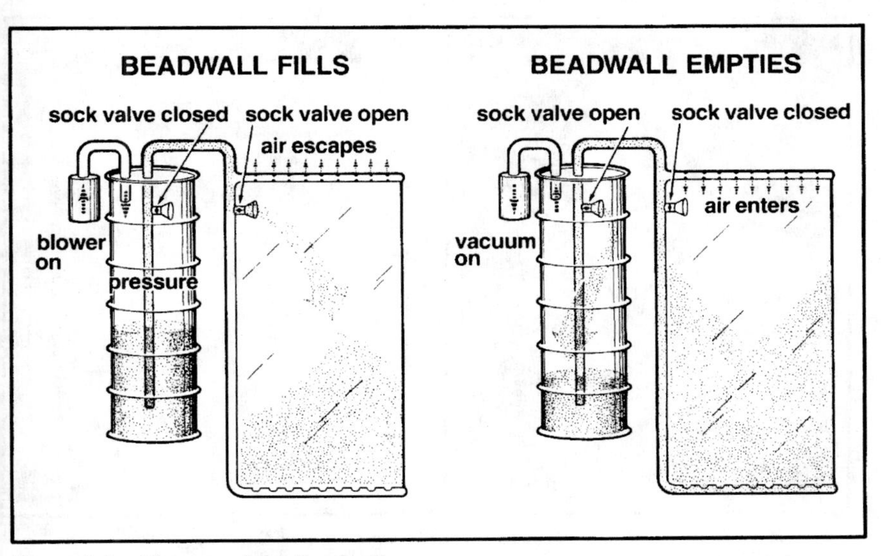

Figure 9-4: Diagram of the Beadwall.

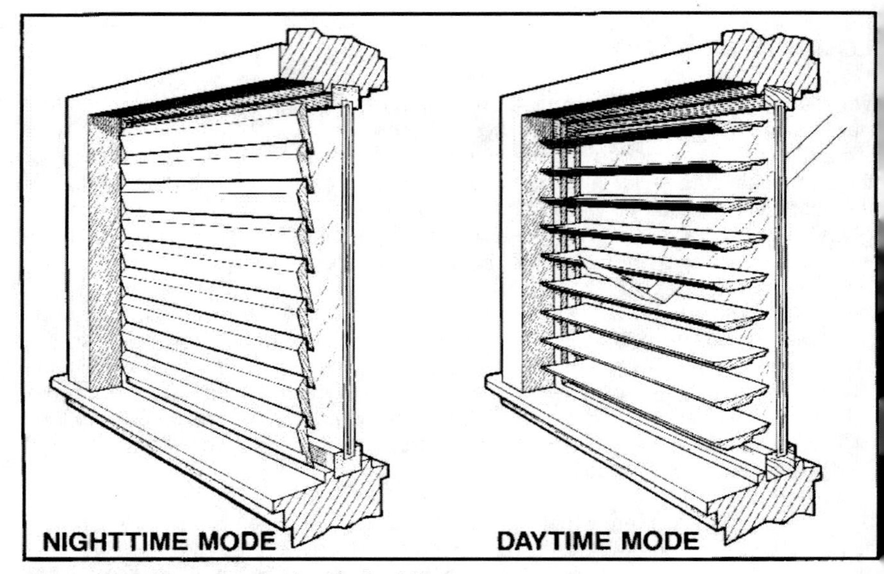

Figure 9-5: The reflective insulating blinds.

Photos 9-4a and 9-4b: Reflective insulating blinds in Oak Ridge, Tennessee—open and closed.

Insulating Blind design strikes a happy medium between thin blinds which do not insulate well and wide louvers which are awkward near floor level.

The RIB (Reflective Insulating Blinds) Design—This design (not a product) was developed by Hanna Shapiro and Paul Barnes at the Solar and Special Studies Section of the Oak Ridge National Laboratory. Cross sections are shown in Figure 9-5. These wooden louvers are about 4 inches wide, with an interlocking edge and weather stripping for an effective seal. The top side of each louver is curved so that all incoming light is reflected to the ceiling and then down onto a work surface. Photos 9-4a and 9-4b show these blinds installed in an office of the Oak Ridge National Laboratory. Blinds similar to these (without the curved face) can be cut from a 1-by-4 at home on a table saw. For more information contact Hanna Shapiro at Oak Ridge National Laboratory, Energy Division, P.O. Box "X," Oak Ridge, TN 37830.

Part III

Movable Insulation and Sunshades on Your Home's Exterior

Outside the dwelling, movable insulation can be positioned during the winter to reflect sunlight into a window, thereby increasing the heat collected by south-facing glass. In the summer, solar heat gain should be stopped before the sunlight penetrates your window because it is difficult to expel. Moisture condensation on window glass, which can be a serious problem with interior systems, is no problem on the exterior. Where interior systems are not practical because of clearances or furniture arrangements, exterior shutters are an advantage because they won't rob your interior of valuable space.

An exterior system, however, must contend with ice, rain, wind, and snow, which in some climates may interfere with operation. Outdoor shutters or shades must be constructed of strong, wear-resistant, water-repellent materials which will withstand exposure to the elements. To operate an exterior system you may need to go outdoors unless it contains some special through-the-wall hardware to operate it from the interior. Bees, trees, bird nests, or ice may all inhibit the operation of an exterior system and regular maintenance may be needed for its continuous use.

The visual impact of exterior shutters on your home will be pronounced, whether for better or worse. A careful study of how a shutter system will alter the appearance of your home is essential. In many cases, exterior insulating shutters make a home much more attractive, providing rhythm and detail to a wall with otherwise bare window openings. Unlike aluminum or vinyl shutters, which—along with false beams made of urethane—are meant to invoke a self-defeating sentimentality about the old times, when craft, ingenuity, and self-reliance were a basic part of our lives, exterior insulating shutters can add a new expression of energy independence and beauty to your home.

Chapter 10

Exterior Hinged and Sliding Shutters

Exterior insulating window shutters can also serve as sun reflectors and thereby increase the amount of heat gained through a south window by as much as 50%. Many new passive solar homes include exterior shutter-reflectors, and they are also being used in home energy retrofitting since they can increase the winter solar heat gain through existing windows without enlarging the glass area.

Reflective insulating shutters are difficult to position in a manner which will bring the most additional sunlight into a room. The task would be easy if the sun were stationary, but the sun's position changes constantly through the day, and its arc in the sky raises and lowers with the seasons. Optimum reflector angles for a given window in a given location can be derived by computer modeling in which the average window gains for a reflective shutter in any position are determined. However, sophisticated analysis of the benefits of reflective window shutters is practical only in large buildings where there are numerous identical windows.

For a shutter-reflector to maximize the daily heat gain through a window, three conditions must be met. Figure 10-1 illustrates the criteria that must meet each of these conditions:

1. There must be a significant quantity of direct sunlight striking the reflecting surface of the shutter.

2. The reflector material must be as mirrorlike as possible.

3. The light from the reflector must reach the window at a steep enough angle to penetrate the glass and not be reflected back by it.

Variations in the first condition are shown in Figure 10-1a. If the sunlight striking the reflector is almost parallel to the shutter, the amount of sunlight striking this surface will be small. As the reflector is rotated to make a larger shadow, it receives more sunlight. However, rotating a window reflector to receive more sunlight is useful only if this sunlight is still directed into a window, and it is likely that turning the reflector fully into the face of the sun will diminish the amount of reflected light reaching the window.

The second condition requires that the reflector material be mirrorlike. Not only must the surface receiving the sunlight be highly reflective, but it should also be a specular reflector (see Figure 10-1b). A specular reflector has a mirrorlike surface where the angle of incidence always equals the angle of reflection. A specular reflector will

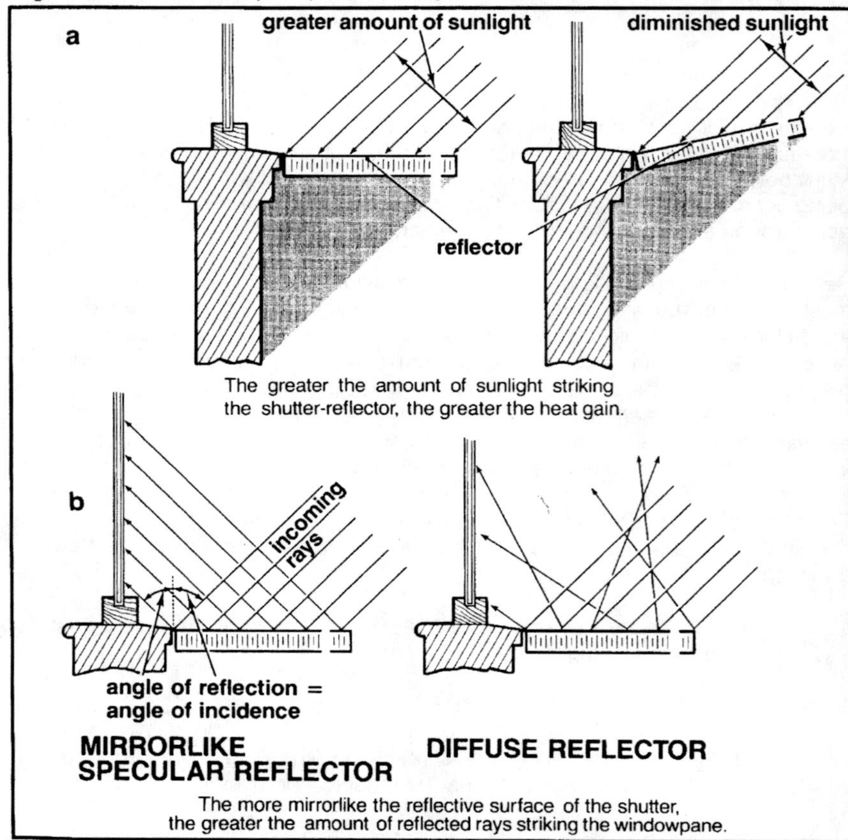

Figure 10-1: Three requirements of an effective shutter-reflector (here and on facing page).

direct parallel sun rays onto a window, while a diffuse reflector will scatter the sunlight in many directions so that a smaller portion of the reflected sunlight actually reaches the window. The way to determine whether a surface is a specular or diffuse reflector is to look for a mirrorlike image on the surface. A highly polished, aluminum surface with a clear mirror image is specular, while a sheet of full mill-grade aluminum has a more diffuse reflection. Several options for specular reflector surfaces are outlined later in this chapter.

The third condition for a useful heat gain from a shutter-reflector is shown in Figure 10-1c. The greater the angle of incidence at which light strikes a glass surface, the

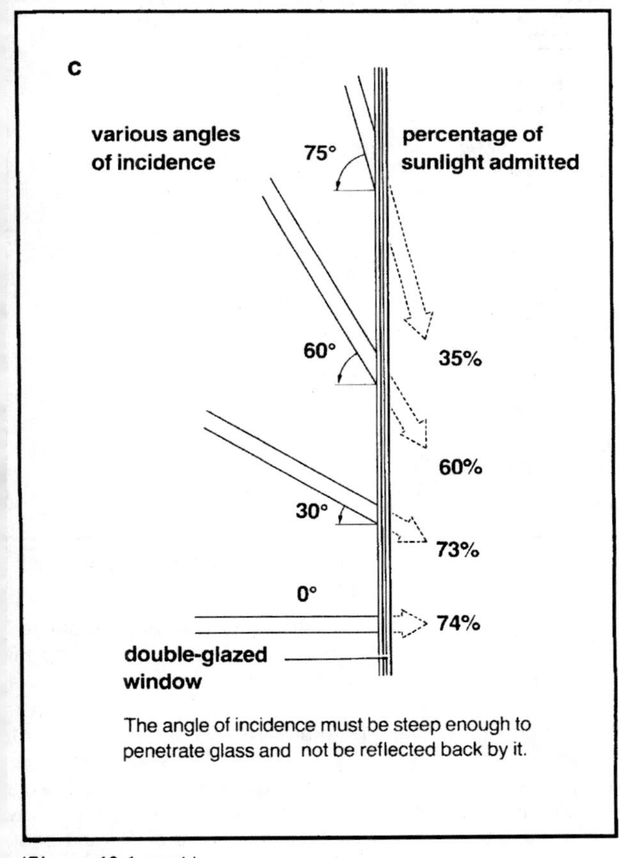

The angle of incidence must be steep enough to penetrate glass and not be reflected back by it.

(Figure 10-1 cont.)

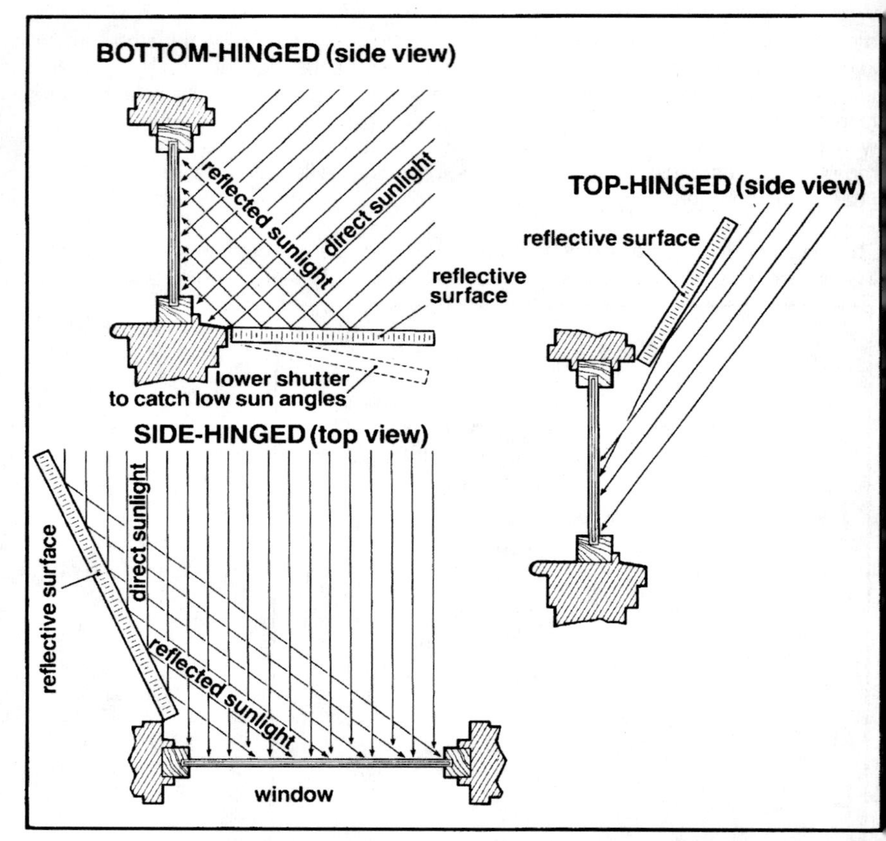

***Figure 10-2:** Three types of hinged shutter-reflectors for increasing winter solar heat gains: bottom-hinged, side-hinged, and top-hinged.*

lower the transmittance of light through the glass. The amount of light transmitted through a window decreases gradually until the angle of incidence exceeds 60 degrees, at which point the transmissivity falls off very rapidly.

Figure 10-2 shows three types of exterior shutters with hinges either on the bottom, sides, or top. The bottom-hinged shutter opens to a horizontal position on a due-south window. It yields a significant increase in solar heat gained because it reflects sunlight into the window at an angle that can penetrate the glass during the midday hours of most-intense sun. This shutter meets the conditions in Figure 10-1 the best.

During December when the sun is very low in the sky, the performance of this shutter can be increased slightly by lowering it 5 or 10 degrees since this increases the amount of sunlight it reflects into the window.

The side-hinged shutter also reflects a significant amount of sunlight into some windows during certain hours of the day as the sun moves across the sky. This shutter is most effective on windows facing southeast or southwest. Its applications are further described in Figures 10-10 and 10-11. The side-hinged shutter adapts very well to retrofit situations where homes usually do not have a due-south orientation.

The top-hinged shutter-reflector is not generally as effective for increasing solar heat gains on vertical windows because the conditions in Figure 10-1 are not adequately met. Less sunlight strikes the surface of a top-hinged shutter, and the light reflected from it onto a vertical window is at too great an angle for adequate penetration of the glass. This type of shutter performs best in northern latitudes where the sun is very low in the sky during the winter.

Bottom-Hinged Shutters

Photos 10-1 and 10-2 show swing-down shutters on two homes in New Mexico—the Steve Baer residence in Corrales, and the First Village House #4 in Santa Fe. Both

***Photo 10-1:** The Steve Baer residence in Corrales, New Mexico. These shutters hinge up to cover a south-facing glass wall with water-filled drums behind it which store heat. For more discussion on movable insulation with heat-storage walls see Chapter 17.*

***Photo 10-2:** The First Village House #4. The operation of this home is similar to the Baer residence, except a concrete wall with water-filled channels is used instead of water-filled drums behind the glass.*

***Photo 10-3:* Detail of a 4-inch-thick shutter for the science building, Marlboro College, Marlboro, Vermont.**

shutters cover a heat-storage wall on the south side. The Baer residence was one of the first homes in the country to use a reflective, insulating shutter on a south-facing heat-storage wall. The shutter is constructed of a 3-inch-thick, honeycomb material filled with urethane foam and faced on each side with 0.024-inch satin-anodized aluminum.

The shutters on the First Village House #4, constructed by Susan and Wayne Nichols, sandwich Styrofoam insulation between an aluminum sheet and plywood. Figure 10-3a shows a cross section. They are constructed with a 2-by-4 frame turned flat and 1½-inch-thick foam core. The bottom of the shutter is faced with ½-inch rough-sawn plywood and the aluminum facing on the top wraps around the edges of the frame to keep out moisture. A shutter with a similar design installed on the science building at Marlboro College in Marlboro, Vermont, is shown in Photo 10-3. In this shutter the 2-by-4 vertical frame members are on edge and contain 3½ inches of foam insulation in response to the severe Vermont winters.

Alcoa Aluminum manufactures an anodized, polished aluminum sheet with good specular reflecting characteristics. It can be used on shutters but is expensive and often hard to find locally. A cheaper one can be made by attaching Mylar-type films to the face of the shutter. Figure 10-3b shows a shutter constructed in this manner. Martin Processing, Inc., P.O. Box 5068, Martinsville, VA 24112, makes a Mylar-type film called Llumar that can be applied to a shutter panel with 3M Scotch Grip Plastic Adhesive 4693. Parsec, Inc., P.O. Box 38534, Dallas, TX 75328, makes a Mylar-type film with a self-adhesive backing. A couple of other reflective films are listed in the *SUN Catalog* from Solar Usage Now, Box 306, Bascom, OH 44809.

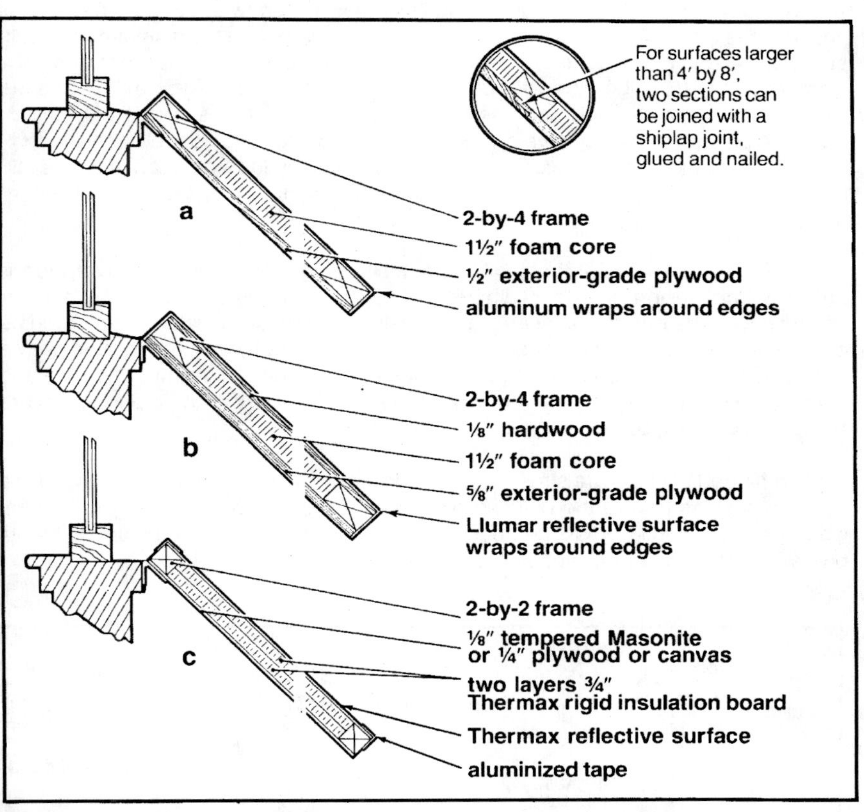

Figure 10-3: Cross-sectional details of several bottom-hinged shutters.

Plywood works well on the outside or bottom of the shutters. Rough-sawn plywood is attractive on homes with a natural wood exterior. If your home is painted, the plywood can be painted to match. An exterior-grade plywood should be used in all cases. One-half- or ⅝-inch-thick plywood is often used to match the siding on a home. This adds considerable weight when ⅜- or ¼-inch plywood is adequate for strength. The plywood should be as thin as possible to minimize weight and should be both nailed to the frame and glued with a waterproof glue such as resorcinol. If the panel needs to be wider than one sheet of plywood, the plywood should be joined and glued over a continuous framing member. Pressed hardboard treated to resist moisture can also be used for the bottom face of a shutter.

A very lightweight shutter design for small windows is shown in Figure 10-3c. This shutter uses the bright foil surface that comes on isocyanurate foam sheathings as its specular reflector. The edge frame is of 2-by-2s with a pressed hardboard or plywood bottom. The edges are wrapped with an aluminized foil tape, which seals the frame against moisture. Brands of isocyanurate sheathing, such as High R by Owens-Corning or R-Max, have a washable ink and should be used on this type of shutter. Thermax prints product information on the shiny foil side in an indelible ink. Substituting a canvas facing on the bottom of this shutter to replace the hardboard makes it even lighter.

Minimizing weight while at the same time maximizing the resistance to warpage is critical in the design of exterior shutters. The wooden frame in a hinged exterior shutter should be cut from a clear-grain, moisture- and warp-resistant wood such as clear heart redwood or clear cedar. Spruce, fir, or ponderosa pine treated with a dose of Cuprinol or Wood Life are other options. By all means, stay away from treated yellow pine that warps every which way. Frames of aluminum or thin-gauge steel are being explored by some design groups.

Bottom-hinged shutters are stressed by the drawcord at the top and tend to warp in a concave fashion toward the building as shown in Figure 10-4. One way to combat warpage is to counterstress the open shutter by supporting the midspan so that its own weight bends it in a slightly convex fashion.

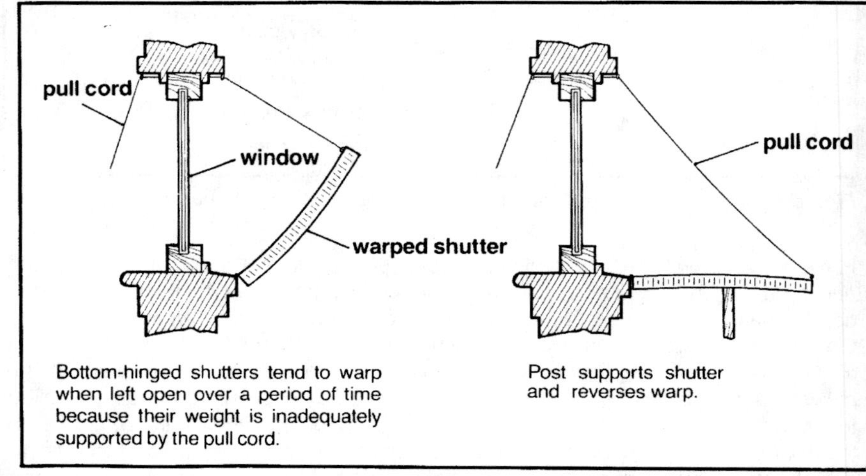

Figure 10-4: Combating warping stresses on bottom-hinged shutters.

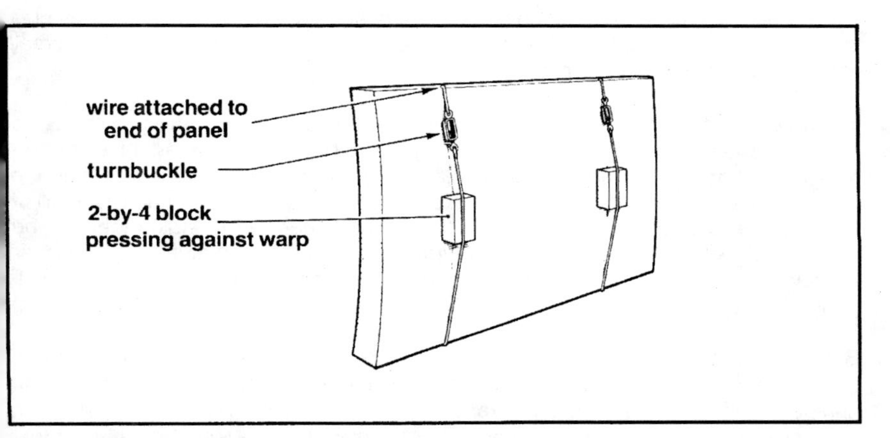

Figure 10-5: Removing the warpage from a shutter.

Another way to combat warpage, either before it takes place or after the fact, is shown in Figure 10-5. To correct a warped shutter, glue and screw a 2-by-4- or a

Figure 10-6: Supporting the bottom-hinged shutter when open.

4-by-4-inch block of wood to each edge of the bottom side of the shutter and fasten a steel wire across these blocks to the top and bottom. Tighten the wire with a small turnbuckle to remove any warpage.

The optimal, reflecting daytime position for a south-facing, bottom-hinged shutter is near horizontal, give or take a slope of 5 degrees to drain off rainwater. If the shutter-reflector is added to a large window near the ground, block, clay tile, or wooden pedestals can be placed on the ground for the shutter to rest upon when opened. If the shutter is for a smaller window where the sill is not near the ground, braces can be constructed with 2-by-2s as shown in Figure 10-6. A large shutter over an upper window can rest on a wooden pole that is anchored to the ground, or held by small-gauge diagonal chains.

Bottom-hinged shutters are raised with a single cord at the center of the shutter. Larger, heavier shutters generally require a winch but a simple pull cord and cleat can be used for very light shutters over small windows. With any heavy shutter which rests near the ground when open, you must consider the safety of small children, particularly when other children might have access to the winch. A safety lock or chain to prevent tampering with the winch is recommended if children are around. Counterweights can also be added for safety. A small low-cost, electrically driven winch would make these shutters practical for those who are unable to operate a

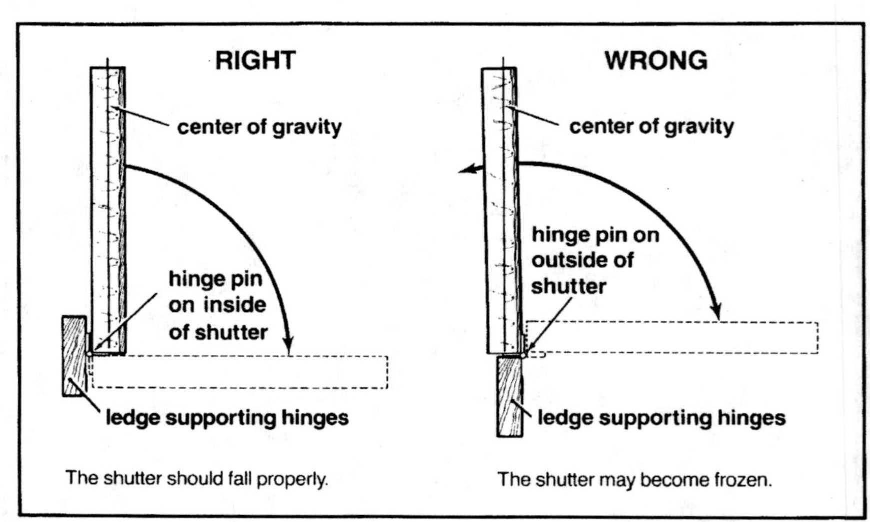

Figure 10-7: Proper location of hinge pins.

hand-driven crank. With the proliferation of small, electrically driven appliances in the home, it is surprising that no low-cost winch motors for lightweight shutters can be found, although several designers have looked hard for this component.

When bottom-hinged shutters are closed, they sometimes become frozen in a vertical position without enough leaning weight for them to fall, particularly if wind is blowing against them. If constructed too close to a vertical position, sometimes two people are required to lower the shutter—one outside who must pull it out until it leans and one inside to lower the winch or rope.

If a shutter will not fall by its own weight, you can hang extra weights on the outside face of the shutter, near the top. A small compression spring at the top of the shutter will also help it fall when released. Compression tube-type weather stripping should be installed on all four sides of the window frame to make a good edge seal. If the shutter closes into a recessed window opening, the weather stripping should be installed on a face or trim piece so that it pushes against the shutter, helping it to open. Installing the weather stripping in a manner that wedges the shutter closed may make it difficult to reopen. The hinges on the shutter should also be placed so they help the shutter to open by gravity. Figure 10-7 shows a right and wrong way to hinge the shutter. The hinge pins should be on the inside face of the shutter when it is closed to help the shutter fall by its own weight. These hinges, no more than 2 feet on center, should be of galvanized steel to resist rusting, and should anchor into a solid 2-by-4 or 2-by-6 ledger strip.

Side-Hinged Shutters

Window shutters are traditionally hinged to the sides of windows. These shutters were at one time closed over windows to provide additional protection during storms. Access could be gained only by opening the window sash and leaning out to reach the shutter. This practice was at times dangerous and invariably brought a chilling blast of cold air into the house. Today with storm windows and fixed window sashes, this thermally wasteful practice is no longer employed, but opening and closing of side-hinged exterior shutters is the greatest barrier to their use as window insulation.

Where windows can be reached from the ground or an exterior deck, they can be opened and closed from the outside (see Figure 10-8). This is practical in some situations, but it requires you to go outside morning and night in the worst winter weather, each time bringing back into your home a puddle of ice, snow, and cold. There must be a better way!

Casement window hardware used to operate window sashes by a hand crank from the inside can sometimes be adapted to operate small shutters. Blaine Window

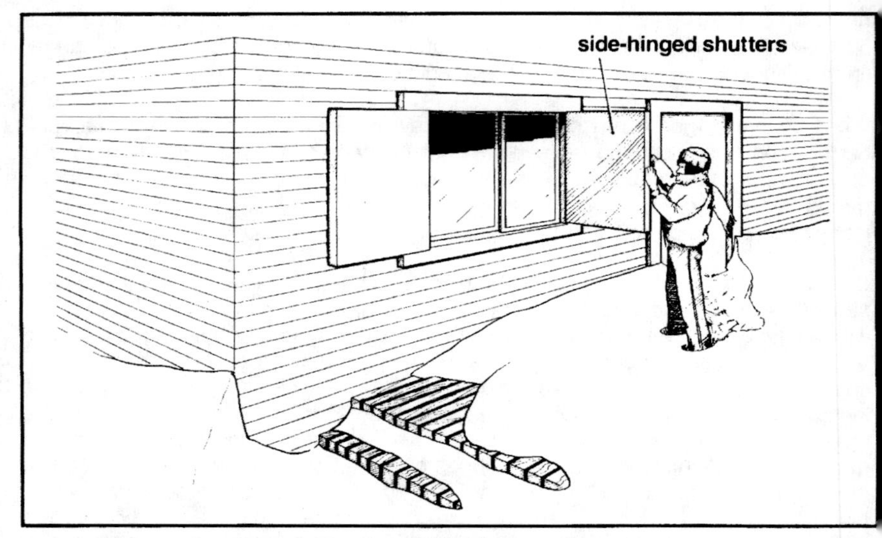

***Figure 10-8:* Swinging shutters opened from the outside.**

Hardware, Inc., 1919 Blaine Drive, RD4, Hagerstown, MD 21740, stocks and ships by mail order a tremendous variety of window hardware items, including casement operators. However, most casement window operators don't rotate a full 180 degrees and many don't turn past 90 degrees. A crank-and-worm-gear pivot with a 180-degree rotation specifically designed for window shutters is a much-needed product. The type of device required (Figure 10-9) has a swing-arm that attaches to the shutter. A semicircular gear is driven by a worm gear connected to a rotary hand crank. The arm-and-gear assembly must be strong enough to resist heavy wind loads and must seal the shutter tightly when closed. In addition, it must be worked into the wall around the jack studs that support the roof above a window opening. Most casement windows also have a locking lever to draw the window tight when it is closed.

Thermafold Shutters—Shutters, Inc., 110 East Fifth Street, Hastings, MN 55033, produces a pair of hinged, folding shutters for 6-foot by 6-foot, 8-inch patio doors. This very clever system has two bifolding panels (see Figure 10-10) that hinge to each side by means of a crank-and-worm-gear system.

These panels are 1½ inches thick, including 1 inch of urethane faced with ⅛-inch tempered hardboard, and three hinges along each vertical joint. One side

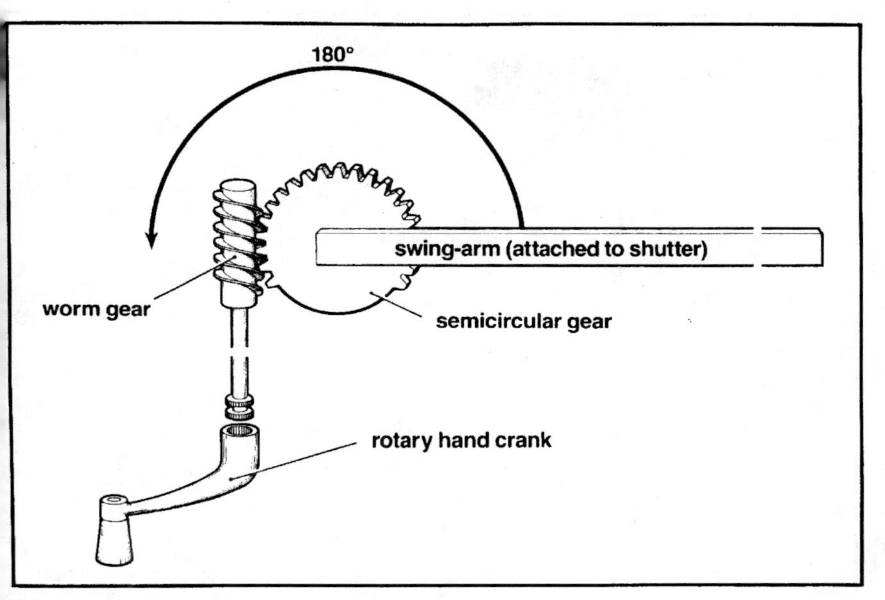

Figure 10-9: Through-the-wall crank operator.

of the bifold is wider than the other so it can be attached with small nylon sliders to a track that runs along the top and bottom. The wider panel allows the shutter to fold flat against an adjacent wall and still remain in the tracks, keeping the shutter from breaking away in a strong crosswind.

Side-hinged shutters can serve as reflectors to boost solar heat gains on winter days. However, they are only useful a couple of hours each day because of constant change in the orientation of the sun. The additional solar gain from reflective side-hinged shutters will be more than nullified if they shade a window during the main part of the day. These shutters must never block direct winter sunlight during the hours of 9:30 A.M. to 2:30 P.M. A clear opening, 45 degrees each side due south (see Figure 10-11), will insure against shading the window.

An exception to this rule can be made on homesites where another building, a mountain, or dense tree shades a southern wall between 9:30 A.M. and 2:30 P.M. On this type of site, the shutter-reflectors should be angled so that they take full advantage of the available hours of direct sunlight.

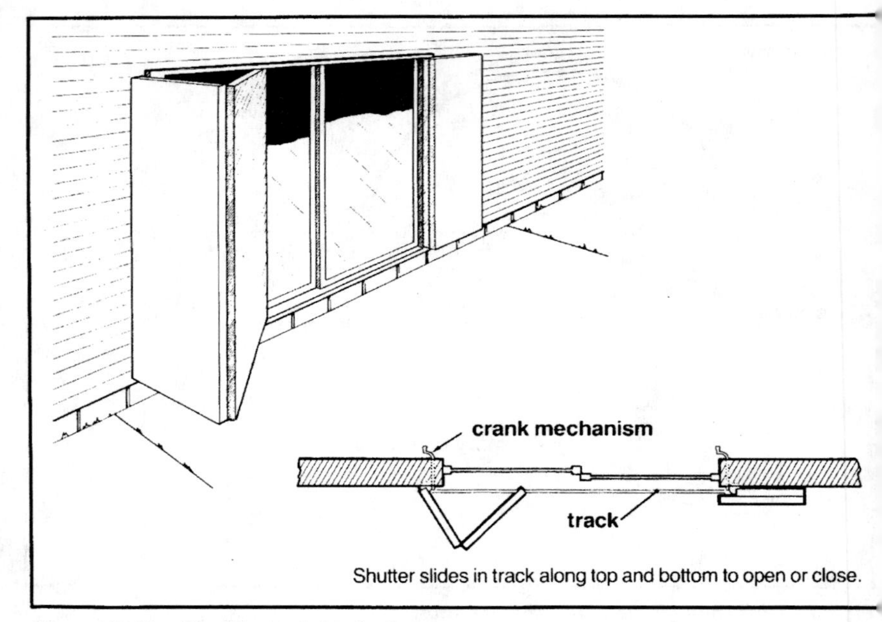

Figure 10-10: The Thermafold shutters.

Side-hinged shutter-reflectors provide a larger increase in solar gains on windows that face southeast or southwest than windows facing due south. In Figure 10-12 a shutter-reflector is shown on a window facing due south and on one facing southeast. Most of the sunlight striking the shutter on the due-south window is either reflected away from the window or strikes the glass at an angle too wide to transmit light through the glass. The shutter on the southeast window significantly boosts the amount of light entering through the window throughout most of the day. Even on due-east or west windows, reflective shutters can boost the sunlight the window receives during part of the morning or afternoon.

Side-hinged shutters are constructed from the same materials as bottom-hinged shutters. However, since these shutters are usually for small- to medium-size windows, the guidelines for lightweight panels can generally be followed. One inch of insulation board is often adequate with ⅛-inch hardboard on each side. Panels with 1½ inches of foam and with a nominal 2-by-2 frame insulate a little better. The same materials can be used for reflective surfaces on these panels as were used on bottom-hinged shutters.

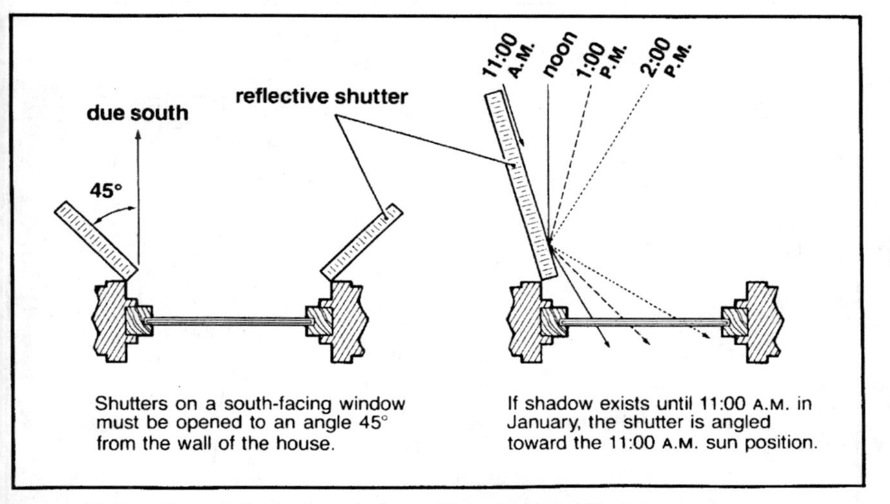

Figure 10-11: ***Do not shade the window with a shutter-reflector.***

Some type of bracing is required for any shutter that doesn't open all the way to an adjacent wall. The loads that crosswinds place on an exterior shutter can be tremendous. To brace a reflective shutter in its optimal position against the wind, a frame

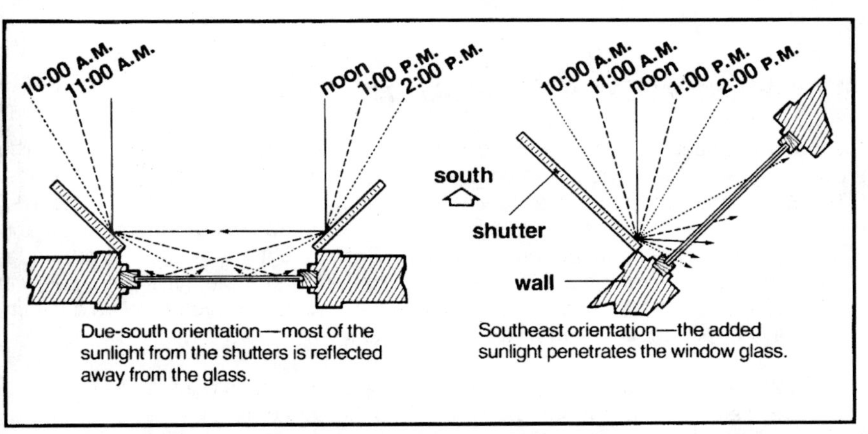

Figure 10-12: ***Comparison of solar-heat gains between shutters facing due south and southeast.***

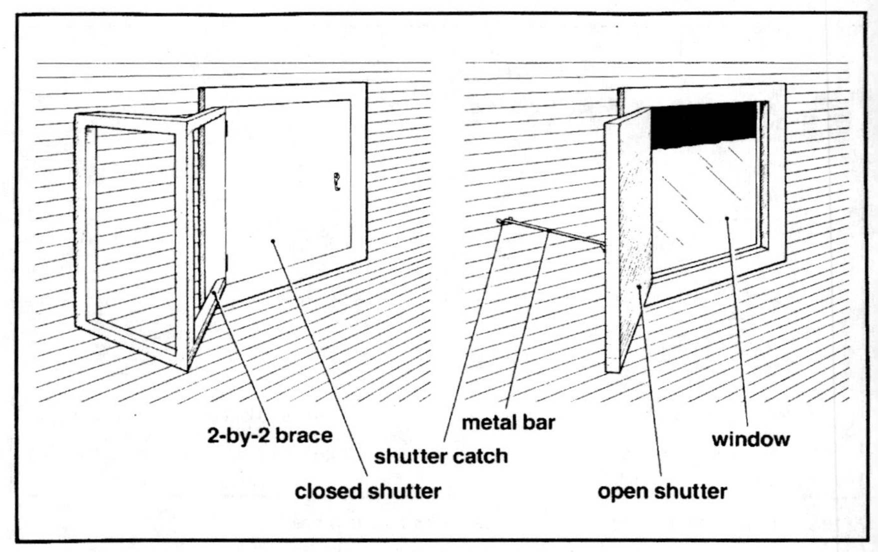

Figure 10-13: Bracing a shutter against wind loads.

can be constructed of 2-by-2s, as in Figure 10-13, and opened to 180 degrees (flat against the wall). A bar can also be used to brace a shutter that is manually operated from the outside.

Top-Hinged Exterior Shutters

Windows beneath a large overhang can employ top-hinged exterior shutters. The overhang above this type of shutter is essential for protection against snow and ice. The Saskatchewan Energy-Conserving Demonstration House shown in Photo 10-4 has 4-inch-thick, aluminum-clad foam shutters which are raised up to an overhang by means of a crank-and-worm-gear system. Even though these shutters are under an overhang, the crank hardware is very strong so that it can lift the shutter with additional ice and wind loads. The shutter can also be lowered partway on hot summer days to protect the window from direct sun.

A simpler approach to raising top-hinged shutters is with a cord and pulley attached to the edge of the roof overhang. However, as the shutter is raised and lowered by this cord it tends not to close entirely without some additional force to draw it tight. The shutter can be sealed tightly with exterior latches but this requires the operator to

Photo 10-4: Saskatchewan Energy-Conserving Demonstration House.

go outside. An excellent remedy to this is described by Bill Shurcliff in his book, *Thermal Shutters and Shades.** Figure 10-14 shows how Shurcliff's design works.

A latching bar is attached to the shutter with yokes, which can simply be steel gate handles. The yokes allow the bar to ride up and down. Notched keepers cut from a hardwood block or bent from ⅛-by-1½-inch steel strips, are bolted to each side of the window. The keeper is beveled on the outside so that when the bar strikes it with some force, it rides up over it and down into a tapered slot to firmly hold the shutter in place. When closing the shutter, the operator swings it vigorously, insuring that the latching bar clears the shutter yokes and falls into place behind them.

The cord which attaches to this bar first runs through a screw eye to keep the bar from sliding sideways. The cord then runs through a couple of pulleys and into the living space. When the cord is drawn, the bar pulls out of the keepers so that the shutter can be raised.

*See Appendix V, Section 1 for book description and complete bibliographical information.

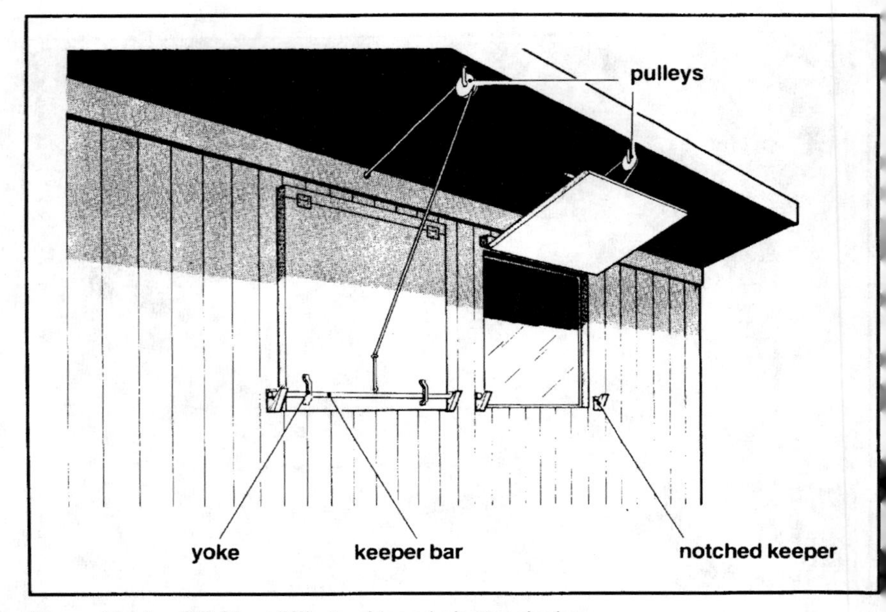

Figure 10-14: Bill Shurcliff's top-hinged shutter design.

A counterweight in the living room is recommended by Shurcliff to help raise the shutter. This weight must be decommissioned by a hook or shelf when the shutter is closed to prevent the weight from unlatching the bar. The counterweight is optional, but it can certainly make handling heavy shutters a lot easier. The shutter can also be raised simply by pulling in the cord and attaching it to a cleat. However, without a counterweight, there is danger of the shutter falling rapidly and hurting someone.

Sliding Exterior Shutters

Sliding exterior shutters cannot reflect sunlight into windows because they remain flat against the wall. However, they do provide good thermal protection for large glass window walls. The best use of exterior sliding shutters is with large panels over large window openings. Tight, durable edge seals are difficult to fabricate for this type of shutter, but if the surface-area-to-perimeter ratio is high, edge seals are less critical. The edge condition should not be ignored, however, even in large panels.

Photo 10-5 shows a pair of barn door-type shutters on the Cooley residence in North Newport, New Hampshire, designed by Clinton Sheerr, an architect in Grantham,

Photo 10-5: Barn door shutters on the Cooley residence.

New Hampshire. These shutters are constructed from 2-by-6 frames laid flat with Thermax insulation inside and ¼-inch plywood skins on each side. An aluminized paint is used on the side of the shutters facing the windows to help reflect heat back to the glass. Each panel is 11½ feet high by 6 feet wide.

Figure 10-15 shows how these shutters are opened and closed. To open one side, a person inside draws a Dacron cord which runs outside and through a series of pulleys. This side is synchronized to the other shutter by another cord and pulley. The shutters slide up an inclined track, and the cord is tied to a cleat to hold both shutters open. To close, the cord is released. An elastic "shock cord" starts the shutter rolling, and the inclined heavy-duty track closes the shutters the rest of the way.

Rollers are mounted on the trailing bottom edge of each shutter to keep it from rubbing against the house and another roller is mounted on the steps outside the bottom of each shutter to keep it tight against the house. Foam-compression weather stripping is used to prevent air infiltration where the shutters meet, and a garage door-type weather strip is applied to the window trim around the entire perimeter of the shutters.

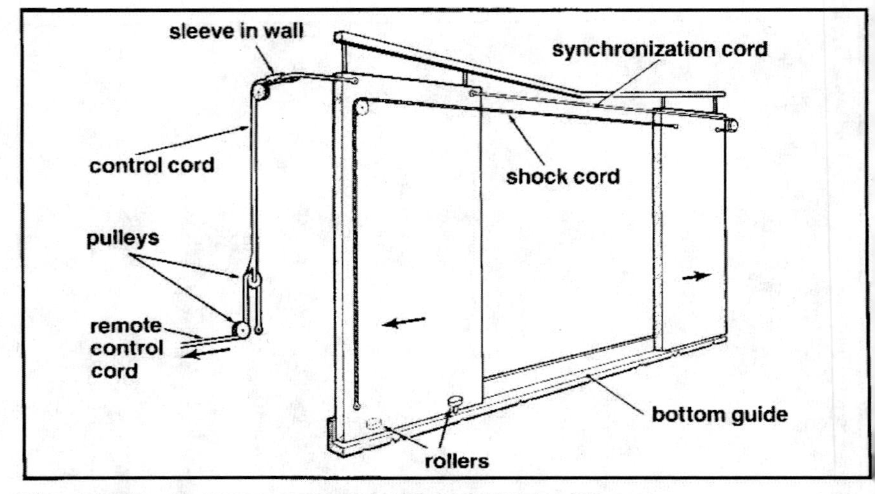

Figure 10-15: Barn door shutters designed by Clinton Sheerr.

Photo 10-6: Shutters on Goosebrook house designed by Total Environment Action.

Another large, sliding exterior shutter was included in the design of the Goosebrook house in Hornsville, New Hampshire. Conceived by Bruce Anderson and Charles Michal of Total Environmental Action, this shutter has two 7-foot-high-by-9-foot-wide panels that slide into the center aisle of a two-car garage (see Photo 10-6). This shutter has 1½-inch polystyrene foam in its core with rough-sawn spruce boards on each side. They hang from a heavy-duty door track with a carpeted lip on each panel that presses against the track to help seal against air flow. An angle iron is attached to the bottom of the shutters so that one edge of the angle is in the space between deck boards to stabilize the shutters and provide a bottom draft seal.

Chapter 11

Sun-Shading Screens

It may come as a surprise to some people to learn that keeping a house air-conditioned in summer can cost more than keeping it heated in winter. It costs you more than twice as much in electricity to extract a unit of heat from your house in the summer as it does to produce a unit of heat with gas or oil during the winter. In most southern areas of the United States the sizing of electrical power generators is based on summer air-conditioning peak loads. In many areas, proper window shading could save enough energy to offset the need for a new power plant. The cost of electricity is also based on peak summer air-conditioning loads, among other things, and is passed on to the residential consumer. (For most utilities a typical peak load occurs on a very hot summer afternoon around 4:00 P.M. when all industrial equipment hasn't shut down yet.) Each square foot of south-facing double-glazed window admits to the interior more than 10 times the heat per hour at high noon than it loses during that hour, and this is true at outside temperatures as low as 30°F.

Because windows collect the sun's heat during both summer and winter, they must be turned off somehow during hot summer months. Exterior window insulation systems are very efficient at shading windows from solar heat gains because they stop the sun before it enters the thermal envelope. Most of the interior window insulation systems in this book (shutters, shades, and curtains) help protect against window heat gains by reflecting much of the entering sunlight back to the outside. However, sun-shading outside window glass that causes the heat to dissipate quickly into the outside air is far more effective than shading inside the house where heat is trapped by the window glass. Several shading options are presented in this chapter for sunny windows in warm climates and for windows where summer heat gains persist, in spite of interior window insulation.

Before applying any type of outside shading, an examination of the hourly location of the sun during each season is necessary. Figure 11-1 shows the sun path for a middle (40-degree) latitude location (Philadelphia, Pennsylvania; Columbus, Ohio;

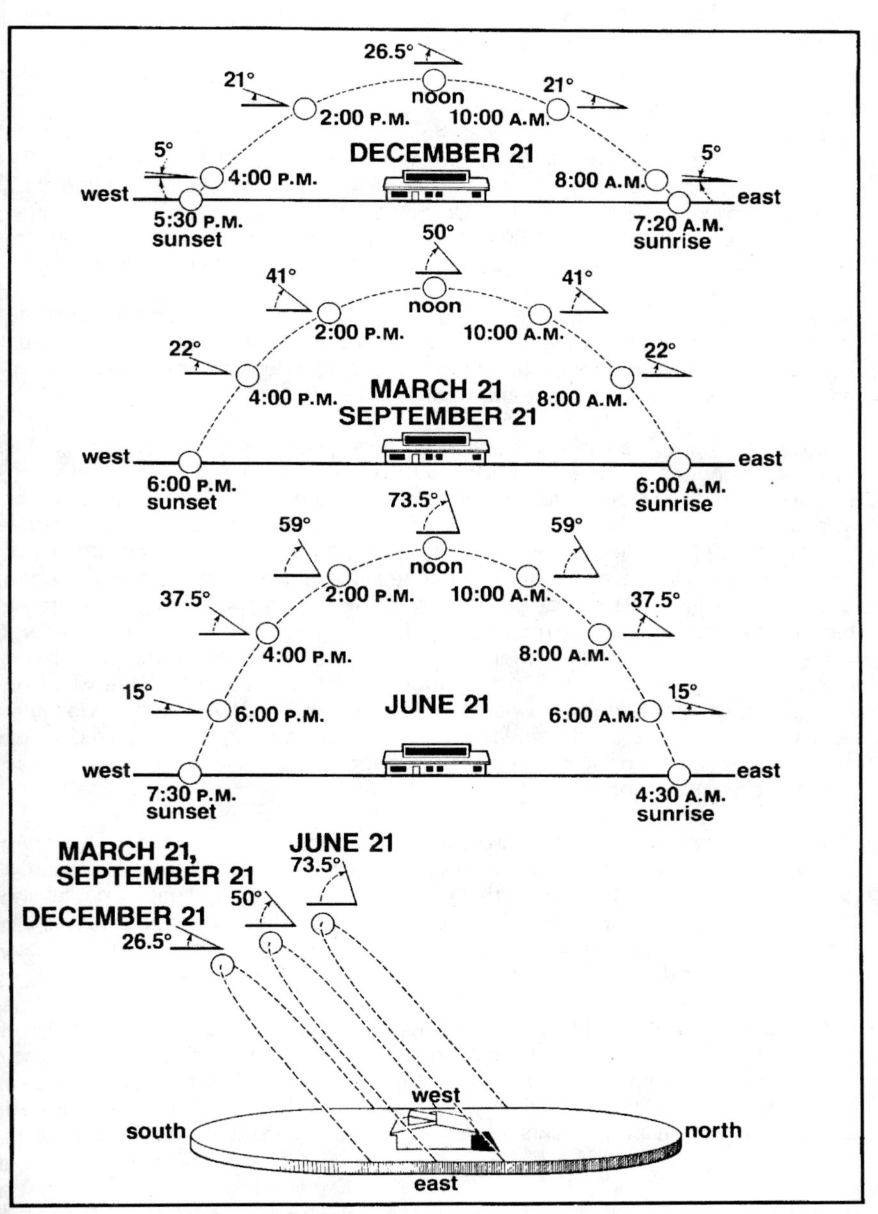

Figure 11-1: Sun angles at 40°N latitude.

Denver, Colorado) on December 21, March 21, September 21, and June 21. The sun rises at 7:30 A.M. on December 21, reaches an altitude of 21 degrees at 10:00 A.M., 26.5 degrees at noon, back to 21 degrees at 2:00 P.M., and sets at 5:30 P.M. On March 21 and also September 21 the sun rises at 6:00 A.M., reaches an altitude of 50 degrees at noon, and sets at 6:00 P.M. On June 21, the sun rises at 4:30 A.M., reaches an altitude of 37.5 degrees by 8:00 A.M., 59 degrees by 10:00 A.M., 73.5 degrees at noon, 59 degrees at 2:00 P.M., 37.5 degrees at 4:00 P.M., and sets at 7:30 P.M.

North of 40-degrees latitude, the sun angles are slightly lower, and south of 40 degrees the sun angles are higher. Note also that on the short days of winter the sun rises and sets south of due east and west, and that on long summer days the sun rises and sets north of due east and west.

Windows facing due south can be shaded during June by a roof overhang. This allows sunlight to penetrate the window during the cold month of December when sun angles are lower. Because the sun is nearly overhead at noon on the summer solstice, the overhang needs to be only 3 or 3½ feet wide to protect a glass window wall extending to the floor. However, fixed overhangs do not adjust the amount of window shading to the seasonal temperature lag and allow either too much sun in late summer or too little sun during mid-spring. In my experience as an architect I have found this problem to be more serious than I originally expected. The owner of a home I designed recently expressed his dissatisfaction at having too much shade on his south-facing windows. A 2-foot overhang, designed to shade these windows during August and September when sun protection is needed, does not allow any solar gain during March and April when supplementary heating is still required. Had I shortened this overhang, the overheating problem in August may have been worse than the shading in April.

East, west, southeast, and southwest windows cannot be protected by a simple overhang. As the sun dips toward the western horizon on a hot summer afternoon, it shines in below even a wide overhang long before it loses its heating capability. Adjustable canvas awnings and shade screens, which can be removed during the winter, are very handy here and are particularly useful for retrofitting existing homes which face southwest or southeast.

Solar Control and Shading Devices by Aladar and Victor Olgyay* and *The Passive Solar Energy Book* by Edward Mazria* provide many charts and designs for fixed, exterior sun-control louvers. During midsummer, the east and west walls and roof of a home receive far more solar radiation than a south wall. Shading the glass of east and west walls is critical in maintaining summer comfort. Besides awnings and shade

*See Appendix V, Section 1 for book description and complete bibliographical information.

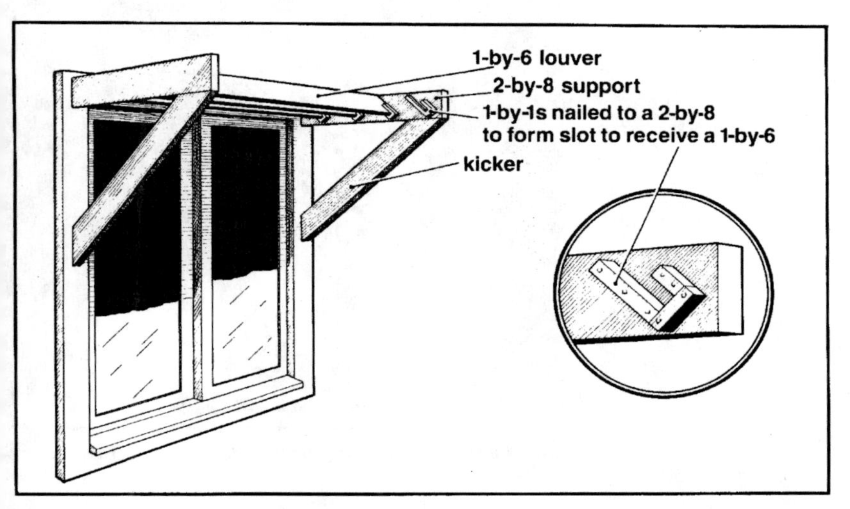

Figure 11-2: Slatted demountable overhang.

screens, carefully designed vertical and horizontal shading louvers can help reduce solar gains on these walls during the summer.

An Adjustable, Wooden Louvered Overhang

An overhang with demountable boards or louvers for shading is shown in Figure 11-2. This adjustable overhang offers a better seasonal performance and greater flexibility than fixed overhangs and can be constructed from 2-by-8 louver support beams which project out over a southern window. One-by-six cedar boards are set at about 60 degrees into 1-by-1 wooden stops which are nailed to the edge of the support beams. These boards are lifted out during the winter and spring to allow full solar exposure to the window, but are replaced when shading is required.

The louver supports should be spaced 6 to 8 feet on center over long window walls or a single support can be mounted on each side of a window where the glass is not continuous. The support beams should be of a weather-resistant wood such as western red cedar, or treated with preservative, and should be supported by 2-by-2 or 2-by-4 kickers whenever added to existing construction. On new construction, the beams can be "scabbed" or nailed onto the sides of the joists that support the ceiling behind the wall.

Photo 11-1: The view through metal louver screen.

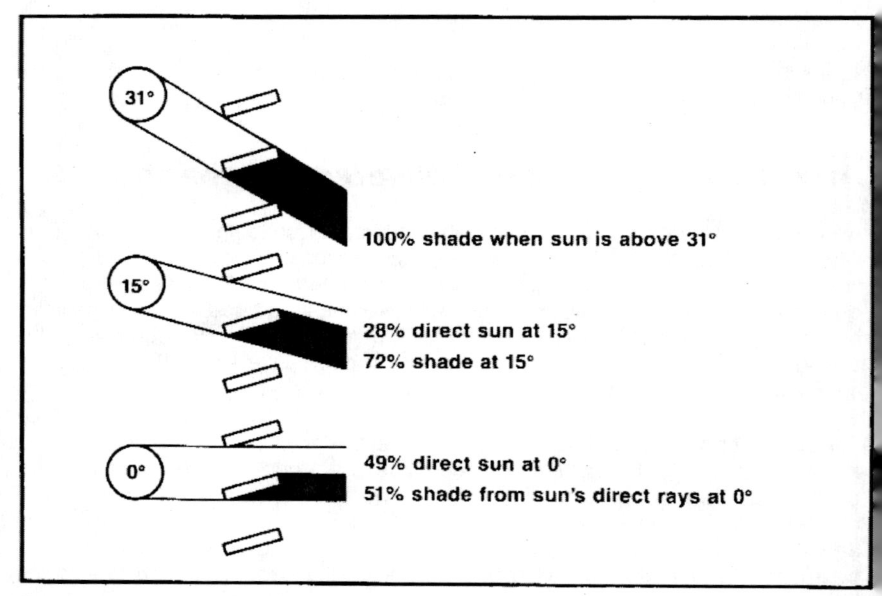

Figure 11-3: Louvered metal screen.

Shading Screens

Sun-shading screens, similar to common insect screens, can be installed over windows to shade direct sunlight as well as prevent insects from getting inside. One type of window shading screen is constructed with tiny metal louvers which are tilted to shade all of the direct sunshine above a certain angle. Photo 11-1 shows a view through a louvered screen. Vision through this screen is surprisingly clear, considering it shades 100% of the direct sunlight.

A less-expensive and less-effective type of shading screen is made from woven fiberglass. It fits into standard aluminum window-screen frames and can be used to replace the existing insect screen.

Metal Louvered Shading Screens

Figure 11-3 shows a cross section through a metal louvered screen. The width of a louver is only about 1/20 inch. Each type of screen has a critical angle determined by the angle and space of these tiny louvers. (In Figure 11-3 it is 31 degrees.) When the sun is at an angle higher than the critical angle, the shade louvers block all direct sunlight. Two types of metal louvered sunscreens are listed below.

KoolShade—Available from KoolShade Corporation, 722 Genevieve Street, P.O. Box 210, Solona Beach, CA 92075. This metal screen contains tiny, fixed louvers woven together with a fine wire. The standard KoolShade screen shades all of the sun's rays from 40 degrees or more above horizontal. The low-sun-angle KoolShade screens out direct sunlight above an angle of 26 degrees.

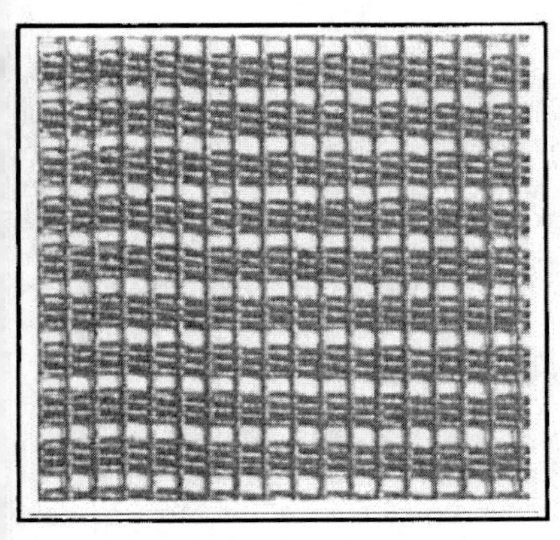

Photo 11-2: A close-up of a fiberglass screen.

Photo 11-3: Fiberglass shade screens are mounted in standard aluminum frames.

Shade Screen—Available from Kaiser Aluminum, 300 Lakeshore Drive, Oakland, CA 94643. This screen shade consists of an array of tiny louvers formed from a sheet of thin aluminum, and can be installed in conventional screen frames. It screens out all direct sunlight above an angle of 31 degrees. It can be mounted against the window in place of a regular insect screen or mounted away from the building to provide free air circulation around the window.

Fiberglass Shading Screens

These shading screens have an appearance very similar to the common fiberglass insect screens but have a special ribbed weave to reduce window heat gains (see Photo 11-2 and Photo 11-3). These screens are low in cost, very easy to install, and block about 75% of the impinging sunlight. Several manufacturers of shading screens are listed here:

Chicopee Manufacturing Company
P.O. Box 47520
Atlanta, GA 30362

Phifer Wire Products Inc.
P.O. Box 1700
Tuscaloosa, AL 35401

J. C. Penney Catalog

Sears Home Improvement Catalog

Vimco Corporation
P.O. Box 212
Laurel, VA 23060

Canvas Awnings

Canvas is a low-cost and very attractive material for exterior sun control on windows. Tents and canopies have been used for many centuries to shade direct sunlight in the hot climates of southern Europe and the Middle East. Canvas awnings, which at one time were commonly used for window shading in the southern United States, are becoming fashionable once again. Awnings provide excellent protection from the sun, even as it lowers in the sky during the late afternoons on hot summer days, and the canvas is easily removed from the frame for full sun penetration at other times of the year. Canvas awnings are light in weight and are available in many attractive colors to add a soft, decorative touch to a home's exterior.

Most awnings are constructed from a few very standard materials: canvas, grommets, ½- or ¾-inch steel pipes, pipe tees and elbows, and nylon cord. Most cities and even moderate-size towns have canvas awning fabricators who custom-make awnings. Some fabricators also sell their materials, which range from all-cotton to acrylic fibers, for home fabrication. All materials gradually fade in direct sun and must eventually be replaced, but if a canvas awning is properly cared for, it should last many years. Acrylics and woven cotton fabrics which are dyed before weaving hold their color the best. Painted or coated fabrics will heat up in the summer sun since they are not porous and cannot breathe. Under extreme heat these coated fabrics deteriorate more rapidly than the woven ones.

Canvas can be hung in a variety of configurations. Photo 11-4 shows an awning hung on a standard, steel pipe frame. These frames are sometimes stationary but often hinge where they mount to the wall as shown in Figure 11-4. This type of awning shades the late-afternoon summer sun very well (see Figure 11-1 for afternoon sun angles) and allows you to raise it with a pull cord when more sun is required. When this awning is raised, a trough which can collect rainwater is formed, so to prevent rotting you should lower the awning after each rainstorm or add grommets to the canvas to allow rainwater to drain out.

A simple, horizontal canvas awning for a patio is shown in Photo 11-5. This awning is constructed from a flat piece of canvas with grommets added along the edges. The canvas is stretched with nylon cord onto a wooden frame.

Another way to use canvas for shading is to construct a roll-down shade or curtain as shown in Photo 11-6. This shade rolls up and down on a floating roller by raising or lowering a pull cord. Details on this roller system are shown in Figure 7-11.

Photo 11-4: Standard canvas awning.

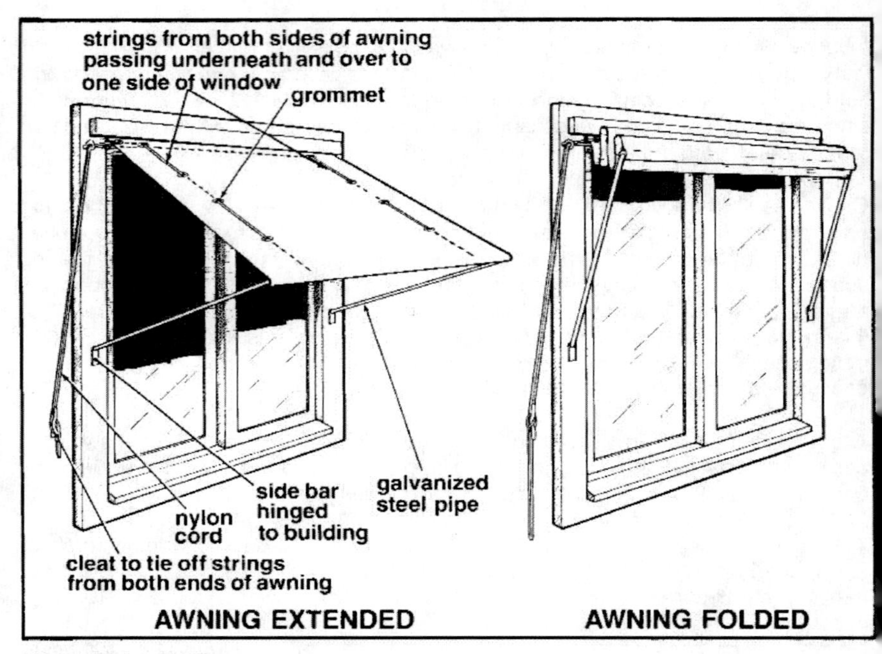

Figure 11-4: Fold-up canvas awning.

Photo 11-5: Flat horizontal canvas awning.

Photo 11-6: Roll-down curtain.

Photo 11-7: Lateral arm awning.

A more expensive type of awning is shown in Photo 11-7. The awning is kept taut as it changes length. Called a "lateral arm" awning, the mechanism holding the outer edge of the fabric telescopes out at the same time the awning is released from a roller. A new, European version of this is a product which has a photoelectric cell and motor to extend the aluminum frame automatically when the sun is shining. The motor is turned off during the colder months to prevent shading glass areas when heat is needed. This type of awning is useful on passive solar homes with large glass areas.

These are just a few options for adjustable exterior window shading on your home. You can also shade your home with careful landscaping; deciduous trees and vines produce leaves when you want shade and drop them as the days get cooler and the sun becomes less intense. Whether you choose natural shades from plants or man-made ones, what is important is that you don't ignore summer shading of windows. You should never be caught with direct sunlight shining into your home at the same time your air conditioner is operating. The personal and environmental costs of this common wasteful practice are much too heavy.

Chapter 12

Exterior Roll Shutters

Exterior roll shutters fall into two general categories. One has numerous, narrow, horizontal slats and is commonly referred to as the "rolladen." It has been used widely in Europe for many years. The slats in this type of shutter are around 1½ to 2 inches wide, allowing them to roll onto a small overhead reel. The other type of exterior roll shutter is borrowed from a popular American product, integral with our love for the automobile—the roll-up garage door. Even though it is not generally recognized as a product that can be adapted for insulation over large glass areas, with only minor modifications, the garage door can do just that. A fully automatic garage-door system, including the motor operator, can be purchased for about $5 per square foot.

Rolladen Shutters

The use of the rolladen shutter dates back to the early 1800s. This window shutter enjoys such outstanding popularity in Europe today that it is quite surprising there are not more of them in the United States. The sales of these shutters during 1976 in Germany alone exceeded $600 million. It is as common in Europe for controlling window lighting and providing privacy as interior shades and draperies are in this country. However, they have always been an expensive item here, one factor which has attributed to their limited use. They sell for as much as $8 to $12 per square foot, and that's without the electric-drive motor!

The first rolladen was constructed from wooden slats which were hung from strips of canvas and wound onto a roller at the top of the window. Modern slats are usually hollow strips of extruded aluminum or PVC, and provide better security, in addition to lighting and heat control (see Figure 12-1). They shield against wind damage and unwanted traffic noise and offer firm resistance to intruders.

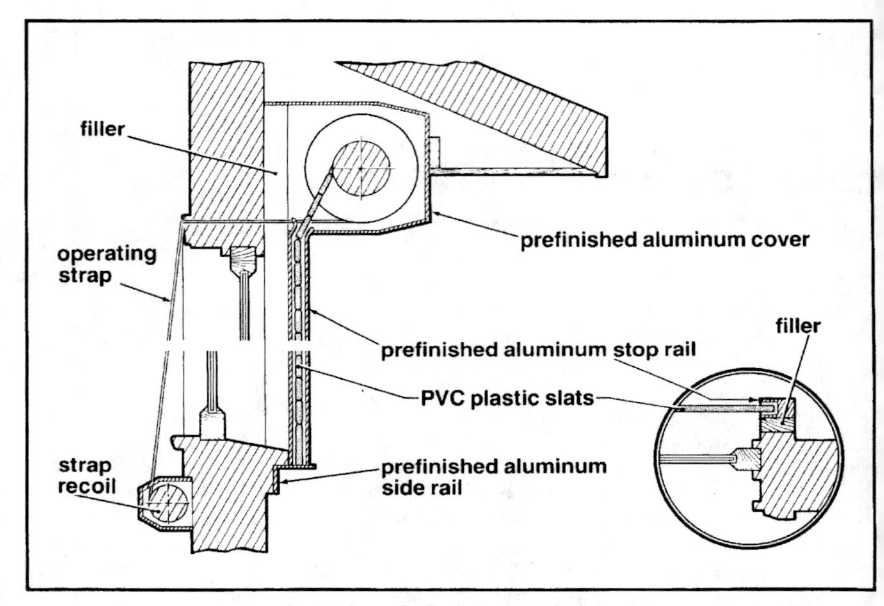

Figure 12-1: Cross section of a Rolladen-type shutter.

The slats on most rolladen shutters have slots along one edge that recess into the adjoining section when the shutter is completely lowered. They open when the upper part of the shutter is raised slightly so that each slat hangs from the one above

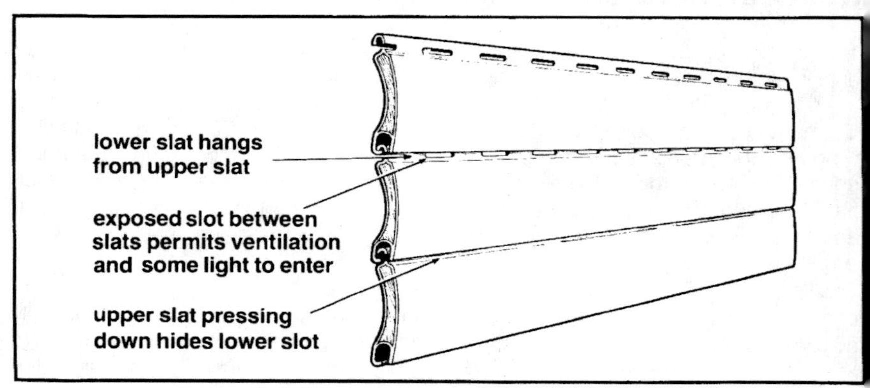

Figure 12-2: Slots open when slats hang.

instead of resting on the one below (see Figure 12-2). This feature is ideal for controlling heat gains in hot climates. The slats are outside the window where they can intercept the sun's rays before they enter the home's weathertight. envelope. By raising the upper portion of the shutter slightly to open the slats, ventilation through the slats prevents a heat buildup between the shutter and the window glass. In addition, a small, controlled amount of sunlight can enter through the slots for daylighting purposes.

The use of exterior slatted shutters for night insulation has only recently become a major concern. The R values of most of the systems currently available are low; this is partly due to the kind of material used for the slats. Hollow aluminum slats, even when filled with urethane foam, have a high rate of conductance since the heat flows quickly through the aluminum around the edges of each slat. The insulation provided by PVC slats is superior to aluminum, but the thermal performance of them is still disappointing. Tests of a hollow, PVC slatted shutter yield an R value of only 1.6. Over half of this additional R value can be attributed to the air space created between the shutter and the window glass; the thermal resistance of the shutter itself accounts for the rest. Foam-filled PVC slats are the most effective insulators, but what is really needed is a slat thicker than the ⅜-inch thickness common to many shutter slats.

The size of the slats is limited to some extent by the need to keep the diameter of the shutter at a reasonable dimension when rolled up. However, this design trade-off must be reexamined in order that a more effective shutter can be developed for use in colder climates. A ¾-inch-thick shutter of foam-filled PVC slats with airtight joints between the slats could provide a reasonable R value for night insulation in all but extremely cold climates. As the state-of-the-art now stands, currently available rol-

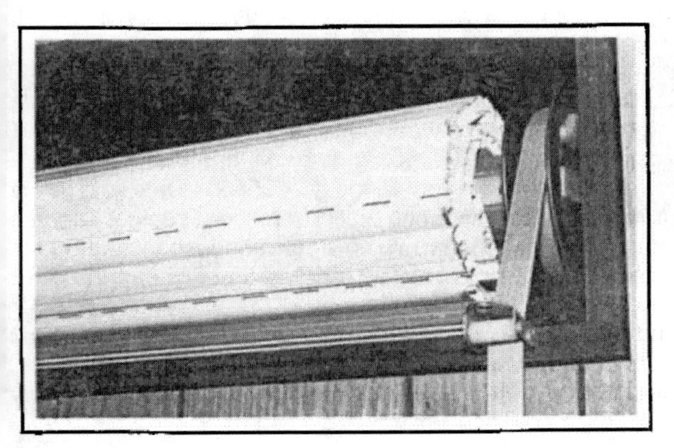

***Photo 12-1:** Manual strap operator. This photo of a display unit shows the wall above the window cut away to reveal the roller. In a real unit the roller wouldn't normally be seen from inside the home.*

laden shutters are only effective in warm climates with mild winters where cooling is a greater problem than heating.

A diagram of a typical, slatted exterior roll shutter is shown in Figure 12-1. The shutter slats are usually about ⅜ inch thick and 1½ inches wide. They can be cut to any length and therefore can be made to fit any window. There is a roller at the top of the unit to raise the slats. In new construction the roller cover can sometimes be omitted because the entire roller assembly is recessed into the eave soffit. Side rails are attached to the sides of the window to hold the slats in place when lowered and to provide an air seal at the edges. A strip of wood or metal serves as a stop to catch the shutter when in its lowered position.

The most common way to open and close them is manually, with a strap (see Photo 12-1) which runs through the wall and down the interior edge of the window to a spring-loaded recoil unit that draws the strap in and removes any slack. The recoil unit has a spring like the one in a carpenter's measuring tape. To raise the shutter you simply pull down on the strap which is attached to a pulley at the end of the shutter roller. To lower the shutter the strap is released. A crank mechanism is available on some systems, as are electric motor-driven rollers.

Pease Rolling Shutter—Available from The Pease Company, Ever-Straight Division, 7100 Dixie Highway, Fairfield, OH 45023. This shutter is formed from interlocking, hollow PVC slats. They fit a wide variety of window sizes and operate by a strap-pulley system from the inside of the home. Slats for the Pease Shutter are available in a standard V size and in a smaller VM size. The V-size slat has a superior R value to the smaller one.

Most of the rolladen shutters available in the United States are imported from Europe. The Pease Rolling Shutter is one of the only shutters of this kind manufactured in this country.

Rolladen Shutter—Carrying the European generic term as its American trade name, this shutter is distributed by American German Industries, Inc., 14601 North Scottsdale Road, Scottsdale, AZ 85260. Another distributor for it is Future Image, Inc., 4348 Jetway Court, North Highlands, CA 95660. The Rolladen is available in two slat profiles: a hollow PVC slat (called "the honeycomb") and an aluminum slat filled with polyurethane foam. You may also choose the method of operation: an electric motor operator, a manual crank, or a pulley strap.

Serrande Shutter—This roll shutter, marketed by Serrande of Italy, P.O. Box 1034, West Sacramento, CA 95691, is available with both wood and PVC slats. With more than ½-inch slats, it was the first energy-conserving shutter to qualify for the Energy Conservation Ordinance of Davis, California. The system can either be operated manually by a pulley-strap system, or by an electric motor operator.

Metrovox Roll Shutter—Metrovox, Inc., 1308 Gresham Road, Silver Spring, MD 20904, markets a PVC roll shutter similar to the other rolladen shutters.

Garage Door-Type Shutters

Overhead garage doors today are as much a common part of the American scene as are superhighways and hamburger stands. Like so many other things around us, they are taken for granted. At face value, the garage door is just something to protect the car from bad weather and vandals. It is rarely thought of as a way to insulate an unusually large area of glass.

The creative mind does not only see things at their ordinary value, but constantly seeks the origins of things to rearrange them in new and more interesting or effective forms. In looking for ways to get solar heat directly into the home through windows, ways to store the heat in the living space around you, and ways to trap this heat with movable insulation panels, passive solar designers are rearranging traditional building components into totally new configurations. The use of the overhead garage door to insulate large, south-facing areas of glass at night is one of the most significant design concepts to come out of the passive solar design movement. As the movement grows and becomes more sophisticated there will probably be improvements on the garage door that will increase its insulating value. See the end of this chapter for a modified door soon to be on the market.

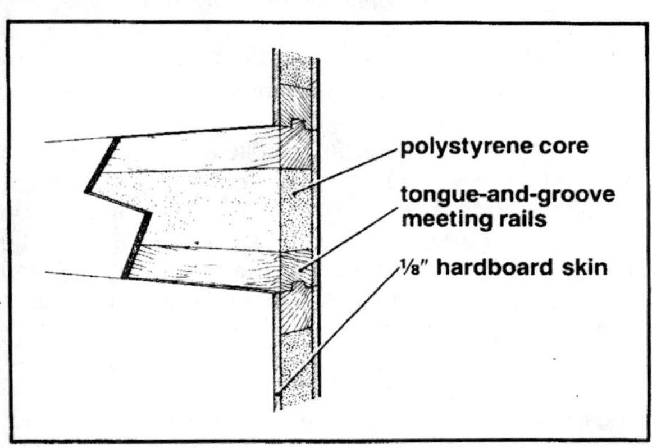

***Figure 12-3:** Standard foam-core garage door panel.*

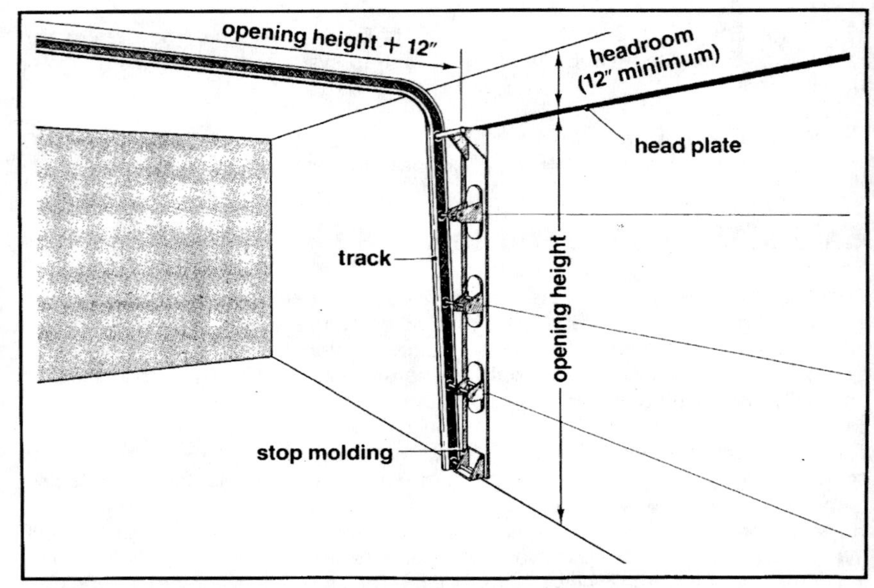

Figure 12-4: The wedge action of the door roller-hinges.

Figure 12-3 shows a section through a traditional garage door manufactured by the Overhead Door Corporation. The door shown in this figure is the Dura 1 125 series, which has 1⅛-inch expanded polystyrene foam between two ⅛-inch-thick exterior hardboard skins. The panels have tongue-and-groove edges which interlock as the door is lowered. Each panel is 1 foot, 9 inches high, in 8-foot, 9-foot, 16-foot, and 18-foot lengths. Four panels make a 7-foot-high door assembly.

At the edge of each door panel is a heavy-duty hinge attached to a steel ball-bearing roller, which runs up and down in a steel door track. On a standard track system the door rolls up overhead into a horizontal position as in Figure 12-4. The vertical section of the track is inclined ¼ inch to help the door seal when closed. Starting at the bottom panel and moving up each successive panel joint toward the top, you find that the rollers are mounted farther out from the hinges. This configuration wedges the door tightly against the stop molding as the door is lowered.

To use this standard door for movable insulation, care must be given to insure proper edge seals. The stop moldings must be carefully aligned and weather-stripped. Because this system is outside your wall, heat warmed from the window area will rise and a top seal is very important. With some finesse, a rubber flap can make a good top seal. Any other cracks between panels should fit snugly or they should be weather-stripped.

Tim Maloney at One Design (see below) has been adapting garage door hardware to movable insulation panels for several years. The overhead track usually must be tilted instead of horizontal if the door is to raise into an attic space above the ceiling insulation. This entails some delicate counterbalancing of the door with heavy springs. Good edge seals, particularly with the top or "throat" seal are difficult to achieve without a great deal of experimentation. The biggest shortcoming of the standard, foam-filled garage door is its low R value, which reduces its cost-effectiveness. The standard door panels have too much wooden frame area, and the type of foam insulation used is an inferior grade. A door system with better insulating value will soon be available from One Design; it is called the Roldoor.

> Roldoor—From One Design, Inc., Mountain Falls Route, Winchester, VA 22601, this door system will be available either individually or in combination with the Water Wall Modules, also by One Design. The Roldoor panels have a vacuum-

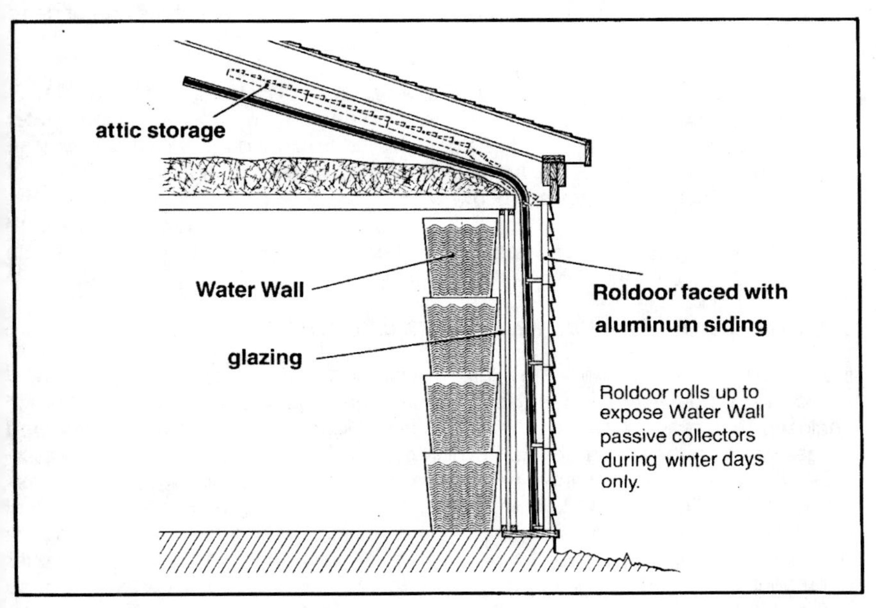

Figure 12-5: The Roldoor system.

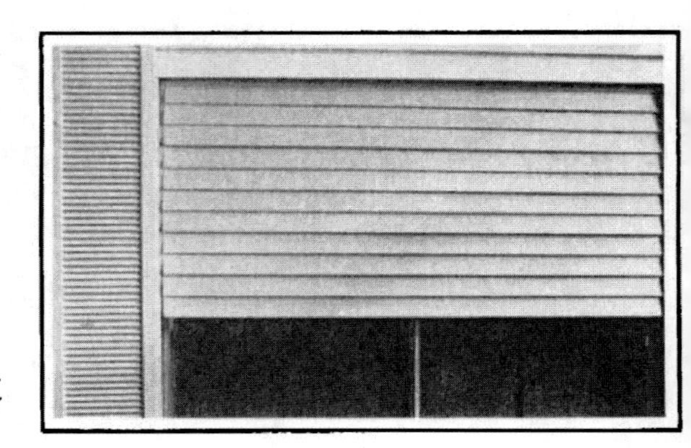

Photo 12-2: Roldoor panels closing over wall.

formed PVC exterior and are filled with 2 inches of isophenol (R-5 per inch) foam. The track is modified so the doors roll up at a 20-degree angle instead of dead horizontal.

Figure 12-5 shows a cross section of a Department of Housing and Urban Development (HUD) passive solar, award-winning design in which this Roldoor-Water Wall combination was used. The door is rolled up during the daytime to expose the Water Wall modules to direct sunlight. It is lowered at night to trap the heat gained. Photo 12-2 shows a home with these panels about to be closed. The panels have a vinyl siding to match the home. When the panels are up, a bold solar heating system is revealed, but when the panels are lowered, the home looks like every other home on the block.

Garage Doors to Form a Winter Enclosure

Most applications of movable insulation described in this book provide a thermal shield over windows and other glass areas at night. The seasonal use of shutters to enclose breezeways should not be overlooked. Figure 12-6 shows a sun-tempered breezeway designed by architect Fuller Moore of Oxford, Ohio. The breezeway in this HUD passive solar, award-winning home has transparent, corrugated fiberglass on the roof to allow sunlight to enter during the winter. Garage doors at each end of the breezeway are closed during the heating season, making this an ideal space to dry clothes, or firewood, or just an extra space for children to play in on sunny winter afternoons. During the summer the garage doors are opened to block the overhead sunlight, and keep this space cool.

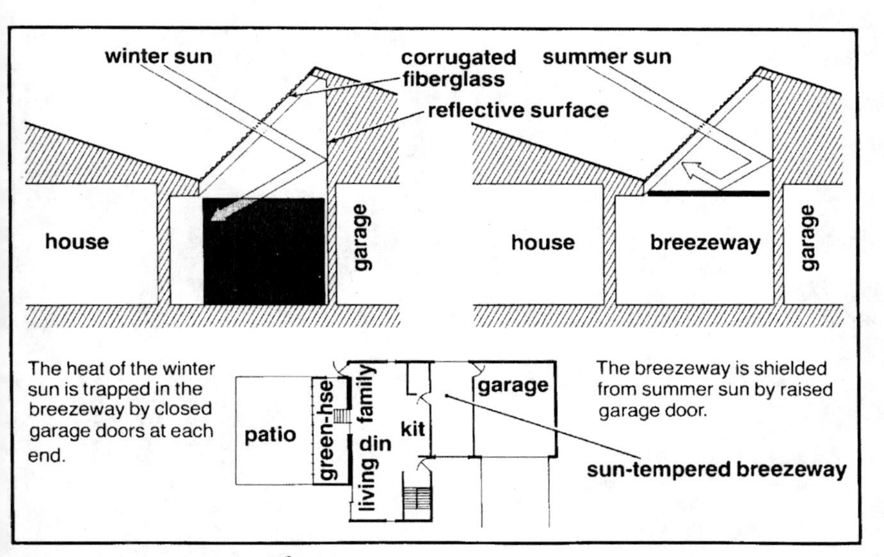

Figure 12-6: Sun-tempered breezeway.

Part IV

Movable Insulation for Skylights and Clerestories

Clerestories and skylights are vulnerable to higher rates of heat loss because the air temperatures near the top of rooms with high ceilings are often 10° to 20°F. warmer than those at floor level. Triple glazing reduces these losses, but not nearly as well as movable insulation.

The design of window insulation for clerestories and skylights requires more sophistication than for windows at standard wall heights. Although there are ample swing clearances in the upper reaches of the room, the insulation system must be operated by remote chains, pulleys, or other similar hardware. The construction of most of the insulating panels in this part is similar to those in Part II, but the requirements for operator hardware are more complex. Only with very careful planning and installation will a remotely operated system seal properly when closed.

Skylights and clerestories, at one time a rare item in home construction, have become more common in the last 25 years. The placement of window glass in a roof opening or high on a wall has long been recognized as an optimal source for natural lighting and has lately been shown to be quite beneficial in passive solar heating applications. Light entering from skylights and clerestories directly illuminates a work surface below while light from lower windows or room lighting fixtures often must be reflected off the ceiling to be useful to the occupants. When properly located in the home, overhead windows can also serve to bring the sun's heat into the northern portions of a home, allowing a more even distribution of passive solar heat. However, the usefulness of skylights and clerestories in space heating is coupled with their potential for high heat losses and thus they are prime candidates for movable insulation applications. In fact, because the insulation is resisting a higher rate of heat loss, it is actually providing a greater benefit than insulation systems used with vertical windows of the same size.

Chapter 13

Shutters for Skylights

As lighting units, skylights are handy on north-facing roofs because they can replace the need for vertical north-facing windows for daylighting. Although north-facing skylights provide good daylighting, they do not provide a useful heat gain during the winter and therefore should be small to minimize heat losses at night. To further reduce heat losses, north skylights should include a night insulation or should at least be constructed with triple glazing. They do not need to be large to be effective because they provide light directly from above. In fact if they are too large, they can bring in excessive, unwanted heat during warm weather seasons. A couple of small skylights on the north side of a home supply as much useful daylight as windows several times their size.

The use of large, sloping skylights in a south-facing roof is a new concept in residential design. They are sometimes called "roof apertures" by passive solar home designers since, like a camera shutter, they open and close the roof to direct sunlight. The south-facing skylight is a very effective way to heat up a masonry or water wall and store heat for nighttime use (see Figure 13-1 and Photo 13-1).

Because a south-facing skylight can collect several times as much heat on a typical summer day as it does on a day in January, it is absolutely essential that some type of insulation be used to prevent overheating by direct summer sunlight. Without summer insulation the benefits of winter sunlight and heat are nullified by warm season discomfort and increased air conditioning costs.

Sliding Skylight Shutters

Photos 13-2a, 13-2b, and 13-3 show sliding skylight shutters in the Village Homes development in Davis, California. On winter days, these skylights heat up water-filled

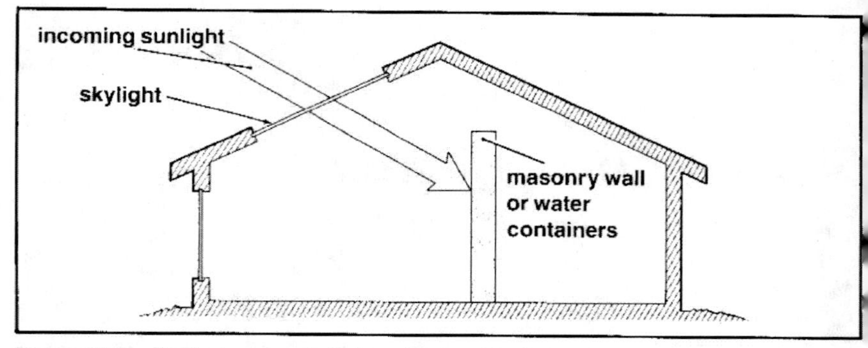

Figure 13-1: Roof aperture skylight collects solar heat.

Photo 13-1: Roof aperture skylights in Village Homes.

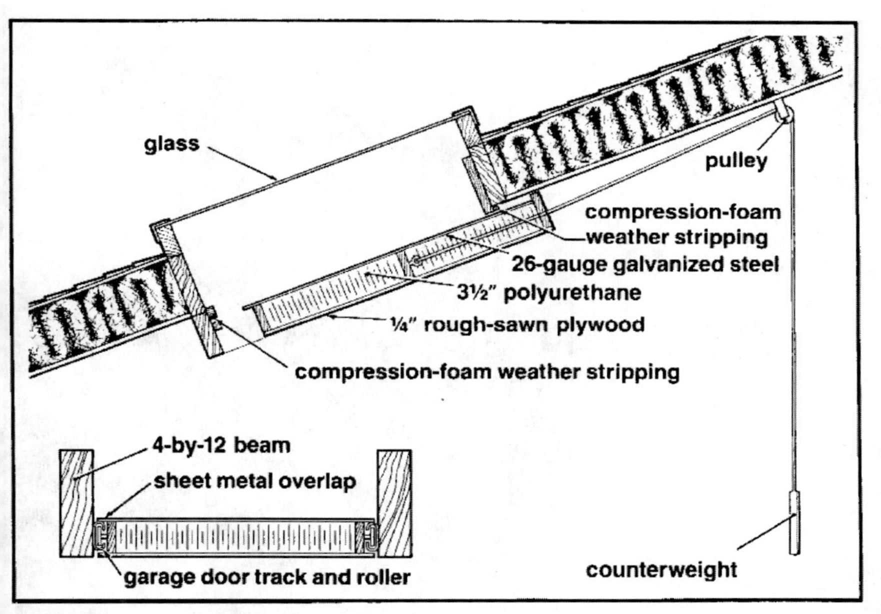

Figure 13-2: The Village Homes shutter.

culverts at the rear of the living room. They are fabricated on the construction site and are considered an integral part of each home. Each shutter is raised by simply pulling down on a counterweight.

Figure 13-2 shows details for the sliding shutter in Davis, designed by Mike Corbett and David Norton. This shutter has a 1-by-4 frame that is filled with 3½ inches of

Photos 13-2a and 13-2b: Interior views of skylights with shutters.

Photo 13-3: Raising the shutter.

polyurethane foam to achieve an R value of about 21. The bottom is faced with ¼-inch rough-sawn plywood and the top is faced with 26-gauge galvanized steel. The panel slides on standard garage door tracks mounted on the sides of roof beams 4 feet apart.

Devising a method to seal the edges on sliding skylight shutters is no easy task. Contact is required, but too much friction will prevent the shutter from lowering into place. In this design the steel sheet overlaps the frame and rests snugly against the top of the garage door track, sealing the sides. The bottom lip of the steel sheet slides into a wooden slot with a compression weather stripping. The top of the shutter has a wooden lip that butts against a weather-stripped wooden lip on the skylight frame. For the sides to seal correctly, the garage door roller-mounts that are bolted to the plywood facing must be precisely the right length or shimmed as required. For the top and bottom to seal correctly, the length of the skylight opening must match the length of the shutter.

Photo 13-4: Pulleys between shutter and counterweight.

The cord for this shutter is 3/16-inch vinyl-coated steel wire which is attached to a 1-by-4 in the center of the panel with a cable tie and an eye bolt. The cord runs up along the ceiling to a pulley, then either straight down to a counterweight, or horizontally along the rear wall and then down as shown in Photo 13-4. This cord sometimes runs through a collar in a side wall to a counterweight in an adjacent room. Counterweights can be made from either a steel or plastic tube that is filled with lead weights. The weights should be added to each tube after the shutter is installed, with the amount of weight in each tube balanced to the load and friction of each shutter. The tubes should be filled until they have just enough weight to prevent each shutter from closing automatically.

Although these shutters are very effective at trapping winter heat and blocking the summer sun, they are quite expensive. The expense could be reduced somewhat by using a thinner section, maybe 2 inches of foam instead of 3½ inches.

Some skylight shutters recede into pockets as shown in Figure 13-3. These shutters, however, reduce the thickness of the insulation in the roof. This is generally not

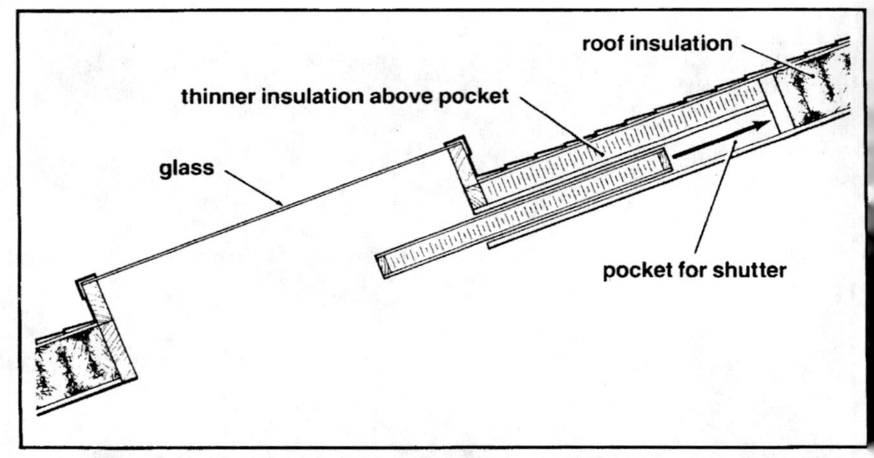

Figure 13-3: Pocket shutters can rob the roof of needed insulation.

recommended since adequate roof insulation is very important, and sloped roofs usually have a minimum amount of insulation to begin with.

Louvered Skylight Shutters

Photo 13-5 shows a louvered skylight shutter designed by Steve Merdler of Santa Fe. The panels are constructed from ⅛-inch luan door paneling on a ¾-by-2-inch frame with 2 inches of polystyrene foam insulation inside. They pivot on a ¼-inch steel pin and strap as shown in Figure 13-4.

These shutters are opened and closed by means of a cord that is attached to the bottom edge. There is very little force exerted from the cord to make the shutters seal tightly, since the operating cord is pulled parallel to their closed position. In spite of this fact, they can be closed reasonably tight.

This type of shutter can be improved by adding tie bars to each end, forcing the louvers to pivot in unison (see Figure 13-5). A tension spring used on screen doors can be mounted on one of the louvers to pull the entire assembly closed, while the top cord and pulley are used to pull the louvers open.

The edges of these shutters have a small lip so that they join together when closed. For an improved performance, the edges can be beveled and weather-stripped, or better yet, they should interlock like the Skylid (see below). The ends of the louvered

Photo 13-5: Louvered skylight shutters.

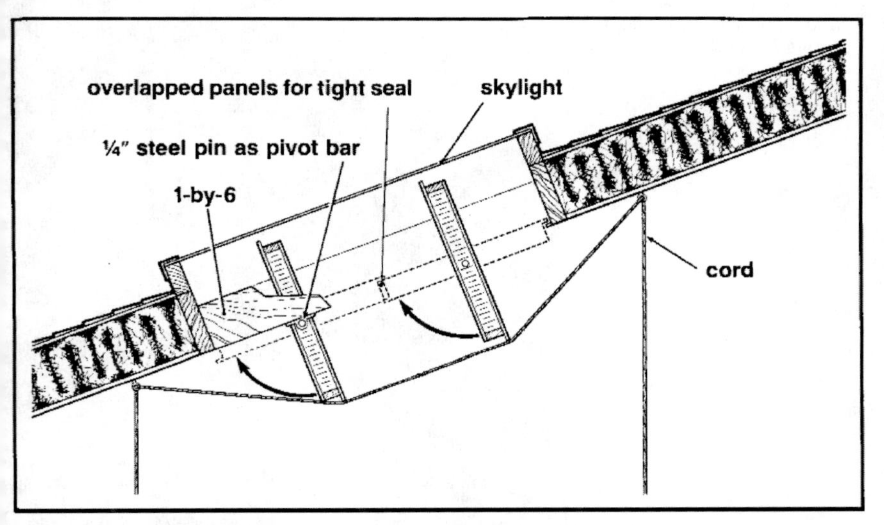

Figure 13-4: Pivoting louvered shutters, designed by Steve Merdler.

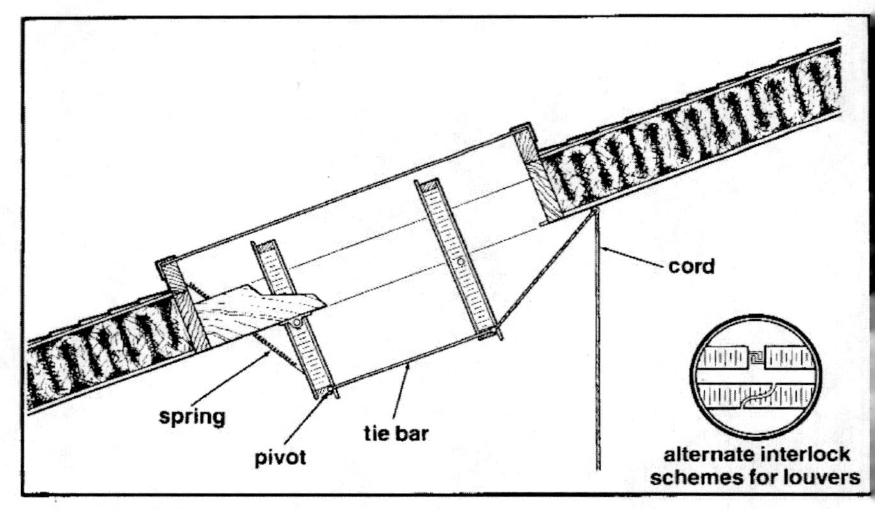

Figure 13-5: Modification of the Merdler design.

Photo 13-6: Skylids in Pitkin County Airport just outside Aspen, Colorado. Note Beadwall panels in background.

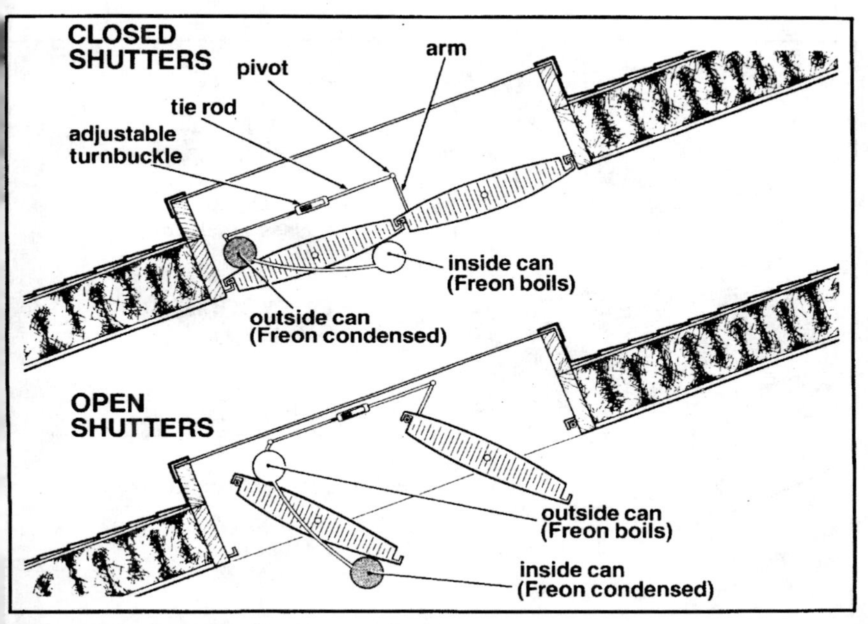

Figure 13-6: Freon gravity drivers in Skylids.

panels are difficult to seal as they must be allowed to pivot freely; a soft rubber or silicone cloth strip helps here.

The Skylid—Zomeworks Corporation, P.O. Box 712, Albuquerque, NM 87103, manufactures and sells a very unique skylight shutter called the Skylid. Photo 13-6 shows Skylids installed in the Pitkin County Airport, just outside Aspen, Colorado. These louvers open and close automatically from the shifting weight of Freon between two canisters, one located on the sun-facing side and the other on the shaded room side. When the sun comes out, the Freon in the sunny canister boils (see Figure 13-6) and moves to the shady side where it condenses and collects until the increased weight of the inboard canister tips the Skylid open. When the sun retreats during the late afternoon, Freon reenters the outside canister, which tips the shutter closed. To prevent the Skylid from opening on hot summer days, a manual override chain is provided.

Steve Baer at Zomeworks has been working with the gravity drivers on Skylids for a number of years. An excellent paper describing the subtleties of using the weight of Freon canisters to open and close shutters is available from Zome-

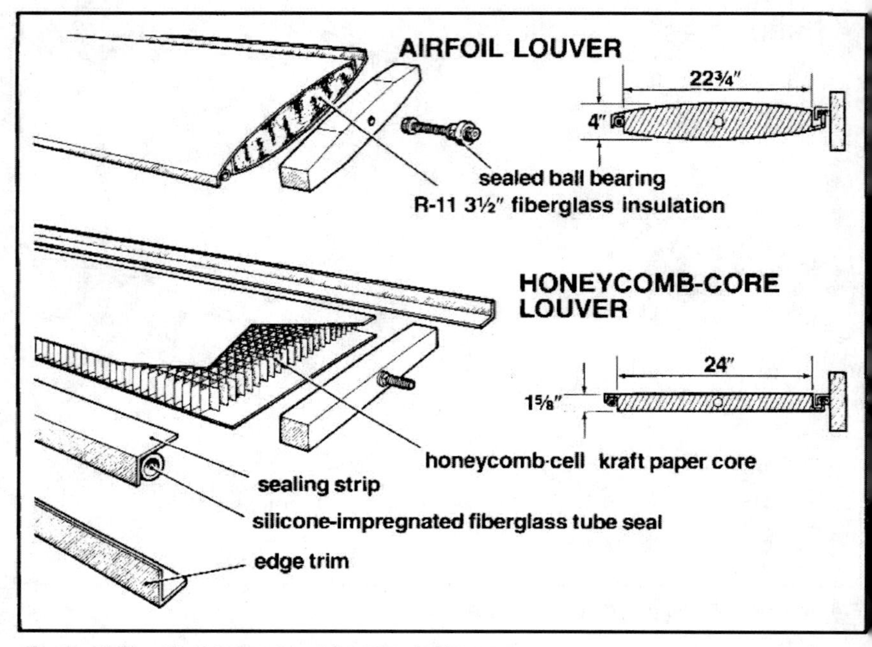

Figure 13-7: Skylid panels — two standard types.

Photo 13-7: Horizontal Skylid at the Benedictine Monastery, Pecos, New Mexico.

works, entitled "Gravity Drivers." Zomeworks is now producing canisters with 15 pounds of Freon while the earlier models had only 5 pounds. Having more force in the canister's shifting weight improves the edge seals in Skylids and offers much more versatility than gravity-driven designs.

Two types of louvers are produced by Zomeworks, as shown in Figure 13-7. The airfoil type consists of two pans of 0.04-inch shop-grade aluminum, curved over wooden ribs to form a cross section 4 inches high at the center. There are 3½ inches of fiberglass inside and the unit has interlocking edges. A flat louver is also available with a 1⅝-inch core of kraft paper honeycomb sandwiched between sheets of aluminum.

Skylids are available in two- and three-louver standard models ranging from approximately 4-by-4 feet to 6-by-10 feet. Variations on the standard sizes are available in the flat honeycomb core louver on a special-order basis.

The standard Skylid works best when mounted at an angle of 15 to 75 degrees, which allows the weight of the Freon canisters to fully open and close the shutters. Special canister mounts are available for vertical installations.

Another type of Skylid shutter that folds out of the way, against the sides of the light well, is shown in Photo 13-7. Note the extension arms on the Freon canisters which allow the Freon gravity drivers to effectively tip the shutters open and closed. This shutter is also available upon request.

Hinged Skylight Shutters

Figure 13-8 shows a detail for a skylight shutter I have used on 45-degree-pitch south-sloping skylights in homes I have designed. This panel contains 1-inch-thick Thermax board set into a frame made of 2-by-2s. The back side is faced with ⅛-inch textured hardboard and the sun-facing side is covered with colored Astrolon. (Astrolon is softer in appearance than the foil face on the insulation but is almost as reflective.) The edges are weather-stripped, and two magnetic cabinet latches hold the panel over the skylight when closed. A snap-in cabinet catch holds the shutter against the opposite ceiling when open. The shutter is operated with a pole that has a hook on one end.

Hinged shutters can also be very useful in skylight wells. Figure 13-9 shows a shutter design that hinges up by means of a pole and hook (a pull cord and pulley can also be used). The shutter folds down onto a 1½-inch ledge strip at the base of the well.

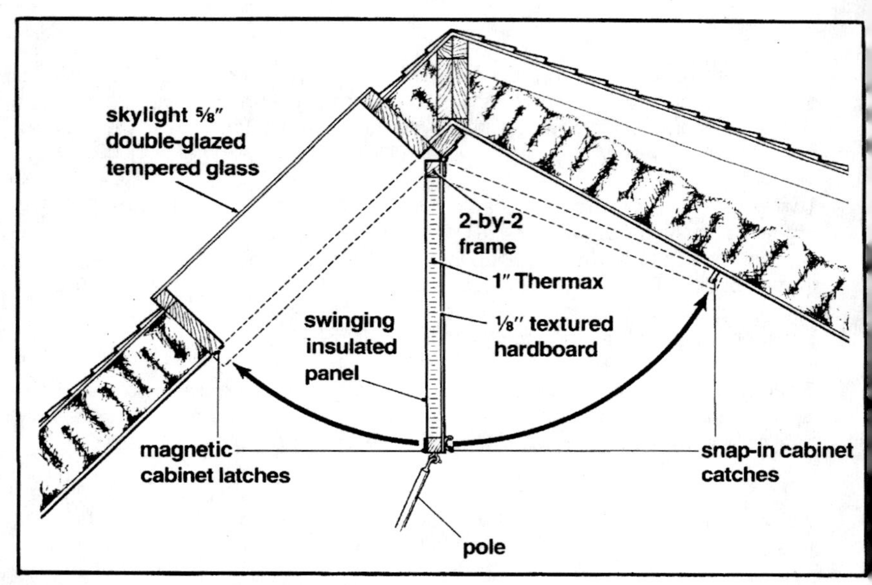

Figure 13-8: Top-hinged skylight shutter.

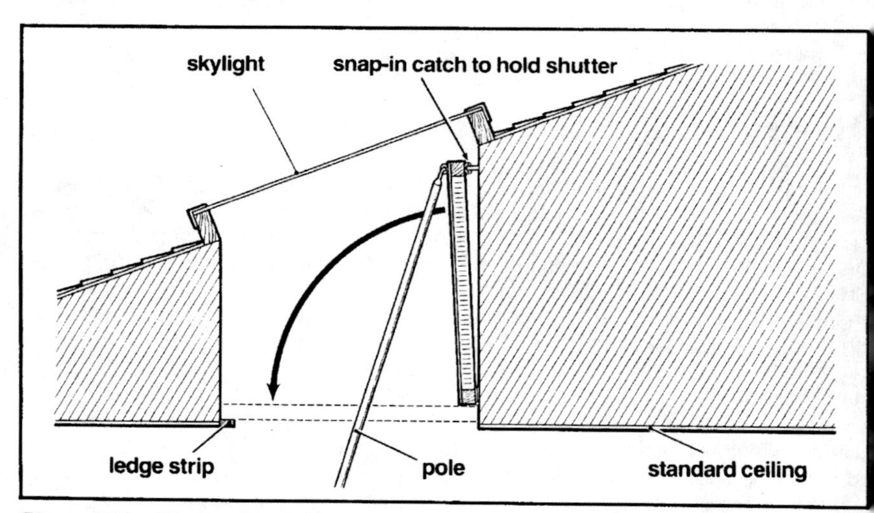

Figure 13-9: Shutter for skylight well.

This type of shutter is normally constructed with 1 to 2 inches of rigid insulation board again mounted inside a frame made from 2-by-2s and faced with thin door paneling, similar to the shutter designs in Chapter 8. A translucent panel can also be made by putting a thin layer of fiberfill between two layers of translucent cloth and stretching it over the shutter frame.

Chapter 14

Shutters for Clerestory Windows

The term "clerestory" originated in the medieval church as the name for the upper windows and walls of the central nave (see Figure 14-1). The clerestory rises above the roof that covers the side aisles and showers light into the central place of worship. Recently, the clerestory window has become a common feature in modern homes. Two sections of roof are separated by a short, vertical window wall that is commonly about 3 or 4 feet high, often running the entire length of the house.

Residential clerestories have often been oriented toward the north because of an outdated, window planning guideline stating that north-facing windows provide the most even quality of light. While there is some rationale to this claim, it is also very shortsighted, and a great deal of direct solar heating potential is lost by this practice. The south-facing clerestory is an excellent way to provide natural lighting for a home while bringing winter solar heat into the living space. Photo 14-1 shows a clerestory on a house in Davis, California.

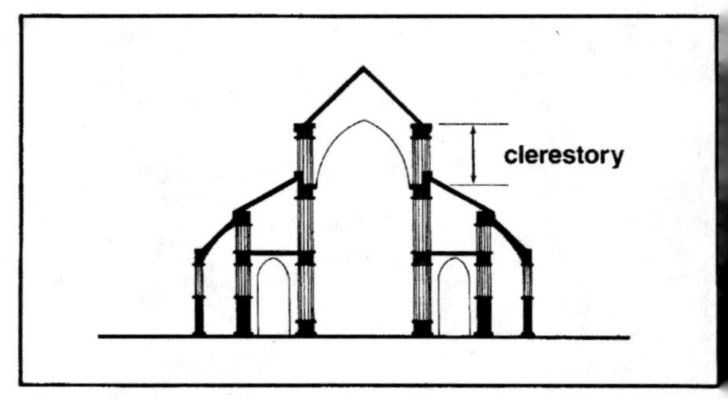

Figure 14-1: Clerestory in a medieval church.

Photo 14-1: A clerestory on a modern residence in Davis, California.

Both skylights and clerestories bring light into a home from above. However, unlike the skylight, which is a year-round solar furnace, the south-facing clerestory can be shaded against the intense summer sun by an overhang since it is located in a vertical wall. Clerestories are still vulnerable, however, to the same elevated rates of heat loss that all lofty windows experience and in northern climates should include night insulation.

Existing homes that are undergoing remodeling can sometimes be adapted to include clerestory windows. Figure 14-2 shows a before-and-after cross section of a bungalow in San Luis Obispo, California, which has been retrofitted with a clerestory. The owners are Kenneth Haggard, a professor of architecture at California Polytechnic State University, and Polly Cooper of San Luis Solar Group. First the existing ceiling was removed, exposing the rafters. The rafters were then removed one by

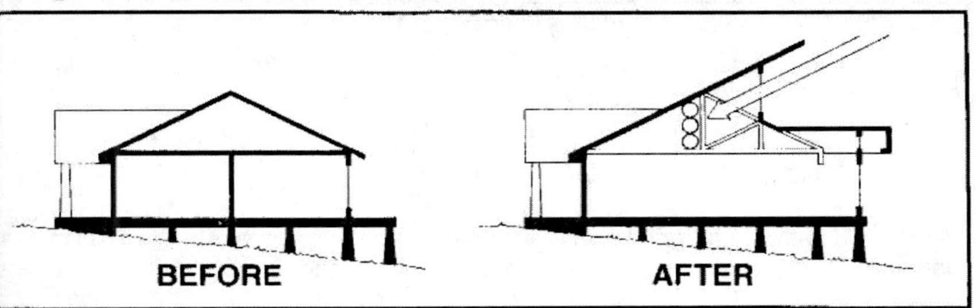

Figure 14-2: Solar clerestory retrofit.

Photo 14-2: The Cooper-Haggard solar retrofit.

one, and every other one was replaced by a truss fabricated from the rafters. Photo 14-2 shows the trusses as they appear from the second floor bedroom loft. Clerestory windows were then added above the southern roof. To complete this project, water-filled tubes will soon be suspended within the trusses to store the direct-gain heat. Night insulation was not necessary in this project due to very mild winter temperatures in San Luis Obispo, but the owners recommend the addition of hinged insulation panels between the roof trusses for homes in colder climates. This very ambitious project shows what can be done by those bold enough to depart from the conventional. However, before you start removing rafters, fabricating trusses, and hanging heavy water tubes from them, by all means consult a structural engineer. This is not a project for beginners!

Top-Hinged Clerestory Shutters

Photo 14-3 shows a row of clerestory windows from inside the Maeden-Nittler residence in Davis, California. A reflective foil surface on the lower side of this shutter directs sunlight onto steel culverts which are filled with water to store heat. Top-hinged clerestory shutters are constructed in a similar manner to the various hinged

Photo 14-3: Top-hinged shutter in the Maeden-Nittler residence.

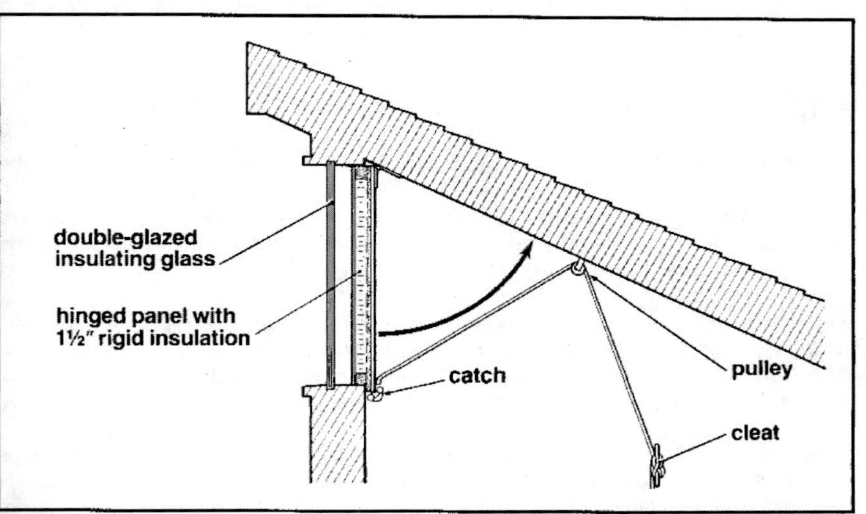

Figure 14-3: A top-hinged clerestory shutter.

shutters in Chapter 8. A frame made from 2-by-2s is cut and fastened together at the corners, and the space inside the core is filled with 1½ inches of polystyrene foam board. Facings are then applied to the panel before it is mounted over the clerestory as shown in Figure 14-3.

Several options are available for facing these shutters. The side which faces the room when closed can be covered with either a decorative fabric, ⅛-inch hardboard, or ⅛-inch luan door paneling. The side that faces the window should be covered with a reflective foil if the clerestory receives direct sunlight during the winter. If you object to a metallic surface on the shutter, a white surface is almost as effective in increasing the amount of light and heat delivered. The primary function of this surface is to reflect the heat coming in the clerestory down to the floor of the living space or into a heat-storage mass, because if it is allowed to build up in the clerestory space, it quickly dissipates back outside through the window glass.

Top-hinged clerestory shutters are commonly raised by using a pulley, cord, and cleat. Unfortunately, top-hinged shutters usually do not lean with enough weight against a vertical window to form a good air seal. However, they can be released to swing with some force into a tightly closed position. A cabinet door latch should be attached at the bottom center of each shutter to grab the shutter and firmly hold it until it is released by the pull cord. Weather stripping should also be applied around the perimeter of each window opening to insure a good seal. Use either a compression-type foam strip or the polypropylene spring V strip (3M type 2743) described in Chapter 2.

Flip-Down Clerestory Shutters

Clerestory windows are usually located in troughlike pockets that are higher than the standard 8-foot ceiling. Warm air rises from the main living space to fill this pocket, increasing nighttime heat losses. Clerestories located in these elevated pockets are best insulated with bottom-hinged or sliding panels that reduce the volume of the space which must be heated during the night.

Flip-down shutters with hinges below the clerestory can be constructed with 2-by-2s and 1½-inch foam insulation board as shown in Figure 14-4. A 2-by-2 ledge strip with its top edge weather-stripped supports the shutter when closed and seals off air flow. A continuous hinge on the back stops air convection along this edge.

This type of shutter must not only have a seal along the leading edge, but the sides should also rest on beams or a ledge strip, and complete shutter array should block the entire opening between the clerestory pocket and the living space below. Sometimes an opening must be boxed in at the edges to form a hatch rim so that the

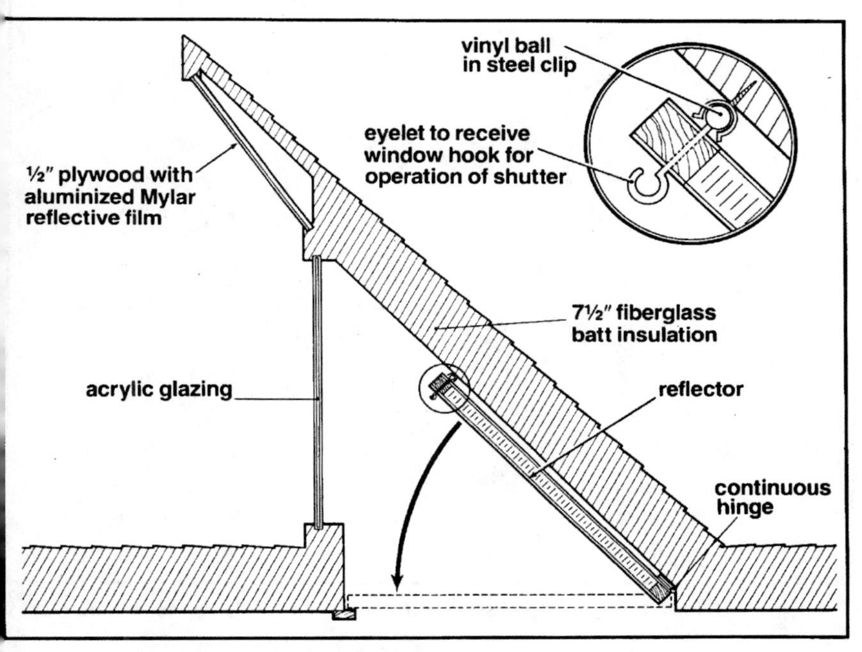

Figure 14-4: A bottom-hinged clerestory shutter.

shutter will block all air flow when closed. Beams are also sometimes added to fill the space between multiple shutters when they close.

Flip-down clerestory shutters can be raised with a pole hook to a simple catch on the ceiling above. A ball-and-clip catch holds this type of shutter well. It can also be raised with pulley and cord. However, the pulley usually hangs down a couple of inches and only allows you to raise the shutter to within 4 or 5 inches of the ceiling. If the shutter cannot be fully opened, it may block some of the light coming in through a clerestory. One potential remedy here is to recess the pulley and cord into the ceiling. If you do this, consider the fact that a jammed cord and pulley which are recessed into the ceiling are difficult to repair.

Another bottom-hinged clerestory shutter, shown in Figure 14-5, is included in a sun-tempered home design I developed several years ago. This shutter rests on beams that brace the clerestory at a 45-degree angle. It works quite well except that it is heavy and unwieldy. Feeling a need to comply with the fine print in the state building code, the architect I was working for at the time included ½-inch gypsum

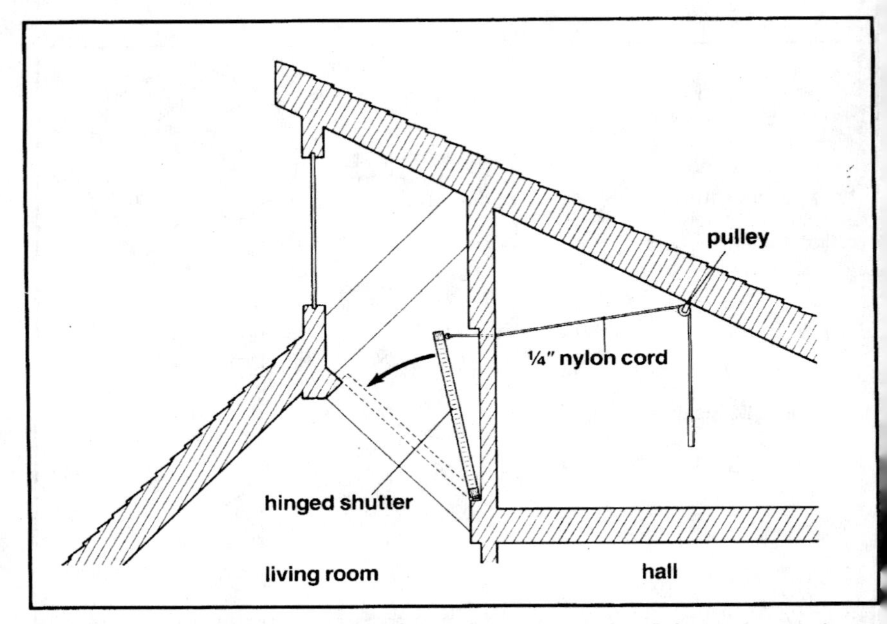

Figure 14-5: A bottom-hinged shutter that rests on the 45-degree beams that brace the clerestory.

board on the interior-facing side of the panels. This provided extra protection against fire, but with a layer of plywood on the other side, these panels are so heavy they can barely be raised by one person.

Local code regulations for the use of foam plastics are less of a problem to the homeowner installing window shutters than to the architect who draws them, since the homeowner can call these items "furnishings." (If an architect draws window shutters on the house plans, they will fall under the building inspector's scrutiny as a construction practice.) Fortunately, new foam insulation types, such as the isocyanurate foams, have a lower rate of flame spread and use restrictions are more relaxed.

A Sliding Clerestory Shutter

Photo 14-4 shows a pocket shutter which recesses into the attic space above the kitchen and dining area in a HUD, passive solar award-winning design by Steve Merdler. The shutter panels slide horizontally on wooden runners when pushed by a pole and hook. They are constructed from 2-inch, rigid Thermax foam sheathing, set

Photo 14-4: A sliding shutter for clerestory window.

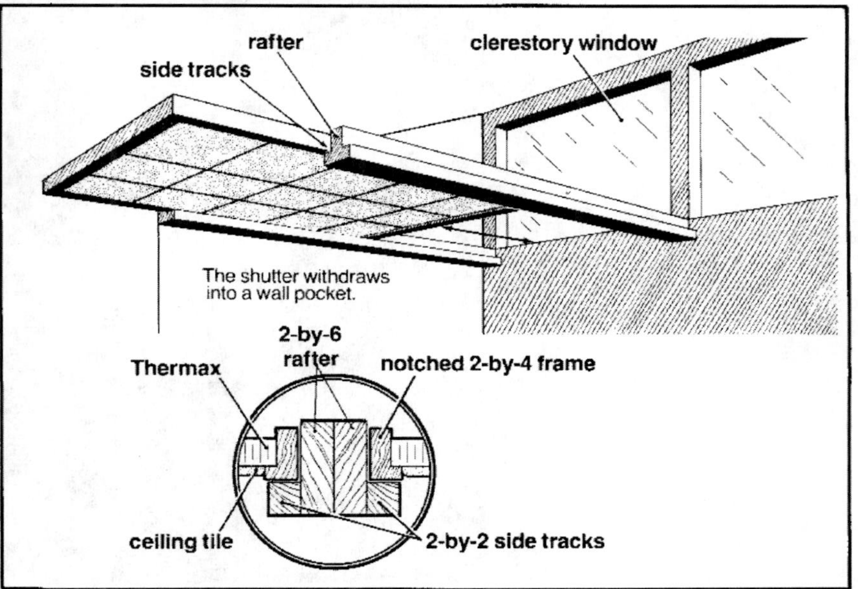

Figure 14-6: A sliding clerestory shutter.

into a wooden frame as shown in Figure 14-6, and faced with acoustical tile on the bottom for appearance. They are lightweight and slide freely on wooden runners, which also serve to seal the sides. When closed, each panel rests on a wooden rafter on all four sides, effectively isolating the high clerestory space above from the room below to reduce heat losses.

Protecting Other Types of High Vertical Windows

Ways to insulate almost any window in the home have been presented in the preceding chapters of this book. One type of window which is conspicuously missing in the treatment so far is the window wall that runs all the way up to a sloped roof. Insulating such glass areas, particularly the triangular windows that meet the roof line, is difficult since most shades and shutters do not work on them.

Photo 14-5: Renee Reiche's pop-in panels.

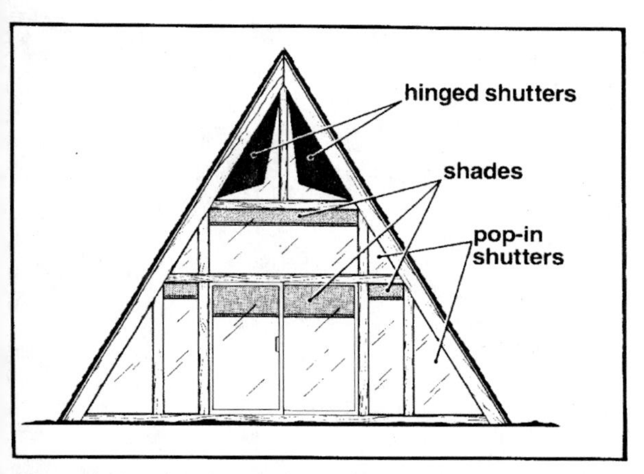

Figure 14-7: Insulating the angular windows of an A-frame.

One solution is to fill these windows each fall with Styrofoam pop-in panels, removing them each spring. Renee Reich of Asheville, North Carolina, did just this. She cut beadboard polystyrene panels to fit snugly in place without magnetic clips or mechanical fasteners. Photo 14-5 shows Renee mounting them over windows which extend up to an A-frame roof in her home. You can see that these panels are somewhat translucent and do allow some light to come through them.

The seasonal pop-in panels leave a lot to be desired since they block the view for months on end. Figure 14-7 illustrates how a combination of shades, pop-in shutters, and hinged, triangular shutters can be used to fill this wall. Triangular shutters can be constructed and mounted in a similar manner to the top-hinged ones in Figure 14-3. However, catches and seals are more delicate in this configuration since the shutter doesn't have a very forceful swing to latch it.

A high, rectangular window can be protected by a thermal shade that rolls into a compact shade and does not block the window when it is open. Some shade products such as the ATC Window Quilt and the SECO Thermo-Shade can be operated remotely from an extended pull cord. Many of the home-fabricated designs, described in Chapter 7, can also be adapted to fit high windows.

Part V

Movable Insulation for Solar Greenhouses

The attached greenhouse or solarium is becoming an attractive, low-cost route to solar space heating for a growing number of Americans. In both new and retrofit situations, a glazed enclosure on the south side of a home can provide indoor space for winter gardening, clothes drying, and a variety of afternoon and evening activities. At times the enclosure can lose heat rapidly to the outdoors such as during winter nights and cold, cloudy days, but with its tremendous gain of solar heat during the daytime, a greenhouse is usually a net provider of heat to the living space it adjoins. For many, winter gardening is the most important function of an attached greenhouse space. The food produced can help sustain a family with fresh vegetables throughout the entire winter. And the dollar savings from food production is often even greater than savings from heat generated.

The role that movable insulation plays in an attached food-producing greenhouse is more critical than in uncultivated solariums or sun spaces that are allowed to freeze at night. Most plants require constant, above-freezing temperatures in order to survive. Purchased heat must be supplied on occasion to most greenhouses without night insulation, but many of these same attached greenhouses can go the entire winter without any auxiliary heat input if they have adequate window insulation.

Figure V-1 shows a simulation of the heat losses or gains in an attached greenhouse in Burlington, Vermont, made by Doug Taff of Parallax, Inc. The greenhouse represented by the middle curve (B) has no movable insulation, only double glazing that is tightly sealed against air infiltration.

This greenhouse, as well as greenhouses A and C, is also well insulated in the areas that aren't glazed. In the upper curve (C), the heating load is increased 30% due to greater air infiltration or due to a portion of the greenhouse having single glazing. Both these curves fall well below the zero line in December and January, losing more heat to the outside at night during these months than they gain during the day. Only

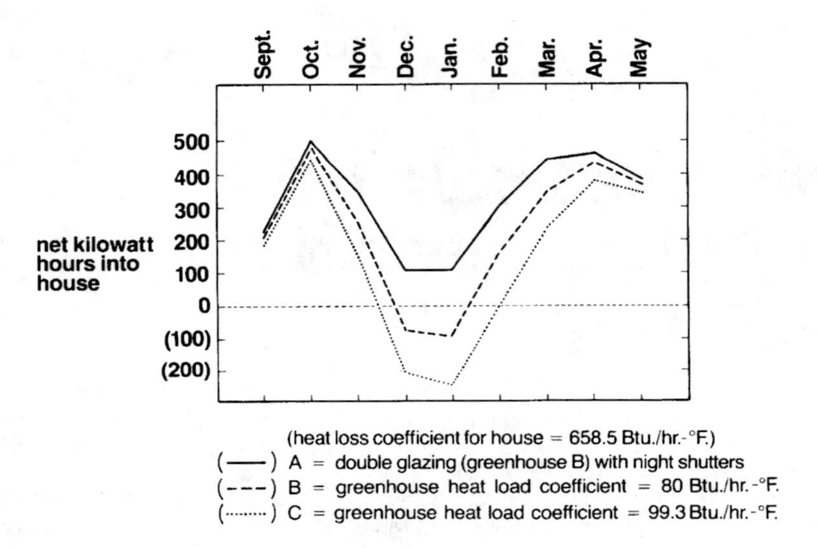

Figure V-1: Simulated monthly performance of an attached greenhouse in Burlington, Vermont.

with double glazing and movable insulation, shown in curve A, does the greenhouse provide heat throughout the winter.

Movable insulation for greenhouses generates not only dollar heat savings, but improves the quality of the growing environment for plants. Warmer temperatures are generally sustained in the greenhouse with movable insulation, which increases the rate of plant growth. Most plants are also sensitive to the thermal shock of rapid temperature changes. Movable insulation slows the rate of heat loss during the evening, and in most climates it can virtually eliminate the need for supplemental heat at night.

With these factors taken into consideration it is surprising how few movable insulation systems have been developed for greenhouses and how sparsely they are used in the hundreds of greenhouses and sunspaces that have been built in the last few years. In an attempt to compensate for the lack of useful design information, I show a number of conceptual designs and design details for systems, some of which have not yet been actually employed. I caution you about the untested nature of some designs but still encourage you to proceed as your knowledge and skill permit.

Chapter 15

Interior Greenhouse Insulation Systems

Movable insulation used inside a greenhouse must not interfere with growing plants. Some plants have delicate foliage that is easily broken by the regular moving of insulation panels. If the glazing extends all the way to the floor of the greenhouse, planting tables must be set back several inches from the glass to allow room for installation and removal. Some clearance is essential even if there is no window insulation, since plants should never come in direct contact with the extreme temperatures (hot and cold) of greenhouse glass.

Insulating shades or blankets are often the most effective when they fill an entire wall or ceiling. Large, one-piece thermal barriers are ideal because they minimize the total perimeter of an insulating system and therefore minimize heat loss by air convection. If a long wall shade is used over glazing or in front of tiered plant beds, as shown in Figure 15-1, the beds must be set back not only from the glass, but also from the vertical posts or window mullions to provide clearance for the shade. Several other window insulation systems can be employed successfully but they must be tailored so they can be maneuvered into place without difficulty. Reaching over a 3-foot-wide plant bed is not too difficult, but reaching under one without disturbing the plants beneath it is very difficult. If shades are used, they should be rigged with pulleys and a cord so their daily operation does not require special gymnastics.

Most package greenhouses are designed with roof support beams that are lighter than those in ordinary structures because the glass shell of a greenhouse tends to melt ice and snow, thus preventing the buildup of ice which can collapse the structure. If you add movable insulation, especially one with a high R value to an existing greenhouse, little heat will escape to melt ice and snow. Care should be taken not to overload the roof beams with snow. However, the danger of the roof collapsing can be avoided if the insulation system is properly managed.

South-facing greenhouses heat up rapidly when the sun shines into them, and this heat will help to melt ice or snow resting on their exteriors. It is very important, how-

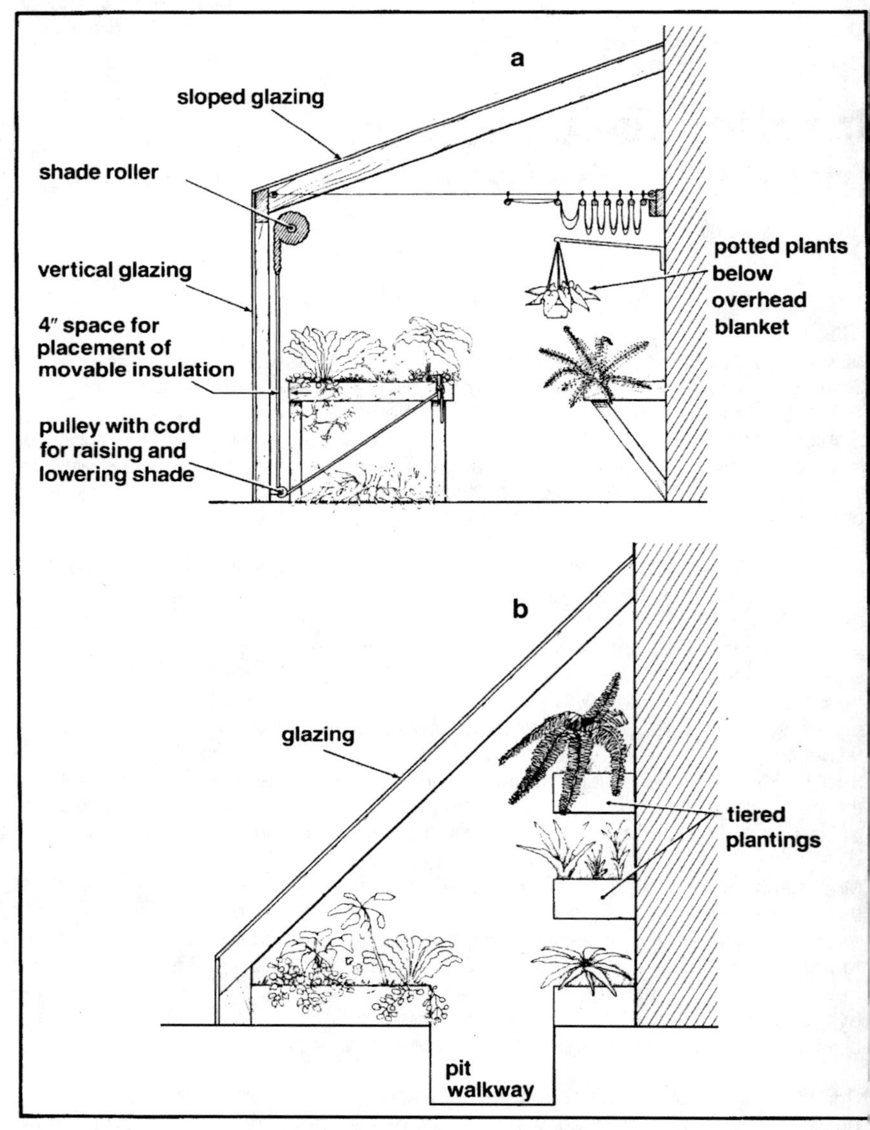

Figure 15-1: Greenhouse insulation slopes and clearances.

ever, that this sunlight is not blocked by snow on the outside or by the insulation on the inside. Any buildup of snow on the outside should be gently brushed, shoveled, or raked aside each morning, and the greenhouse insulation system should be opened regularly. While this is adequate protection for most greenhouses, as an added precaution you might want to leave the insulation open on nights when exceptionally heavy snowfall is anticipated (more than 12 inches). The blanket of snow on the exterior of your greenhouse should also provide some insulation against extreme heat losses and protect the glass from harsh winds.

Plants are often hung from the roof beams in residential greenhouses to make the best use of all the space. In many cases, however, they interfere with an overhead ceiling blanket spanning the entire length of a greenhouse. To solve this problem brackets to support hanging plants can be hung on the walls of the greenhouse below the thermal blanket (see Figure 15-1a).

The operation and maintenance of greenhouse window insulation is simplest if the glazing on the south wall begins above the growing beds. In the pit greenhouse shown in Figure 15-1b, tiered beds are raised along the north wall to make excellent use of the narrow, compact growing space. However, there are many different shapes and sizes of greenhouses designed to meet a range of sites, growing needs, and home types. A variety of insulation systems, or combinations of systems, are required to fulfill each greenhouse insulation need.

Pop-In Shutters

The pop-in shutter described in Chapter 5 is one of the simplest ways to insulate glazing on the interior of a greenhouse. The end (east-west) walls on greenhouses are invariably triangular or irregular at the top, making it difficult to use shades or other vertical-hanging systems on them. Pop-in shutters can easily be cut to fit odd-shaped sections of glass.

Figure 15-2* shows how to make a panel that slides behind a plant bed up to 3 feet wide. Make the panel from two sections and join them together at a point 1 to 3 inches above the top of the planting bed to fill an entire glazing panel. Hinge both sections at this joint with a high-quality cloth tape, such as duct tape. Wrap the edges with a couple of layers of tape to protect them against the considerable wear that results from sliding them in and out of position behind the bed. Add stops along the glazing mullions or posts to form a slot for the panel. Apply Nightwall magnetic clips to the

*The system in Figure 15-2 was designed by the author. Since it has not yet been constructed, there may be some unforeseen problems in the design.

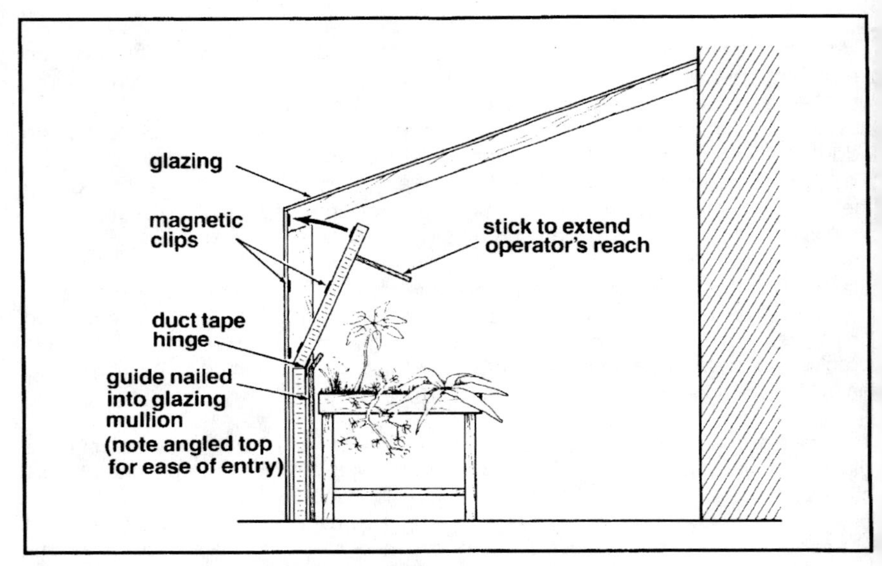

Figure 15-2: A pop-in shutter for greenhouses.

upper portion of the panel to hold the shutter tight against the glazing. If needed, you can tape a small wooden furring strip to the upper portion of the panel to make it easier to open and close the shutter.

To use this panel, simply lower it into the slot and press it against the window. When not needed it should be removed and stored in a convenient place.

Roll Shades

A number of roll-up shade designs were presented in Chapter 7, and they can be readily adapted to vertical or near-vertical greenhouse glazing. Fuller Moore, an architect in Oxford, Ohio, has developed a greenhouse thermal shade system, patent pending, shown in Figure 15-3. A single layer of Astrolon III, a heat-reflecting fabric, is rolled up and down on a 4-inch-diameter PVC pipe at a slope of 60 degrees to match that of the glazing. This roll-down shade is supported by a 2-by-2 at each end of the greenhouse and hung by wires 6 inches behind each glazing mullion. Weights are suspended at the bottom to keep it at an even tension so that warm greenhouse temperatures will not cause it to sag. The shade is raised and lowered by a ⅛-inch nylon cord that eases the shade and roller up and down the wire sup-

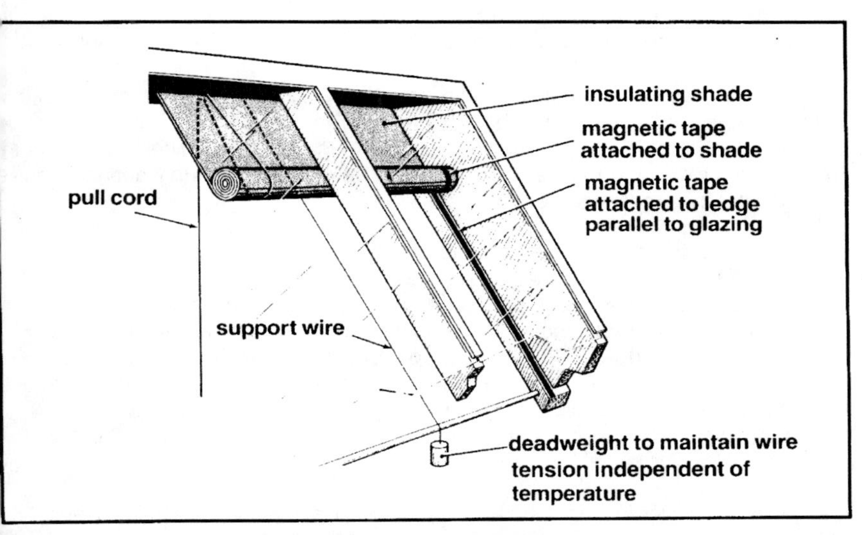

Figure 15-3: The Fuller Moore roll-down greenhouse shade.

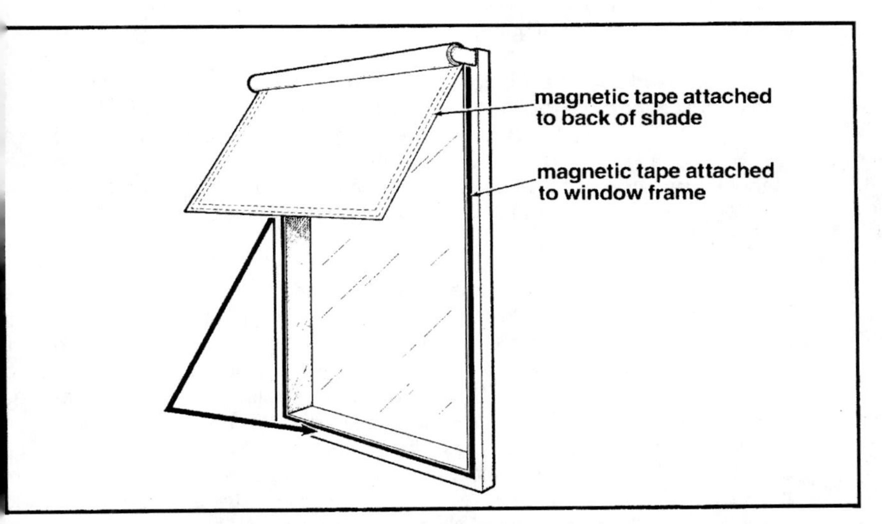

Figure 15-4: A similar, magnetically sealed pull-down shade.

ports. The edges are sealed with a strip of 3M Plastiform magnetic tape attached to both the shade and the 2-by-2. This double strip insures proper alignment.

Fuller Moore has done a considerable amount of research on the use of flexible magnetic strips to seal window shades. Figure 15-4 shows how a pull-down shade can also be operated and sealed with flexible magnetic strips. The following is an excerpt from a paper on these seals which he delivered to the 4th National Passive Solar Conference in Kansas City in October 1979:

> Magnetic Perimeter Seal—An alternative to mechanical seals is a magnetic seal. Flexible magnetic tape is available in various thicknesses and widths. In a thickness of 0.03 inch, such a tape is sufficiently flexible to be used as a perimeter seal on roll-down (similar to the familiar bamboo shade) or pull-down (similar to the spring-loaded window shade) insulating shades. In this thickness, magnetic attraction is sufficient to effect an airtight seal, while allowing ease of "peel-away" separation for shade movement. The magnets are a permanent type, available with adhesive backing.
>
> Although a single magnetic tape can be used in conjunction with a ferrous strip, there are advantages in using a pair of tapes (one around the glazing frame, and the other attached to the perimeter of the shade material). First, the magnetic attraction is increased. Second, and more important, the tapes become self-aligning. . . .
>
> Roll-Down Shades—The roll-down shade is particularly suitable for insulating applications due to its mechanical simplicity and suitability for large areas (limited only by the structural bending stiffness of its center core) and bulky shade materials. Its configuration makes it ideal for use with a magnetic perimeter seal because (unlike a pull-down shade) the shade itself does not slide over the glazed opening, but instead is "rolled on" and "peeled off.". . . .
>
> Pull-Down Shades—Magnetic seals are also adaptable to the conventional, pull-down window shade configuration in moderate sizes. In this case, the shade is pulled down and out at an angle to a predetermined stop point. (Pulling at an angle prevents the magnetic tapes from sealing until the shade is fully down.) Once in the down position, and the spring-locking mechanism "set," the shade is allowed to fall in against the window frame with the magnetic tapes aligning the shade and completing the seal automatically. To raise, the shade is pulled away at the bottom until the seal is "peeled off," then tugged sharply to activate the spring wind-up mechanism in the conventional manner. . . .

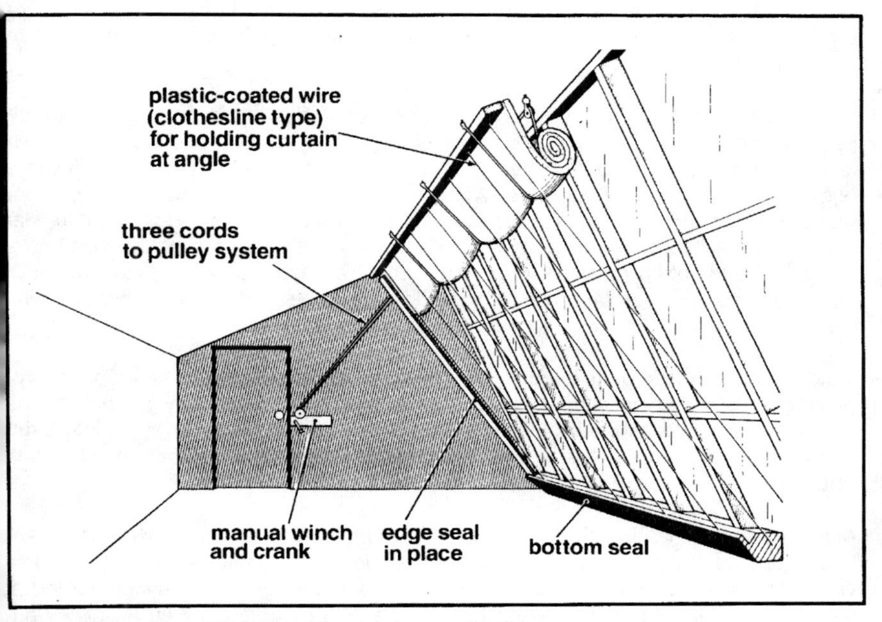

Figure 15-5: The NCAT fiberfill greenhouse shade.

Another thermal shade design developed by Andy Shapiro, at the National Center for Appropriate Technology, is shown in Figure 15-5. This shade rolls up and down on a floating roller and has support wires similar to the ones in the Moore design. The NCAT shade is constructed from a 1-inch-thick layer of PolarGuard between two layers of Duratuff, a vinyl-coated nylon material made by Duracote Corporation, Ravenna, Ohio.

Moisture levels and condensation are high in any greenhouse and it's common to see water dripping from the glazing supports. Thus, greenhouse insulation systems invariably get wet from time to time and must have some means of venting or releasing moisture. In the Shapiro design, 8-inch-wide, semiporous, water-repellent canvas strips were sewn between 54-inch sections of a nonporous vinyl-coated fabric. These canvas strips, arranged vertically so they face the exterior glazing, were intended to ventilate the PolarGuard and help keep it dry. The same vinyl-coated fabric used on the outside face was used to cover the entire side of the shade facing the greenhouse to prevent moisture from penetrating the shade. Due to the high amount of moisture dripping from the interior surfaces of the glazing, the canvas strips retained more moisture than they vented. Adding a series of small holes or grommets along the outside bottom edge of the shade is probably a more effective way to provide this venting.

The Shapiro shade is operated by a boat winch located on a side wall. Three ¼-inch nylon cords run from the winch to pulleys along the top of the shade and down around the shade to raise and lower it. A ¾-inch galvanized-iron pipe is sewn into a pocket at the bottom of the shade. When the shade is lowered, the pipe rests in a V-shaped wooden slot constructed from 1-by-4s joined at a 45-degree angle with holes drilled in it for drainage, this wooden slot provides a bottom seal for the shade. The side seals, often referred to as "ear flaps," are hinged 1-by-6s covered with carpet scraps. The ear flaps are wedged tightly against the shade with short metal rods. Spring-loaded hinges might be more convenient.

All in all, this shade is quite successful. The materials cost only $1 per square foot of glazing area. A weakness in this design, though, is that it blocks sunlight through the top 18 inches of glazing, a problem characteristic of bulky fiberfill shades. However, reduction in lighting by this type of shade is not a problem in greenhouses with opaque roofs.

Most of the commercial shades listed in Chapter 7 perform well on vertical greenhouse glazing, but how well they perform on sloped glazing varies. The ATC Window Quilt can be operated on windows 45 degrees or steeper. On windows mounted at less than 45 degrees, friction along the side tracks prevents the Window Quilt from falling properly. The IS High R cannot be used at all on sloped glazing. The spacers between the reflective Mylar layers collapse on sloped applications.

One of the best commercial systems for greenhouses is the Thermo-Shade by Solar Energy Components, Inc. This system can be drawn back on a horizontal or sloped track to the rear of a greenhouse. A curved track allows the slatted shade to wrap continuously around any dual-pitched greenhouse wall-roof combination. A 12-volt, motor-drive system is also available with this shade upon request.

Overhead Blankets

Overhead blankets or drapes are becoming a standard way to reduce nighttime heat losses in commercial greenhouses. Photo 15-1 shows an installation of an insulating overhead blanket at Pennsylvania State University. Under the direction of Professor John White, many materials commonly used in insulating blankets were tested. The greatest heat loss reductions were found to be from reflective layers of fabric like Foylon or Tyvek, but even black polyethylene helped considerably over single glazing.

Photo 15-1: Commercial overhead thermal blanket at Penn State University.

Symtrac—Symtrac Inc., 8243 North Cristana Drive, Skokie, IL 60076, makes motor-operated tracking systems for overhead blankets used in commercial greenhouses. The blanket is suspended from wires stretched across the greenhouse. The Symtrac system was used in the Penn State experiments.

Generally little attention is given to edge seals used on commercial greenhouse curtains. The vast areas covered by these blankets are small compared to the length of their edges, so the heat lost through convection in commercial greenhouse blankets is relatively insignificant. In small, residential greenhouses, however, where the ratio of edge length to surface area is higher, edge seals are far more important. Clamp strips with spring-loaded hinges can be used in several ways to provide edge seals for residential greenhouse insulation.

Figure 15-6* shows a diagram of an overhead suspended-blanket system for residential greenhouses. Stainless steel wires are stretched from the south to north walls

*The system in Figure 15-6 was designed by the author. Since it has not yet been constructed, there may be some unforeseen problems in the design.

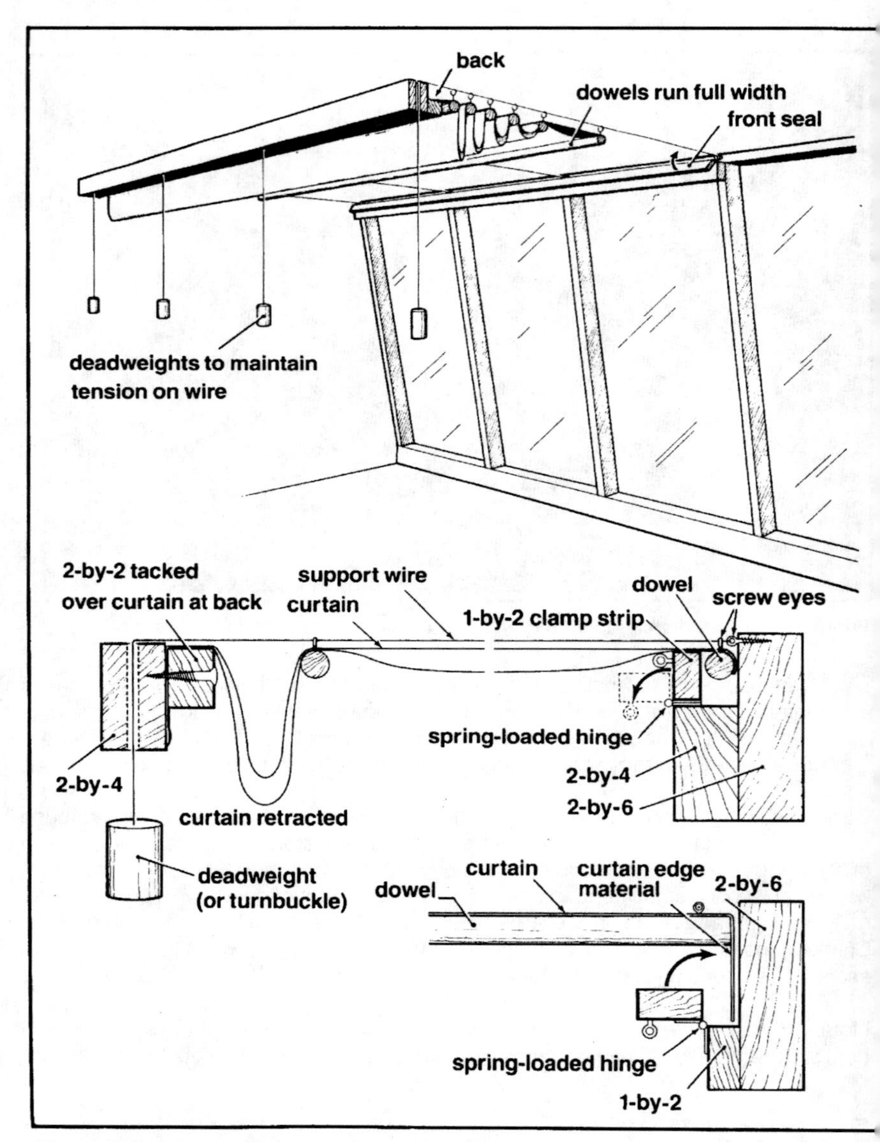

Figure 15-6: A continuous overhead blanket.

parallel to the overhead beams at 3 to 4 feet on center. Each wire runs down through a hole in a 2-by-4 nailed to the north wall. Beneath this 2-by-4, each wire is stretched taut with a heavy weight, a short, heavy-duty spring, or a small turnbuckle. One-inch dowel rods are suspended from the wire by screw eyes at 18- to 24-inch intervals, and they move freely in a north-south direction. Two layers of reflective cloth are sewn to these dowels so that they hang separately as shown in the illustration.

When installing this blanket system, sew the fabric to the dowels first; then attach the screw eyes to the dowels and thread the wires through the screw eyes to hang the system over the greenhouse. About ½ inch of slack should be left in the bottom layer of cloth so there is an air space between the fabrics when the upper layer of fabric is pulled taut. Clamp the north side of the fabric to the 2-by-4 with a 2-by-2. You can open and close this blanket with a short pole-and-hook assembly that hooks into a large screw eye in the leading dowel rod. Sew a 5-inch flap of fabric along the side edges of the blanket, so you can seal these edges with spring-loaded clamp strips. Attach the south or leading edge of the blanket using a clamp strip with spring-loaded hinges. Attach the blanket to the north wall with a 2-by-2. Attach the 2-by-2 on the south wall last, with the hinged clamp strip closed. This will help you determine the length of blanket required for a snug fit when the blanket is in use.

Sometimes a single, overhead blanket is objectionable because it does not allow plants to be hung from individual beams nor does it allow the control of separate areas of the greenhouse for summer shading. Figure 15-7* shows a design similar to the previous one except that it fits between the gently sloping beams that support the glazing.

Instead of wire, a ½-by-¾-inch steel track is attached to the beam to suspend the blanket. A ¾-by-¾-inch wooden piece is attached outside this track to act as a shoe for the clamp strips. Clamp strips with spring-loaded hinges hold the shade tightly against these shoes when closed. The clamp strips are opened and closed by a short pole that hooks into a screw eye in the clamp strip. The dowels here need only be ½ inch in diameter and are terminated short of the wooden shoe to avoid interfering with the clamping. Care must be taken in choosing a tracking system for this design that doesn't bind up under lateral weight of the shade.

The Sun Quilt Thermal Gate—Sun Quilt Corporation, Box 374, Newport, NH 03773, makes a unique thermal blanket of nylon and polyester fiberfill. This system can be drawn horizontally, vertically, or over sloped glazing. Tracks can also be run on dual-pitched glass areas, making it ideal for greenhouses. The

*The system in Figure 15-7 was designed by the author. Since it has not yet been constructed, there may be some unforeseen problems in the design.

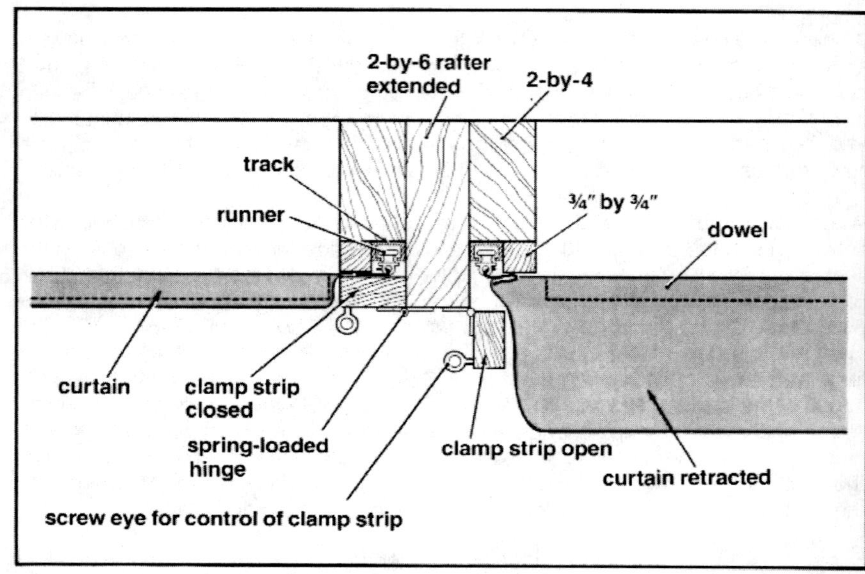

Figure 15-7: An overhead curtain between beams.

Photo 15-2: The Sun Quilt.

Sun Quilt is operated by a small motor and can be activated by a switch, or automatically, by a light sensor. Photo 15-2 shows an installation of the Sun Quilt in a dining room solarium.

Hinged Shutters

The use of rigid shutters inside greenhouses is limited by the large areas of continuous glazing and the availability of convenient places for daytime storage. In some greenhouse designs, however, hinged or sliding shutters perform quite well.

Photo 15-3 shows top-hinged shutters in an A-frame greenhouse in Corrales, New Mexico. Designed by Steve Baer of Zomeworks for a greenhouse addition on the north side of the Stan and Melinda Handmaker residence, these panels are simply swung by means of a pole and hook to cover sloped south glazing. The panels are held in the open or closed position by a heavy, circular magnet, similar to the type used to attach devices onto automobile dashboards.

Shutters can be used with a bottom hinge that allows them to swing down to rest and seal horizontal stops in the upper, inaccessible areas of a tall greenhouse. This

Photo 15-3: Top-hinged shutter for the Handmaker greenhouse.

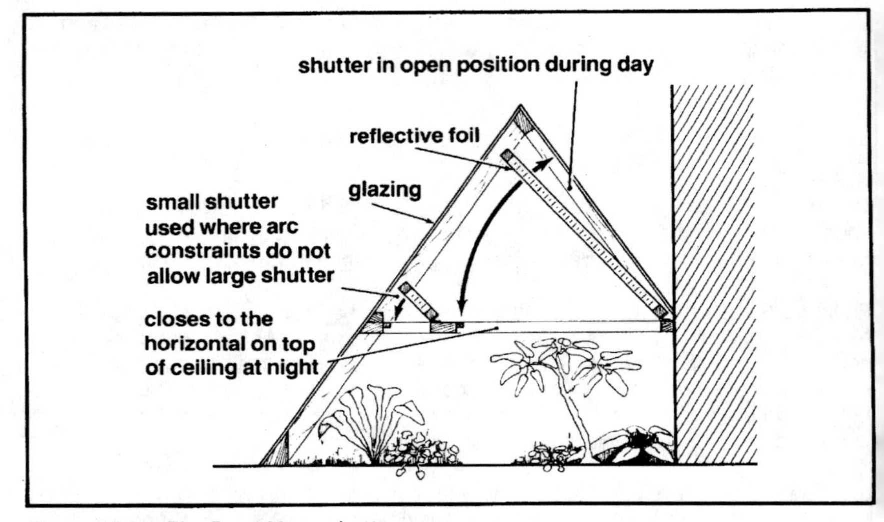

Figure 15-8: The Reed Maes shutter.

arrangement also helps to conserve heat by reducing the volume of the greenhouse space at night. Figure 15-8 shows how reflective insulating shutters were used in the upper part of a greenhouse designed by Reed Maes of Ann Arbor, Michigan. The bottom surface of each shutter is foil-lined to reflect sunlight down into the greenhouse during the daytime. Since these shutters must be trimmed to clear the south wall as they swing closed, they do not cover a portion of the ceiling rafters on the south side. To seal the entire upper space without blocking daytime sunlight, the Maes greenhouse uses a second, smaller hinged shutter, as shown in the figure.

Wing Bifold Shutter

Charlie Wing of Cornerstones has designed a bifold shutter to be used in greenhouses between window mullions that are placed 30 inches apart. These shutters are constructed from Thermoply and 1-by-2s like the Homesworth Sun Saver shutter in Chapter 8. They can be hung from both vertical windows and glazing having a slope of up to 60 degrees. The 60-degree bifold shutter is held in place by a roundhead screw placed in the top of the shutter to slide along a wire at the top of the window. Photo 15-4 shows this shutter in the Davis residence in Brunswick, Maine.

Photo 15-4: The Wing shutter in the Davis residence.

Chapter 16

Exterior Insulation for Greenhouses

In many cases, particularly in smaller greenhouses where space is at a premium, it's wise to use movable insulation on the outside instead of the inside of the glazing. Outdoor locations present the same problems with exposure, and even though exterior insulators are hardy, they no doubt will have a shorter service life than the inside models.

Blankets for Sun-Heated Pits, Cold Frames, and Small Greenhouses

One of the oldest greenhouses in America is the Lyman greenhouse, built in the 1880s in Waltham, Massachusetts. It was sunk into a gentle, south slope to cut the chilling north winds and take full advantage of the insulating value of the earth. This greenhouse inspired the construction of a number of pit greenhouses in the Boston area during the 1930s and 1940s. Most of them were sunk about 30 inches below grade, with glass rising at a 45-degree angle to 4 or 5 feet above the ground. Plant beds were built right into the ground, one on each side of a central walkway. While it was still in operation, the Lyman greenhouse was covered with straw at night to help insulate it, but the pit greenhouses of the 1930s and 1940s were usually insulated with an exterior insulating blanket rolled up and down by hand.

Numerous sun-heated pit greenhouses are described in detail in an excellent book by Kathryn Taylor and Edith Gregg, *Winter Flowers in Greenhouse and Sunheated Pits* (New York: Charles Scribner's Sons, 1976). It describes sunken greenhouses that were almost entirely solar heated. Earth on the walls and dirt floor in these sunny pits stored some of the heat, and with the help of window insulation, the plants were kept safely above freezing temperatures at night.

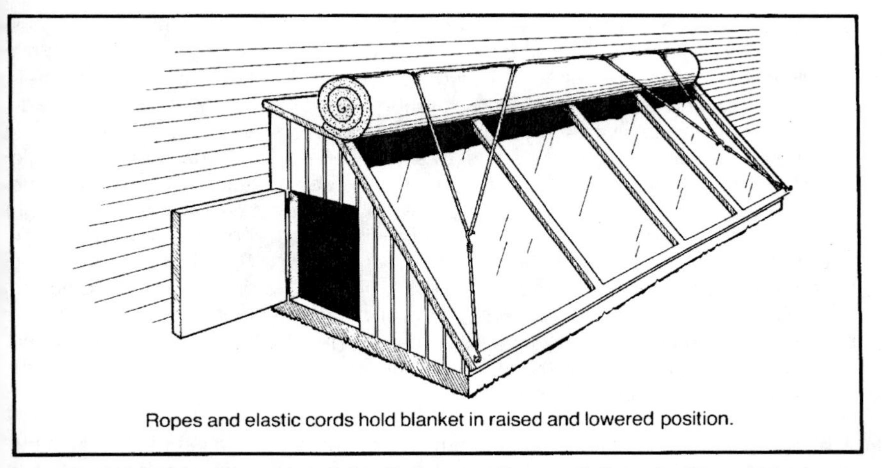

Ropes and elastic cords hold blanket in raised and lowered position.

Figure 16-1:* *Blanket for a small pit greenhouse. (Adapted with permission from* The Complete Greenhouse Book *by Peter Clegg and Derry Watkins.)

Of particular interest in this book is the simplicity of the insulation blankets used over the greenhouse glazing. These were usually burlap bags filled with leaves or waterproof canvas filled with the fiberglass insulation like that used in the walls of homes. Figure 16-1, from *The Complete Greenhouse Book* by Peter Clegg and Derry Watkins,* shows how this type of blanket rolls down over a small pit greenhouse or cold frame. Ropes keep the blanket from flapping in the wind. Elastic shock cords or Bungees can be added on the corners to help keep the ropes taut in both the rolled-up and rolled-down positions. When the blanket is rolled down, it fits snugly over the glass and greatly reduces air infiltration.

The polyester fiberfill used in sleeping bags and outdoor clothing is far superior to fiberglass wall insulation for exterior insulating thermal blankets. Polyester fiber materials are designed to flex, while fiberglass insulation is designed to be stationary and will tend to break down in time. These fiberfill materials are light and resilient and they usually cost less than the fabrics used as coverings for thermal blankets.

The insulating capability of any fibrous insulation material drops significantly when moist and it is virtually nil when the material is waterlogged. To prevent moisture

*See Appendix V, Section 1 for book description and complete bibliographical information.

buildup in an exterior blanket, the top fabric should be a moistureproof, nonporous material. Vinyl-coated canvas, which is sold in most awning shops, is quite effective in preventing moisture penetration. However, even if the entire blanket has a waterproof skin, moisture will eventually enter through the seams so it must also have some way to escape. Using a vinyl-coated canvas on the bottom, with rows of grommets to vent the insulation, is probably the best design. (Generally, porous materials should not be used to vent moisture because they can tend to act like a sponge.) The blanket should be aired periodically on sunny days to help moisture escape through the grommets. One row of grommets, about 8 inches apart, should be installed on the upper and lower edges of the blanket, as shown in Figure 16-2. Another row across the middle of the blanket will also be helpful. However, the grommets should not be placed on either the top or bottom face of the upper quarter of the blanket, which will be exposed to the weather when rolled up. These grommets should also be placed so that they do not lie in obvious areas of puddling or water will seep in when the blanket is rolled down over the glass.

A simpler alternative to the multilayer thermal blanket is to use a couple layers of the closed-cell bubble or foam plastic materials used in swimming pool covers. These materials do not insulate as well as the fiberfill blanket, but they are designed to be placed directly on water and therefore will not become waterlogged. One product

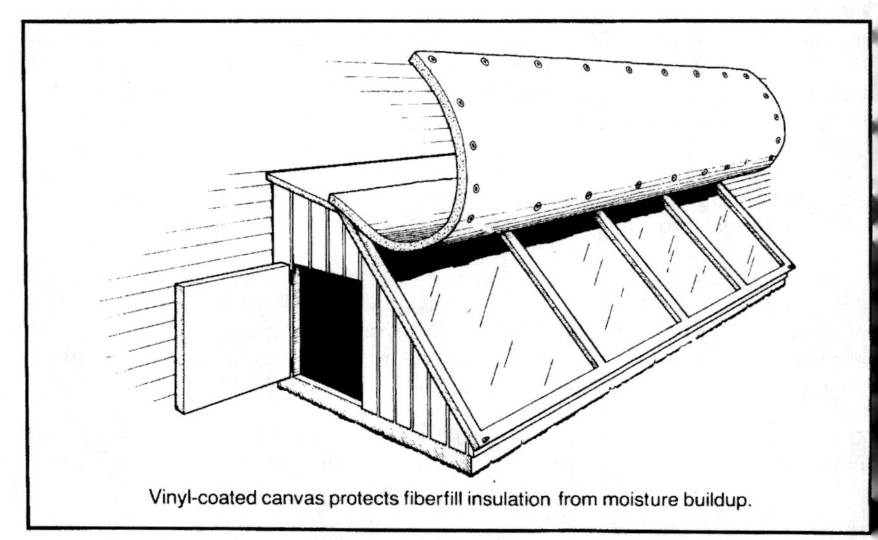

Vinyl-coated canvas protects fiberfill insulation from moisture buildup.

Figure 16-2: Grommets vent moisture that enters through the seams of exterior blanket, keeping its insulating capability high.

that can be used here is a bubble plastic material called Sealed Air Solar Pool Blanket, available from the Sealed Air Corporation, 2015 Saybrook Avenue, Commerce, CA 90040.

Hinged Shutters

While exterior blankets are limited primarily to small greenhouses, hinged shutters can insulate greenhouses, both large and small. Cold frames can be insulated with a small, hinged shutter (Figure 16-3) that can be easily raised by hand. A foil-faced, Thermax-type (isocyanurate foam) insulation board can be applied to the inside face to both insulate at night and reflect up to 50% more sunlight onto the growing beds. The additional sunlight will warm the soil, help plants to germinate earlier, and extend the growing season.

Another insulating shutter design for cold frames is illustrated in Figure 16-4. This clever device called the Solar Frame was invented by Leandre Poisson of Solar Survival. The sloped glazing on this cold frame, constructed from two layers of fiberglass glazing 3½ inches apart, is filled with Styrofoam beads to protect the cold

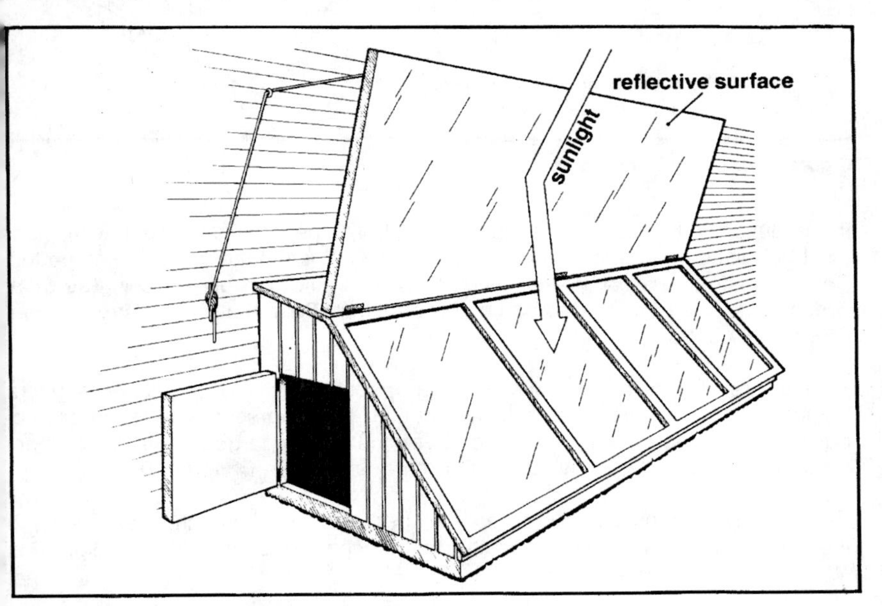

Figure 16-3: Insulated reflecting shutter.

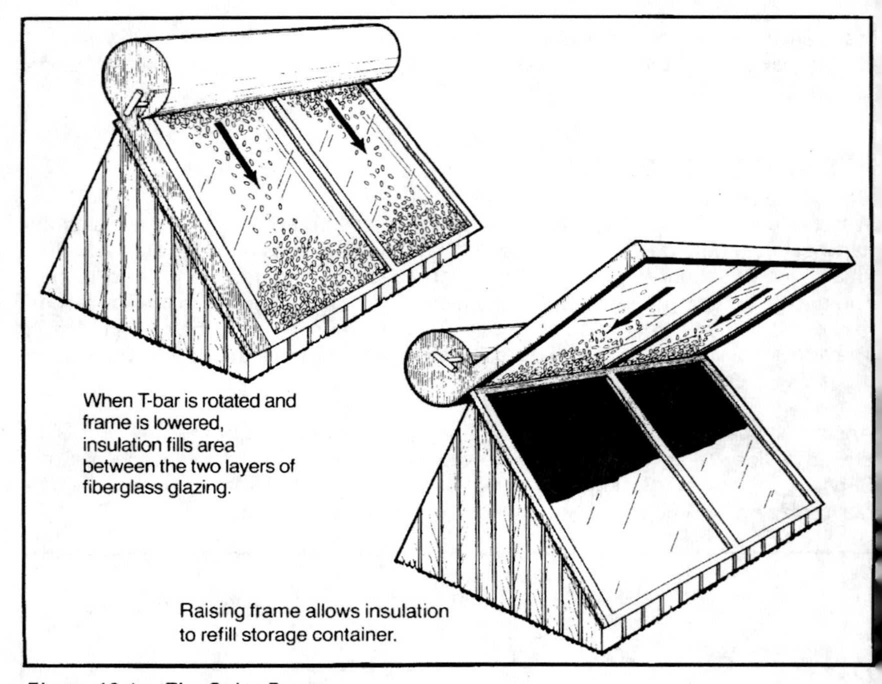

Figure 16-4: The Solar Frame.

frame at night. A flap control releases the beads into the insulating area, but the glass must be raised so the beads can fall by gravity back into the round bin at the top. Detailed instructions on how to build this solar frame can be found in *The Solar Greenhouse Book,** and plans can be ordered from Solar Survival, Cherry Hill, Harrisville, NH 03450.

Vertical greenhouse glazing can also be protected with bottom-hinged shutters. Photo 16-1 shows reflective-insulating shutters on the greenhouse in the science building of Marlboro College, Marlboro, Vermont. These large panels are 4 inches wide, 8 feet tall, and 4 inches thick. For more details see Chapter 10.

Figure 16-5 illustrates two other exterior, hinged shutters for greenhouses. The top of the greenhouse has a reflector-shutter that reflects low-angle winter sunlight onto

*See Appendix V, Section 1 for book description and complete bibliographical information.

Photo 16-1: Shutters on the science building greenhouse, Marlboro College, Marlboro, Vermont.

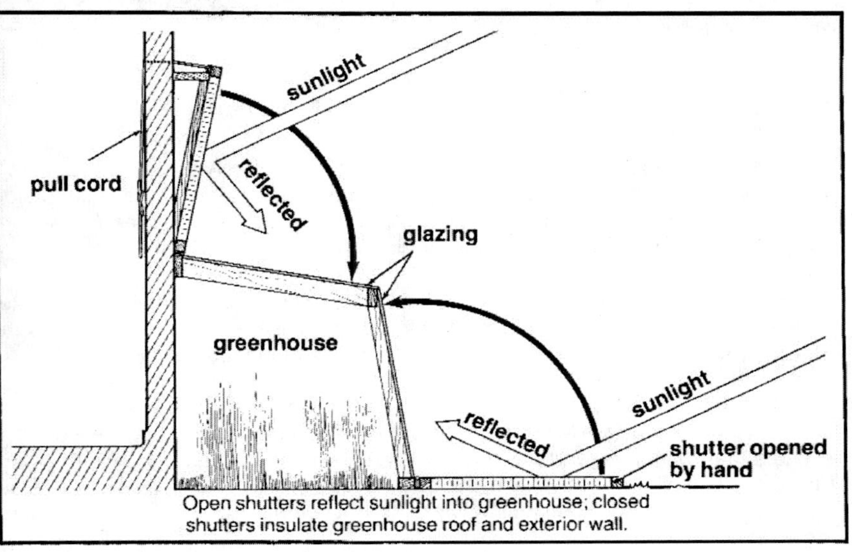

Figure 16-5: Exterior shutters for greenhouses.

Photo 16-2: Shutters on a porch greenhouse in Milwaukee, Wisconsin.

plants. Shutters that swing up like this are exposed to heavy winds and require a sturdy frame behind them. A wooden frame can be braced against the second story of a house or posts can be extended through the top of the greenhouse and anchored to a wall or floor at the rear of the greenhouse.

The sloped, south wall of the greenhouse in Figure 16-5 has a bottom-hinged shutter. Unlike vertical-wall bottom-hinged shutters, which fall from their own weight, this particular shutter must be opened manually, a less-than-ideal solution.

Photo 16-2 shows a greenhouse built onto a porch by architecture students at the University of Wisconsin in Milwaukee. This small greenhouse has reflective-insulating shutters that swing open all three ways—up, down, and sideways. With this dual function "coverall" shutter system, the solar heat gains into small greenhouses like this one are increased substantially, while at night complete protection against freezing is achieved.

Bifolding Shutters

Photo 16-3 shows a bifolding shutter designed and constructed by Bob Forer, Bob Nelson, and Scott Hicks for the Nelson residence in Madison, Wisconsin. This clever design joins two rigid panels so that they swing open and closed like a bifold closet

***Photo 16-3:** Greenhouse shutters on the Nelson residence, Madison, Wisconsin.*

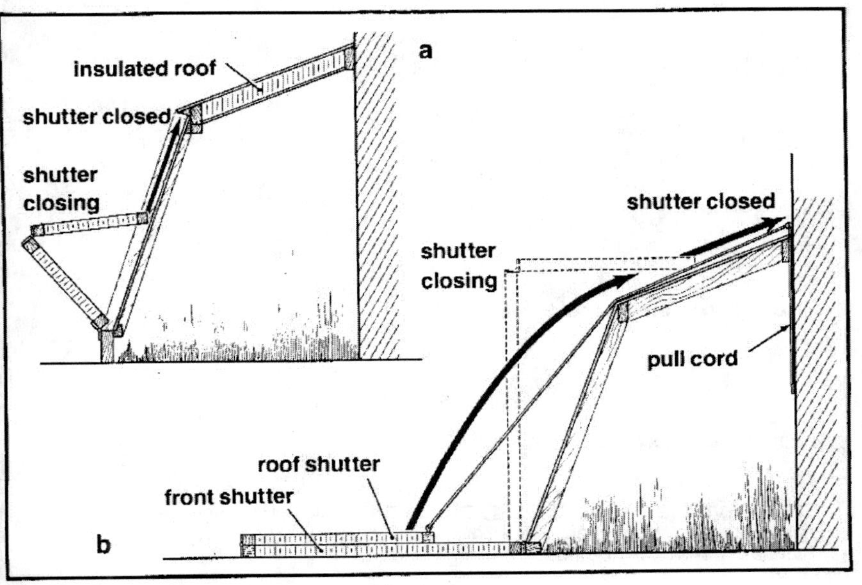

***Figure 16-6:** Bifolding greenhouse shutters.*

door turned on its side. The shutter is constructed from ¾-inch Thermax insulation board recessed into a frame made from 2-by-2s to create an air space between the foil face and the greenhouse glazing. The exterior is made from ⅜-inch exterior grade plywood. The corners of the shutter slide up and down on aluminum flashing attached along the edges of the greenhouse glazing. A tube-type weather stripping made from a black, rubberlike material runs along the edges of the shutter to seal against drafts when closed.

There are two separate bifold shutters for this greenhouse, each of which is raised and lowered by a boat winch inside. The procedure is as follows: The winch is first unwound a little to allow a foot of slack in the line; it is then locked for safety. Next, the bifold shutter is pulled out from the outside so that it begins to fall. Then it is lowered by the winch from inside.

Figure 16-6a shows a section through the Nelson shutter. A variation of this shutter can be applied to a dual-pitched greenhouse shown in Figure 16-6b.

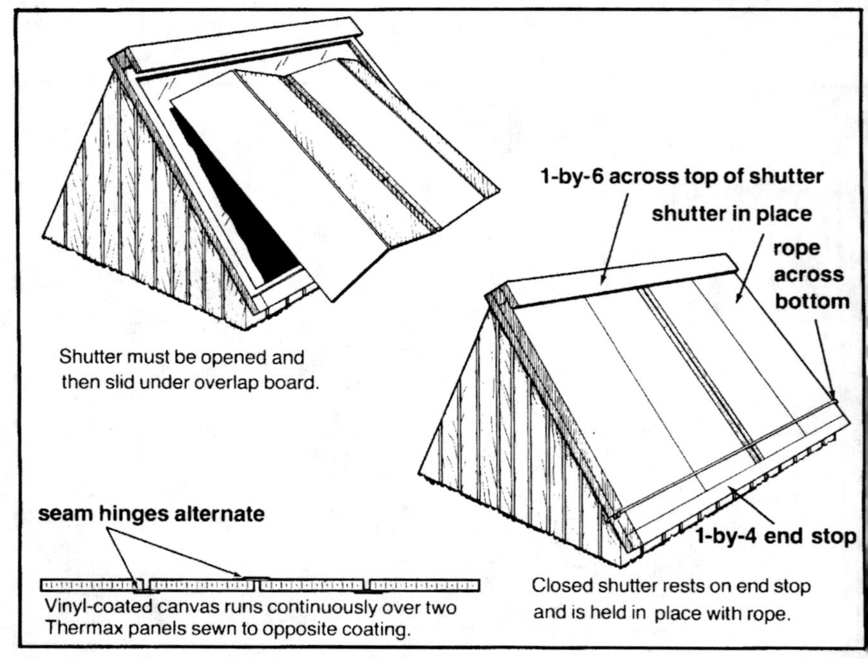

Figure 16-7: Accordion-folding shutters.

Another way to use folding panels is to assemble very lightweight panels that fold accordion-style as shown in Figure 16-7. This design, along with several other concepts for movable greenhouse insulation, is presented in *The Complete Greenhouse Book.** All should be constructed from Thermax insulation board and covered with a vinyl-coated canvas. A lip on the upper portion of the greenhouse holds the panels against crosswinds at night, and the bottom is held in place with a rope and elastic Bungee straps at each end. These shutters work best on small greenhouses where they can be managed easily while standing on the ground.

Combinations of Systems

As you design a system to insulate your greenhouse, you may want to use combinations of interior and exterior systems. Figure 16-8 shows several combinations. The roof is generally easier to insulate from the inside than the outside, and a ceiling blanket is used in this example. The vertical wall above the planting bed is also insulated on the inside with a thermal shade. An exterior shutter-reflector is used to provide extra light and insulation below the planting bed where access is difficult.

*See Appendix V, Section 1 for book description and complete bibliographical information.

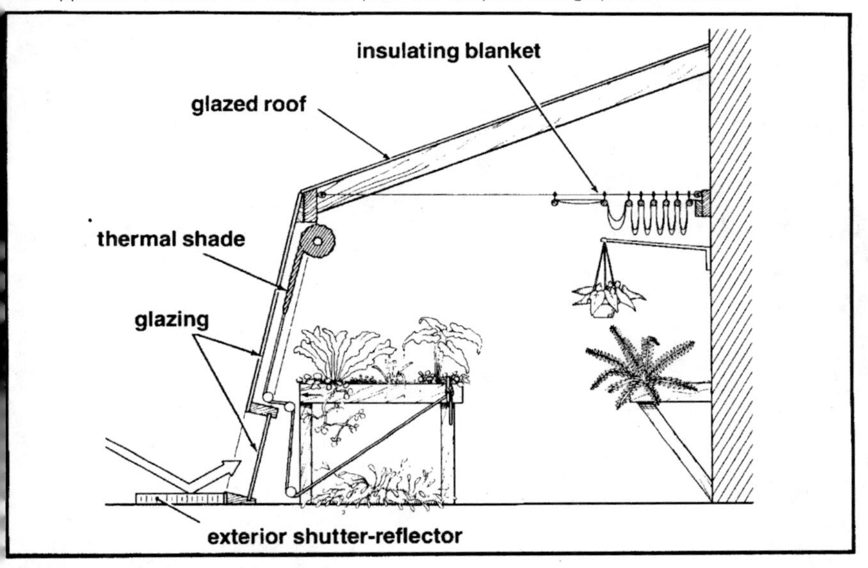

Figure 16-8: Combining systems.

Part VI

Movable Insulation to Assist Solar Heating Systems

The window systems included in this section are especially tailored for passive solar heating systems. "Passive" systems combine south-facing glazing with a heat-storing mass to enable the home to heat itself naturally, without relying heavily on outside energy sources. Movable insulation is a very important component of passive solar heating systems because it allows whole wall areas of the home to regulate heat flow. Movable insulation between the glazing and the heat-storing mass (masonry or water-filled containers) can be opened to receive solar heat during the day and closed at night to trap heat. The release of heat from storage walls can also be better managed by movable insulation, helping the wall to retain heat until it is needed. Even domestic water heating can be greatly simplified with movable insulation.

While these final two chapters focus on movable insulation as an integral component of passive solar space- and water-heating systems, all the forms of window insulation discussed in this book assist the solar heating of homes by reducing heat losses. The thermal envelope of a building is probably the most important facet of any solar heating system because it reduces the amount of expensive equipment and auxiliary energy required to maintain comfortable temperatures. Anyone who installs solar heating in a home without first thoroughly insulating, caulking, and weathersealing in every practical way, is throwing money away on an oversize heat supply system. The most extreme example of squandering energy resources can be found in the mechanical air-temperature control systems in large, commercial buildings. These systems sometimes expend energy to heat a room and at the same time use additional energy to cool the *same space*.

In direct contrast to these wasteful practices is the use of movable insulation as part of a home's passive energy fabric. This passive fabric is truly a domestic form of technology; one where the home occupants skillfully adjust curtains, shades, or shutters to make the best use of natural energy sources. With the passive energy fabric, the dwelling in many ways becomes lifelike, collecting and storing solar energy like plants, and regulating the heat flow through its skin or shell like animals.

Chapter 17

Movable Insulation to Assist Passive Space Heating

Passive solar homes are constructed with heavy or *mass*ive materials located inside a tightly insulated exterior shell so that the sun's heat which enters through south-facing glass is stored inside the home for a reasonable period of time. The thermal mass and envelope of a passive solar home retain heat in a similar manner to a Thermos bottle filled with a hot liquid. However, the windows that charge the mass are also weak areas in its thermal envelope and allow much heat to escape.

Although south-facing double-glazed windows without movable insulation generally gain far more heat during the winter than they lose, windows still often account for about 50% of the heating load in a passive solar home. Because a passive solar home has larger glass areas than a conventional home, movable insulation greatly enhances its performance. Insulating these glass areas not only reduces the overall heating load but also reduces in turn the amount of glass and mass required to supply the home with heat. By focusing your efforts on minimizing the overall heating load, a home can be solar heated with very modest amounts of glass and mass.

There are many ways in which the three elements—insulation, glass, and mass—can be combined into an effective, passive solar heating system. In one approach, south-facing windows allow sunlight to directly enter the home. These windows bring heat into a room virtually as soon as the sun appears. When sunlight enters a home and strikes a lightweight material such as carpeting, it quickly heats the air in the home. Without an adequate mass inside the home, like masonry walls or floors, the interior can overheat by noontime, requiring doors and windows to be opened so that the remainder of the day's heat is vented to the outside—not an efficient operation.

The heat-storing mass in a passive solar home is four times more effective when sunlight strikes it directly than when this mass is warmed by air in a room. To insure that sunlight will shine directly upon it throughout the entire day, a heat-storage wall is often located only inches behind a south-facing wall of glass. This heat-storage wall, often referred to as a Trombe wall after one of its earliest developers, can be constructed from brick, block, stone, adobe, or from water-filled containers. The Trombe wall absorbs the sun's heat before it directly enters the home's interior and stores this heat for use at night.

South-facing windows are commonly referred to as a "direct-gain" component by passive solar designers because they allow sunlight to *directly* enter the living space. The Trombe wall, on the other hand, is referred to as an "indirect-gain" component because heat must *conduct through* 1 foot of masonry or water-filled containers before it can radiate into the room. These two approaches, direct and indirect gain, and their relative merits, are cause for much debate among solar designers. Indeed, they each serve different needs. Windows admit heat early in the day, provide daylight, allow a view to the outside, and help to charge walls on the home's interior with heat. The Trombe wall stores heat that can be used late at night and on into the morning.

For my residential design work in the Asheville area, I often plan for about one-third of the south-facing glazing to be direct-gain windows and two-thirds to be used with indirect-gain systems. Rooms that need heat earlier in the day such as a kitchen or breakfast room have primarily direct-gain windows. Bedrooms, which need heat mostly at night, may have predominantly indirect-gain or Trombe wall-type glazing.

Night Insulation for Trombe Walls

The glass that covers the Trombe wall generally loses more heat to the outside at night than a direct-gain window wall. Heat losses through a glass surface are proportional to the temperature difference between the outside air and the air on the interior side of the glass. Temperatures on the sun-exposed face of a Trombe wall often exceed 140°F. during the day, and though the surface temperature drops rapidly during the late afternoon, it can still remain above 80°F. throughout the night. Thus, twice as much heat can be lost through a square foot of Trombe wall glass as through an equal area of window glass.

Studies have shown that even without night insulation, the Trombe wall gains far more heat during the day than it loses at night. However, these studies have also shown that movable insulation greatly increases the net gain or performance of a Trombe wall. Table 17-1 shows a comparison of the solar fractions (percent solar heating) achieved with and without night insulation from identical homes in 21 locations in the United States and Canada, for both masonry and water Trombe walls. Movable insulation over the masonry wall generally increases the solar fraction by about 18% and over the water wall by 21%, almost doubling the performance of

Table 17-1: Annual Solar Heating Fraction Comparison for Masonry and Water Trombe Walls with and without an R-10 Night Insulation System*

City	Masonry Trombe (with night insulation)	Masonry Trombe (without night insulation)	Water Trombe (with night insulation)	Water Trombe (without night insulation)
Grand Junction, Colo.	64	45	68	46
Washington, D.C.	59	40	61	40
Boise, Idaho	57	39	59	39
Ames, Iowa	48	31	49	29
Manhattan, Kans.	58	39	60	39
Boston, Mass.	52	35	54	33
East Lansing, Mich.	46	29	47	27
Columbia, Mo.	58	39	61	39
Great Falls, Mont.	52	35	54	34
Reno, Nev.	62	43	66	45
Albuquerque, N.M.	75	56	81	61
New York, N.Y.	54	36	57	35
Raleigh, N.C.	70	51	76	55
Cleveland, Ohio	44	28	44	25
Oklahoma City, Okla.	70	50	76	54
Rapid City, S. Dak.	54	37	58	36
Nashville, Tenn.	64	45	67	47
Seattle, Wash.	58	41	59	40
Madison, Wis.	45	29	46	26
Toronto, Ont.	45	28	45	25
Edmonton, Alberta	39	25	39	21

*Comparison is for a load-to-collector ratio of 32, interpolated from tables by J. D. Balcomb and R. D. McFarland, "A Simple Empirical Method for Estimating the Performance of a Passive Solar Heated Building of the Thermal Storage Wall Type." *Proceedings of the 2nd National Passive Solar Conference.* 16–18 March 1978, Philadelphia.

some of the systems! These figures are not too surprising when one realizes that movable insulation can reduce nighttime losses through the glass by 80% or more.

While direct-gain windows can be insulated by any of the techniques described in Parts II or III of this book, Trombe walls must be insulated on the exterior of the wall mass. The exterior shutters detailed in Chapters 10 and 12 serve well to protect the Trombe wall against nighttime heat losses. In addition to these exterior systems, shade designs which operate in the narrow space between the mass wall and glass are discussed in this chapter.

Insulating the Exterior of Trombe Walls

Night insulation on a Trombe wall is no different than exterior night insulation over windows. Photos 10-1 and 10-2 show bottom-hinged shutters that insulate south-facing glass having a heat-storage wall behind it. The Steve and Holly Baer residence in Photo 10-1 has 55-gallon drums of water behind its south-facing glass wall. The First Village House #4 has an 8-inch-thick precast concrete panel filled with water behind the glass. Both of these homes have reflective insulating shutters as described in Chapter 10. The roll-up shutters described in Chapter 12 are another option for night insulation over Trombe walls, as are rolladen shutters and foam-filled, overhead garage door panels.

One Design, Inc., Mountain Falls Route, Winchester, VA 22601, manufactures a roll-up garage door that is integral with a Water Wall Modular system. A cutaway view of this system, called the Roldoor, is illustrated in Figure 17-1. Standard gypsum wallboard is used inside the water wall to give a simple, clean, and traditional appearance to the interior, and also to slow down the rate of heat transfer into the living

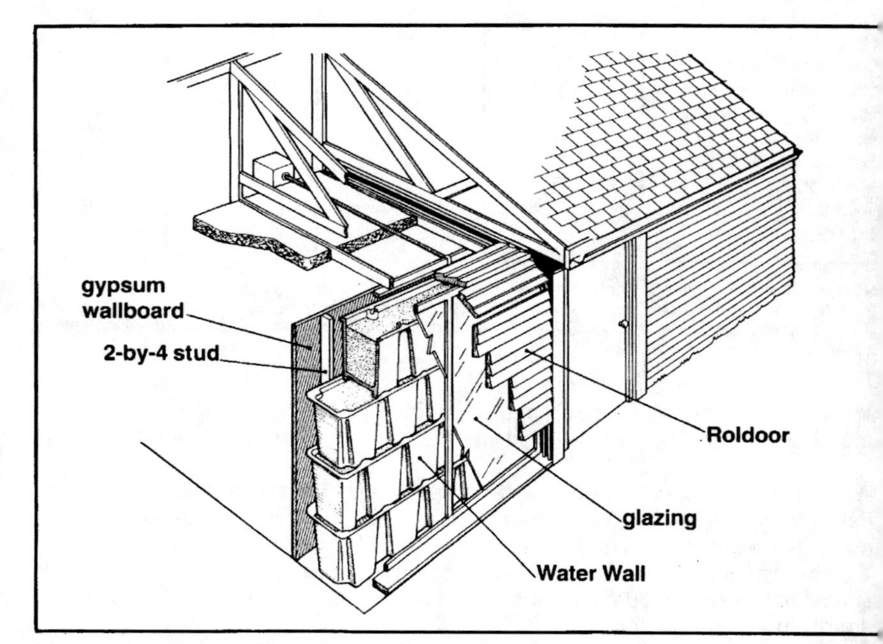

Figure 17-1:* *The Roldoor system with stacking Water Wall modules.

space. This gypsum board has about the same resistance to heat flow as the glass on the outside. However, the temperature difference between the water wall and outside is much greater than between the water wall and inside. With the insulating effect on the interior gypsum board included in this design, the night insulation provided by the Roldoor system becomes very important. Without the Roldoor system, most of the heat stored in the water wall would exit through the glass to the outside instead of gradually entering the interior.

Movable Insulation Between a Trombe Wall and the Glazing

The space or air gap between a masonry heat-storage wall and the glass that covers it can be anywhere from 2 inches to 2 feet wide. The type of insulation used certainly depends on the width of this space and as well on the degree of accessibility to it. Whatever system is applied, easy access should be incorporated into the design in case the system requires repairs. All things considered, thermal shades are generally the most practical for this application.

Figure 17-2 shows a cross section of a two-story, masonry Trombe wall that I recently designed in Asheville. Access is gained into the 18-inch-wide air gap from the living room through removable, clear acrylic panels. These panels form windowlike openings in the masonry wall to allow direct sunlight into the living room and give a view to the south. With the air gap being 18 inches and with the window wall framed with 4-by-6 mullions, it leaves a 12-inch-wide corridor for a person to snake through periodically to adjust the thermal shade and clean the window glass. (The average-size person can squeeze through a 10- to 12-inch space without any difficulty, but if you're building, be sure to design for "occupant girth"!) The shade for this particular space has two layers of Foylon hung from a Mecho Shade roller bar. The ball chain-type operator allows the shade to be raised and lowered manually from inside.

A simple, lightweight, jamfree thermal curtain for Trombe walls has been needed since the advent of this passive solar heat-storage system. Ron Shore of Thermal Technology Corporation, P.O. Box 130, Snowmass, CO 81654, has spent several years developing an effective automatic curtain to reduce Trombe wall heat losses. He had built numerous shade prototypes with multiple layers of reflective Mylar films and experimented with different ways of separating the reflective layers of fabric. One day, by accident, he discovered a very important physical phenomenon. When installing one of these experimental shades over a Trombe wall, the shade wouldn't roll up because the layers of reflective film had inflated with air. When he realized that the curtain could inflate and therefore have more insulating value, he quickly added air inflation slots to the bottom of the shade and started designing ways for the curtain to deflate as it rolls up (see Photo 17-1).

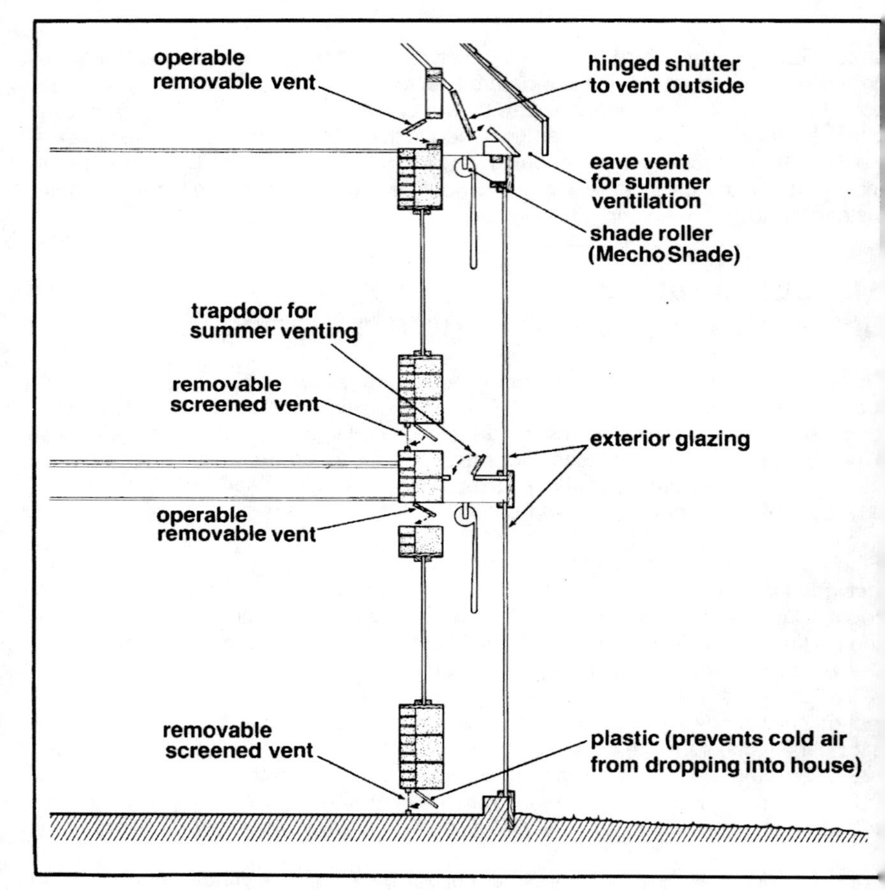

Figure 17-2: Two-story Trombe wall with shade.

Shore's curtain inflates whenever the temperature on one side is significantly different than on the other. As a result, it not only warms during cold weather, but insulates a Trombe wall from the hot, outdoor sun during the summer. Because heat always moves from hot to cold, the heat on the warm side of the curtain gradually warms the air in the curtain chamber next to it. This causes the air in this space to expand and rise to the top of the chamber, inflating the curtain. Each chamber in this curtain warms and inflates the adjacent chamber, forming temperature gradients. For exam

Photo 17-1: Inflating thermal curtain for Trombe walls.

ple, if the temperature on the outside of the curtain is 30°F. and the temperature on the Trombe wall side is 70°F., each chamber inside the curtain is successively warmer—40°, 50°, and 60°F.—as they approach the 70°F. Trombe wall. This dynamic air and heat flow phenomenon is still a puzzlement to the scientific community, but Shore's experience has shown repeatedly that the inflation principle works. A note of warning should be made for those who experiment with this inflation principle. Because the air forces are slight in this shade, the reflective fabrics must be very thin and lightweight in order to inflate properly and prevent collapse.

Following many months of refinement and development, a four-layer shade called the Curtain Wall, specifically designed for the narrow space behind a Trombe wall, is now available from the Thermal Technology Corporation. Figure 17-3 shows details for the use of this shade. The shade comes in a 6-inch-diameter housing for one-story applications and 8-inch housing for two-story applications. Inside the housing is a 3-inch-diameter aluminum roller that attaches to an 18-inch-long motor. The motor mounts inside the aluminum tube. (When installing the unit, an access panel should

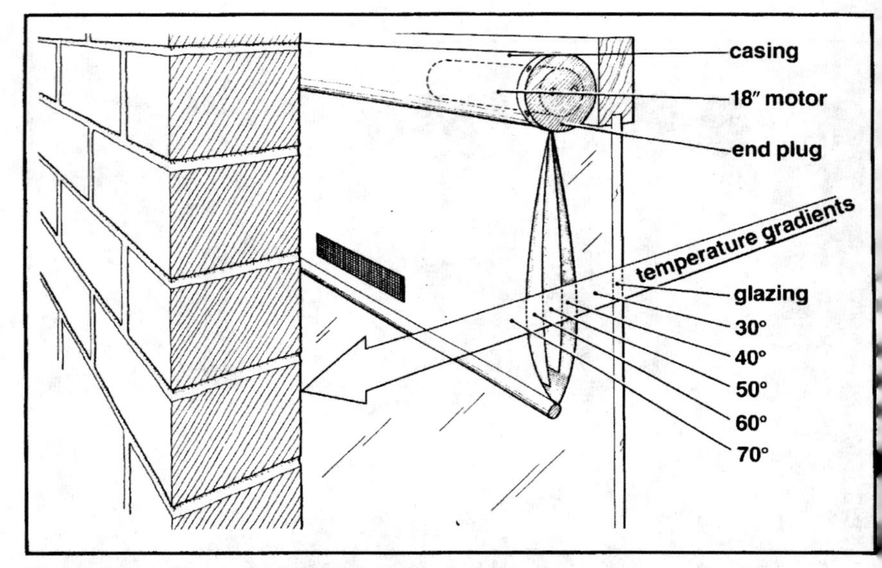

Figure 17-3: Curtain Wall by Thermal Technology Corporation.

be provided at this end to service the motor when required.) The outer layers of fabric for this shade or curtain, called Mirror-Fab 101, are available as a separate item upon request from Thermal Technology Corporation.

Roof Ponds and Movable Insulation

Harold Hay of Skytherm Process Company, 2424 Wilshire Boulevard, Los Angeles, CA 90057, developed movable insulation for a passively heated and cooled roof system many years ago. A vinyl bag filled with 8 to 10 inches of water rests on a steel roof deck. Movable insulation panels above the water-filled bags are opened on winter days to let sun shine on them and to heat the water. At night, the insulation panels are closed, trapping the heat that radiates from the steel roof deck to the space below. During the summer, the process is reversed, opening the insulation at night so the water bags can radiate their heat to the clear. desert sky and closing the panels during the day to protect against the hot, overhead sun. Several homes built with this roof system in Southern California have proven that it can be very successful for both heating and cooling—tne interior temperature varies less than 4 Fahrenheit degrees the year-round.

***Photo 17-2:** Reflecting insulating roof-pond shutters, operated by electrically driven hydraulic cylinders.*

These water-filled roof panels work well for heating in very mild climates, but in colder regions, the winter sun is too low in the sky to reflect adequate heat onto a horizontal surface. To adapt this system to the cooler temperatures in northern California, Jonathan Hammond designed hinged insulation panels which, when opened, reflect the low winter sun onto horizontal, vinyl water bags on a roof. The panels are constructed with a wooden frame and insulation sandwiched between aluminum sheets on the top and foil-faced plywood on the bottom. An electric motor drives hydraulic cylinders that raise and lower the panels as shown in Photo 17-2.

Movable Insulation to Control the Rate of Heat Flow into a Home's Interior

Movable insulation can also be used to control the rate at which heat flows out of a Trombe wall and into the interior of a home. Heat-storage walls can lose their heat too rapidly, especially during the early evening hours when the house is still warm from the daytime sun. The performance of many passive solar homes would be improved with added insulation to control the heat flow on both sides of a Trombe wall. While the insulation outside the wall would be quite substantial to reduce night losses to the cold outside, the inside insulation of this wall need be no more than a reflective curtain.

This curtain would no doubt improve the performance of a mass wall to some degree, since it helps the wall to retain its heat until it is needed later in the evening. A reflective curtain is more necessary with water walls than masonry walls because heat travels quickly through the former while heat moving through a masonry wall has about a 12-hour temperature lag before it begins to radiate into the room. Whether or not significant dollar savings are realized from installing and managing this interior insulation blanket has yet to be determined. However, it is clear that it can help regulate a more gradual release of heat into the home's interior and subsequently provide a more comfortable temperature in the home.

Insulating a Freestanding or Discontinuous Mass Wall

A masonry heat-storage wall made from small and separate units is often preferable to a large (or continuous) wall design. Access to periodically clean, operate, or repair glazing and window insulation behind a continuous wall is difficult. A continuous wall also blocks daylight and a view to the outside. To allow light, view, and access through a Trombe wall, windowlike openings can be added and faced with a removable acrylic glazing. However it is often more practical to break this wall up into distinct segments through which light and people can pass.

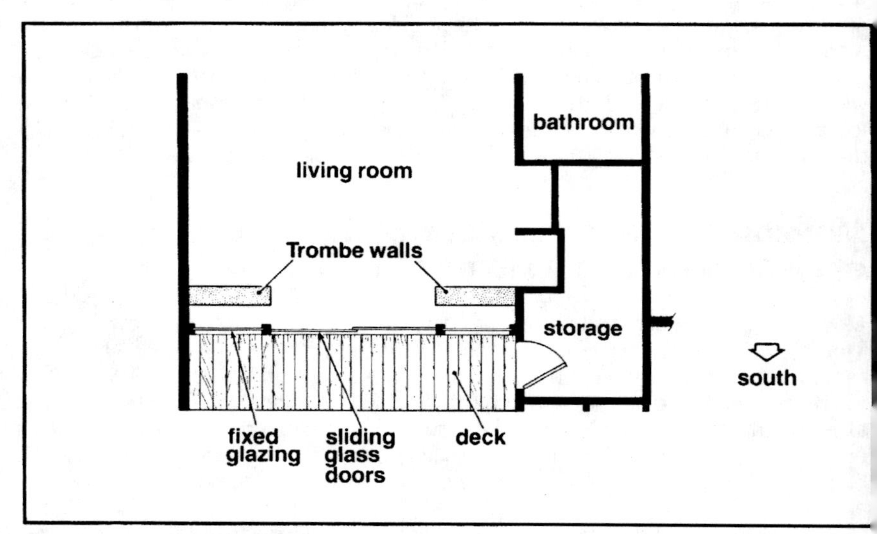

Figure 17-4: Trombe-type wing walls.

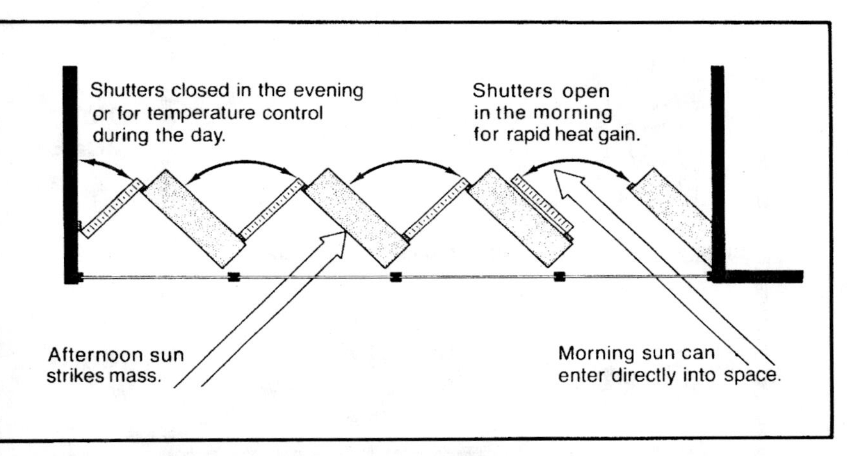

Figure 17-5: Segmented mass wall turned at an angle.

Figure 17-4 shows part of the floor plan of a home that I recently designed in Asheville. The south-facing walls of the bedroom and living room above have 8 feet of patio door exposed in the center, with a 4-foot-wide masonry wall on each side. The glass behind the walls is accessible so that it can be insulated at night. Several methods could be used for insulation; the owners plan to use pop-in shutters.

The masonry heat-storage wall can also be broken into segments that are rotated as shown in Figure 17-5. These short, vertical walls have doors between them that are closed at night to reduce heat losses and help isolate the windows from the main living space. The doors are opened during the morning to let direct sunlight in for a rapid heat gain. During the afternoon, sunlight shines directly on these short walls, charging them with its full heat intensity.

There are many ways to insulate around these walls. The doors in Figure 17-5 are less than ideal since they do not provide a continuous thermal barrier between the heat-storing walls and the glass. They are there primarily to prevent convective air flow between the living spaces and the spaces next to the glazing. However, with the access provided by breaks in the wall, the glass area behind can be insulated by pop-in shutters, hand-operated shades, or a wide range of hinged-shutter combinations. Bill Shurcliff, in *New Inventions in Low-Cost Solar Heating*,* proposes many

*See Appendix V, Section 1 for book description and complete bibliographical information.

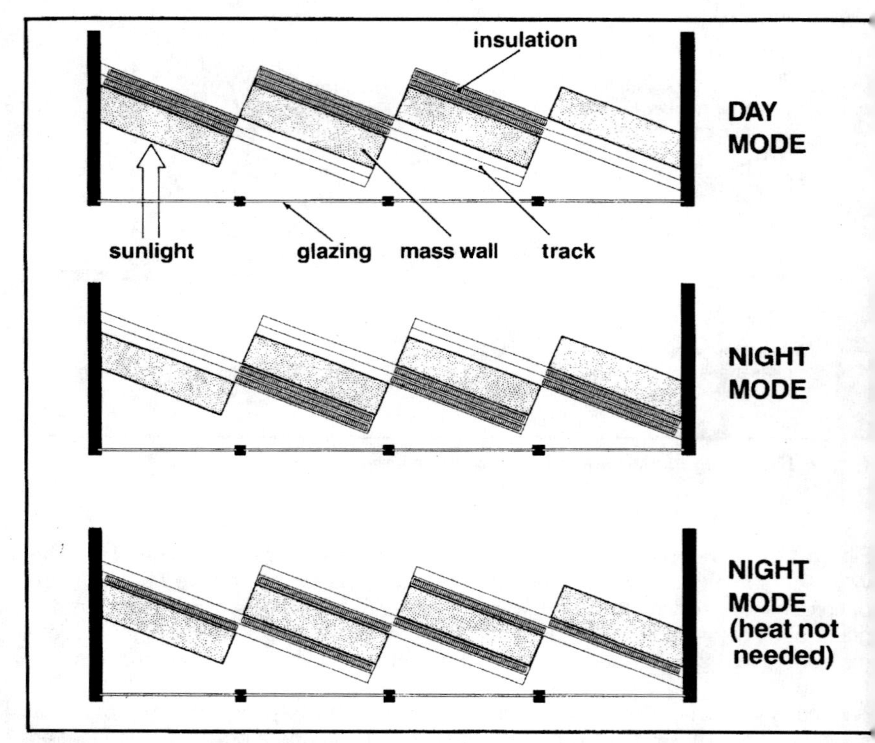

Figure 17-6: Sliding panels for insulating a segmented mass wall.

practical ways to insulate discontinuous mass walls with hinged panels. Among these are shutters that slide from behind one wall to the front of an adjacent wall (see Figure 17-6).

Another design I plan to experiment with is a thermal curtain that operates on a continuous, oval-shaped overhead track (see Figure 17-7). The curtain would have magnetic edges on both ends. The tracks at the top would nearly touch so that the edge of one curtain can join with the edge of the next one. The curtains would be joined on the interior side of the wall during the day and Velcro tabs would be used on the outside of the wall at night.

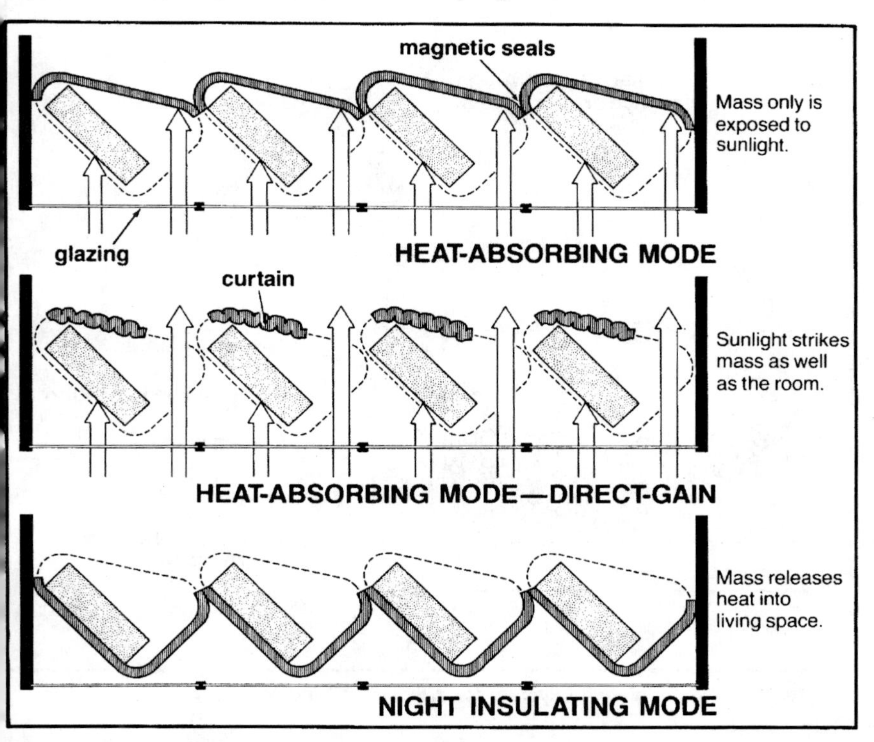

Figure 17-7: Thermal curtains on oval-shaped overhead track.

Window Insulation Which Converts South-Facing Glass into a Hot-Air Collector

Direct-gain, south-facing windows are often preferred to indirect-gain mass walls because they provide a pleasant view and enhance the home's interior. These large windows provide heat on winter mornings until the air inside becomes so warm that ventilation must be provided. As the home is saturated with heat by late morning, much of the heat entering south-facing window areas is often vented to the outside to maintain comfortable temperatures inside, but lowering the efficiency of the system. Two window insulation designs are shown in Figures 17-8 and 17-9 which allow surplus heat to be pumped instead into a rock storage bin.

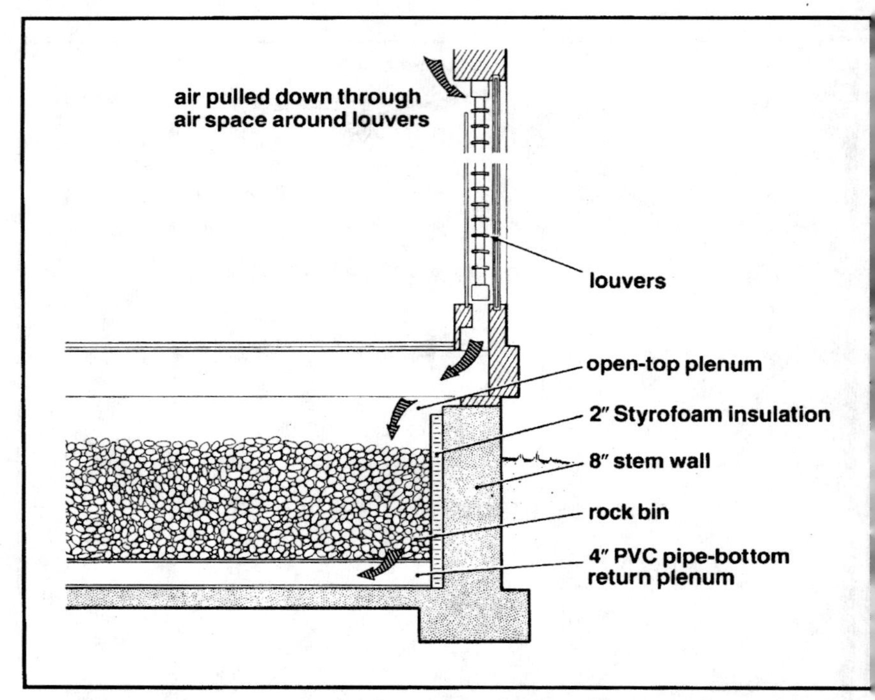

Figure 17-8: Venetian-blind system for direct-gain window/solar collector.

The first design shown in Figure 17-8 has been employed in a number of homes by Peter Calthorpe, a Calfornia architect. This south-facing window has a sealed, double glazing on the outside, a 4-inch air space, and then another layer of glass that is open to the room at the top. This cavity also opens into an air plenum (air space) in the crawl space below the floor. Venetian blinds, with thin slats that are black on one side and reflective on the other, are hung in the air space between inner and outer glazings. On winter mornings the slats are tilted to allow direct sunlight to enter until the home's interior is warm. Then they are tilted so that their black surfaces block all direct sunlight and absorb most of the sun's heat. However, indirect light can still enter for daylighting between the slats. When the blinds are closed in this way, a blower is turned on to pull air down through the air space around the louvers, drawing heat away to be stored in a 20-inch-deep rock bin beneath the floor. The heated air enters this rock bed from a plenum across the top of the rock bin and exits out the bottom of the bin through 4-inch drainpipes that are laid on 16-inch centers.

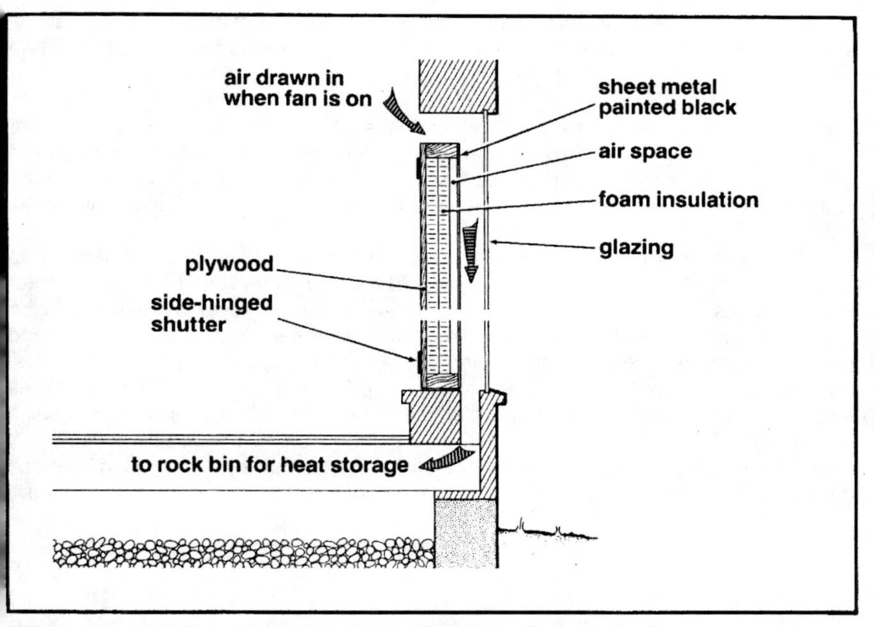

Figure 17-9: Shutter system for direct-gain window/solar collector.

On winter nights, the blinds are shifted so their reflective surfaces face the room, reflecting heat back toward the room and helping to reduce heat losses through the window glass. On hot summer days, these reflective surfaces tilt in the opposite direction to face out so that sunlight is reflected to the outside before it is absorbed and becomes unwanted heat. Finally, on cloudy days throughout the entire year when direct sunlight is no problem and diffuse sunlight is desired, the entire venetian blind can be raised to allow a clear view to the outdoors.

The other design approach, Figure 17-9, is suggested by Nick Nicholson, a Canadian solar home builder, in his book *Harvest the Sun* (Quebec: Ayer's Cliff Centre for Solar Research, 1978). This system is designed more for a northern climate with low winter sun angles and very cold nights. Instead of a venetian blind to collect the sunlight, this system simply uses a series of shutters that are closed tightly over a south-facing window wall during the daytime.

The shutters should have a sheet of galvanized steel or aluminum painted flat black and mounted onto a shutter frame. Behind the sheet metal is an air space and then 1 or 2 inches of foil-faced isocyanurate foam board. A plywood interior facing can be used to prevent the shutter from sagging. These shutters form an air cavity when closed so that a third layer of glass is not necessary.

This system has three modes of operation: In the direct-gain mode, the shutters are open to let in sunlight and allow a clear view to the outside. In the collection-storage mode, the shutters are closed and a blower is turned on to charge a rock bin. In the night insulation mode, the shutters remain closed with the blower turned off.

These are just a few examples of the many ways that a movable panel or fabric can serve as a solar hot-air collector during part of the day, allow direct sunlight to enter at other times, and insulate against heat losses at night. For a window insulation system to serve as a hot-air collector, it must absorb the sunlight which strikes it and readily transfer the sun's heat to moving air. Metal surfaces are best suited for this, but even plastic shades with tracks along the edges, like the Arc-Tic-Seal Inseal-shaid, can successfully perform as solar hot-air collectors. The solar window, which will be common in future homes, may be very different from the common window of today. There are numerous design possibilities for windows that control, in several different modes, the flow of heat and the kinds of daylight which enter the home.

Chapter 18

Movable Insulation to Assist Solar Water Heaters

Solar water heating is rapidly becoming common as a way to heat water for the home. Solar domestic water-heating systems save money year-round, not just in cold weather, and some installed systems have a payback period of less than 5 years. With many commercially available active systems, however, the price is far greater than justified, extending the time it takes to recover the investment to as many as 10 years and more as the system becomes increasingly sophisticated.

Much of the expense in some domestic water-heating systems is incurred in overcoming freezing problems. A single water freezeup in a water-filled solar collector can destroy an investment of well over $1,000. To insure against freezing, a majority of the solar installations in northern climates don't pump water directly through the collector, but instead use an antifreeze solution that circulates through a heat exchanger in a storage tank filled with potable water. However, the heat exchanger is an expensive item and it also causes the collector system to operate at a higher temperature than a water-filled collector. With the collector at a higher temperature, it loses more heat to the outside and is less efficient at collecting the sun's energy. In addition, ethylene glycol or other toxic or corrosive antifreeze heat-exchange fluids are often used in these systems, and they require the use of double-wall heat exchangers as extra protection against contamination of the potable water. A double-wall exchanger further increases costs and impairs the transfer of heat.

Some solar hot-water systems use house-pressure water in the collector plate and avoid the use of antifreeze and heat exchangers by automatically draining whenever the pump is off, or when freezing temperatures are approached. These systems can be tricky to install; moreover they rely on electrical controls to make it all work, which means that Murphy's Law has a perfect in.

Other approaches to the freezing problem include attaching electric resistance wires to the back of the collector to prevent freezing and circulating small amounts of stored hot water back through the collector. However, these approaches waste hea to the outside at night and are all in some ways vulnerable to failure.

Bringing the Collector into the Home's Thermal Envelope

Probably one of the most overlooked approaches to domestic hot-water heating is to bring the collector into the home's thermal envelope and use the heat within the home or in an attached greenhouse to keep the absorber plate from freezing at night. Figure 18-1 shows a design which I am currently developing to simplify solar heating systems and insure against freezing where the collector is separate from the tank.

In this passive system the collector can be located in an attached greenhouse, or better yet, beneath a skylight in the main heated space of the house. Heated water flows by thermosiphoning (no pumps) from the collector into a tank located above it. A thermosiphon system operating without the antifreeze, the pump, controls, heat exchangers, and special valves, is about a third of the cost of the "high-tech" system

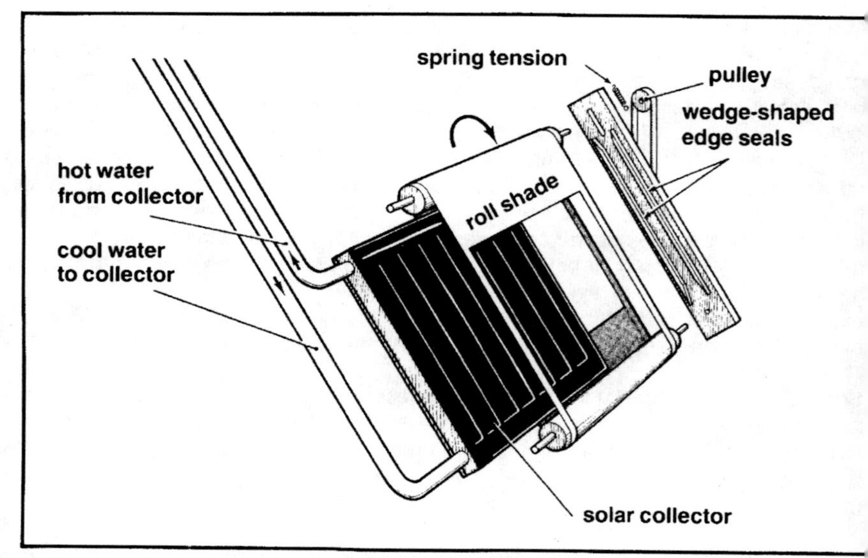

Figure 18-1: A thermosiphon solar water heater designed by the author.

To reduce nighttime heat losses from the indoor collector, a double-layer reflective fabric is mounted on two rollers with connecting straps along each side so that the fabric runs in a continuous loop around the rollers. The insulation blanket is operated by a pulley and strap, similar to those used on draperies. Electric resistance heating wires with an automatic thermostat (sold by plumbing supply houses) are attached to the collector to insure against freezing. However, if the insulating blanket is drawn over this plate at night, the electric resistance wires should never come on. During the summer, the reflective blanket remains under the collector plate to prevent the plate from heating the home.

If you install an elevated storage tank on the second floor of a home or in an attic, be sure to include a galvanized steel or plastic water pan beneath the tank to prevent water damage to the rooms below should the tank leak. This pan should include a drain line to carry off the water spilled in an emergency or during servicing.

The Bread Box—
Eliminating the Collector Altogether

The bread box or batch-type water heater is by far the simplest of all domestic hot-water heating systems. It even eliminates the collector plate by placing the water tank directly in the sun. The bread box water heater is composed of one or more water tanks painted black and mounted in an insulated box, and which has a glass cover facing south. Reflective insulating lids both improve the performance of a bread box and help store heat overnight.

The bread box water heater can be built as a freestanding unit like the one shown in Figure 18-2. It features two reflective insulating lids, one on the top that can be adjusted seasonally for optimum sun reflection angles, and a lid on the front that opens to rest on the ground. Plans to construct this unit are available from Zome-works Corporation, P.O. Box 712, Albuquerque, NM 87103.

The best tank for a bread box is an electric, glass-lined hot-water tank that has been stripped of its outer casing, insulation, and heating elements. Glass-lined tanks will resist inside corrosion better than other tanks and therefore last longer. They can sometimes be picked up very cheaply secondhand. Tanks from discarded gas-fired water heaters are also usable, as are galvanized pressure tanks and some well-site pressure tanks.

The mixing of hot and cold water in the tank of a bread box should be avoided as much as possible. Good designs form layer after layer of progressively hotter water.

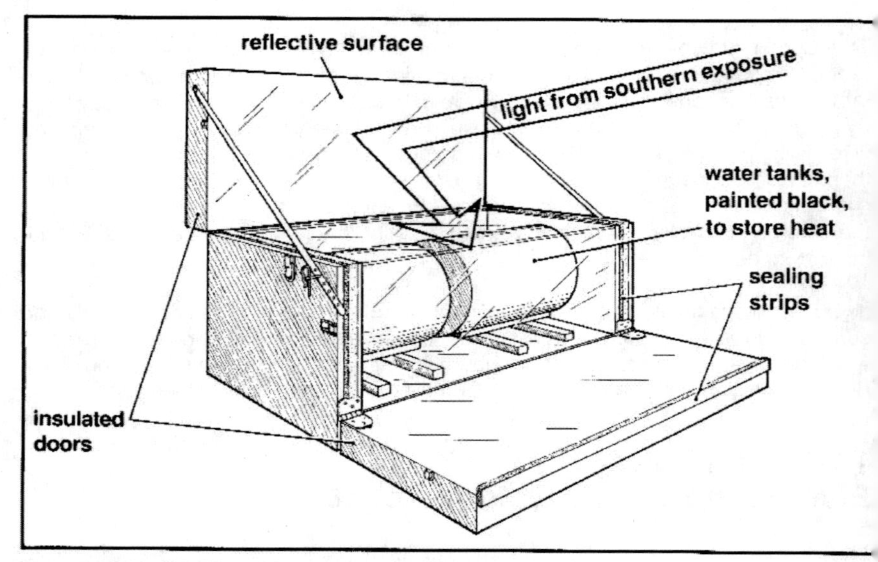

Figure 18-2: The Bread Box design by Zomeworks.

Photo 18-1: Three-tank bread box, Davis, California.

This is accomplished both by connecting tanks in series and by tilting tanks upright to allow temperature stratification. Photo 18-1 shows a three-tank, bread box water-heating system in Davis, California. To receive maximum sunlight, the tanks are upright and tilted back at an angle that equals the latitude of the city of Davis. There is still enough column height to allow the stratification of temperatures within each tank. With three tanks, each developing progressively warmer temperatures, the system is able to provide the entire hot-water needs for a family in Davis for eight or nine months out of the year. And it also serves as a preheater to a small, 20-gallon backup water heater during the remainder of the year.

Night insulation is not required in the Davis locale because winter nighttime temperatures there rarely drop below 30°F., but in colder climates an insulating lid is necessary to avoid freezing and help the water tanks retain more heat. The better the box and lid are insulated, the warmer the water remains and the less you have to rely on a backup system.

Cold water should be fed into each tank near the bottom, and exit where it is warmest, at the top. Regular water tanks have cold and hot water openings on the top. The fitting for cold water connects to a pipe that runs to the bottom of the tank. This is most desirable for upright tanks because it allows temperature stratification. If you lay a tank on its side, the hot and cold water will mix. In this type of system, a curved pipe should be added to carry the water entering to the bottom tank and another curved pipe to draw the hot water off the top.

About 1 square foot of glazing is required for every 2½ gallons of water in the system. Twenty- to 30-gallon water tanks for each person in a household should be ample. Therefore, a 90-gallon system of three 30-gallon water tanks should supply domestic hot water for a family of four and would require about 40 square feet of glazing. Reflectors can increase the efficiency of a system and reduce the amount of glazing required per gallon.

If a bread box is mounted into a roof, caution should be given both to potential structural and water damage problems. A 90-gallon, three-tank bread box can weigh 1,000 to 1,200 pounds and many roofs aren't designed to take this heavy load. A bearing wall or posts running down through the structure will help support the weight. Consult a structural engineer or experienced carpenter if necessary. If the system is mounted above a roof, leaks are no problem, as far as water damage is concerned. All bread box tanks should have drain lines so they can be emptied for servicing or replacement. Depending on the climate, freezing can be a problem in bread boxes, particularly in the water lines that run between a bread box heater and into a home. These lines should be heavily insulated and electrical heating wires should be wrapped under this insulation where pipes are exposed to the outside in cold climates. However, maintaining above-freezing temperatures in these lines by electric

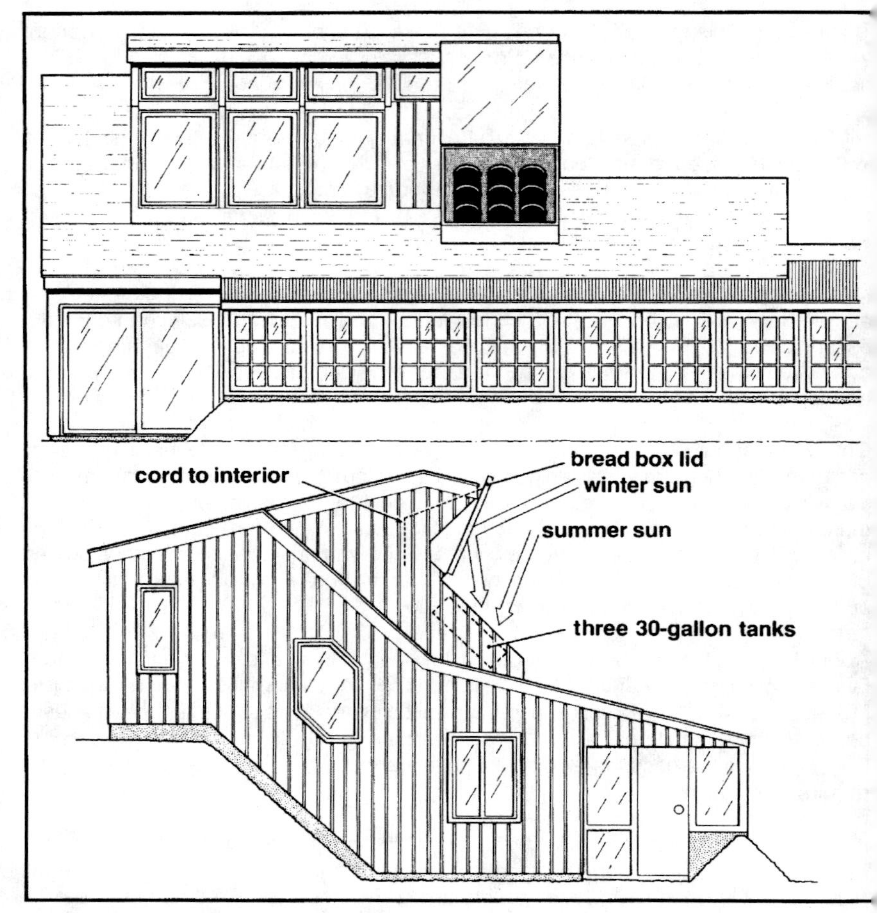

Figure 18-3: A bread box design by the author.

resistance wires can be expensive and wasteful and freezing problems can be best avoided by building the bread box into or up against the south side of a home. This also helps to reduce heat losses from the bread box since temperatures in the house are warmer than the outside. However, the bread box is still much warmer than the home's interior and insulation is still required between the box and the house.

Figure 18-3 shows a bread box I recently designed into a home in Asheville. The insulating lid will be raised from the second story by a pull cord, and secured in place by fastening the cord to a cleat. The lid should raise to an angle equal to the altitude of the noontime sun on the summer solstice. The box frame will be constructed from 2-by-2s with 3 inches of Thermax insulation board on the bottom side and ⅜-inch plywood on the top. When the shutter is opened, the foil-faced Thermax serves as a reflector.

The Capsule Collector

Photo 18-2 shows a very simple, single-tank bread box water heater, which has been named the Capsule Collector by its designer, John Golder of Santa Cruz, California. Designed for economy and using a minimum of materials, this bread box contains an 80-gallon tank and a layer of clear glazing covered by a wraparound fiberglass blanket. The insulation in this design works very efficiently because when it is closed, a minimal surface area is exposed to heat loss. If this tank were in a box with two or three times the exterior surface area, it would lose two or three times more heat.

The fiberglass blanket in this design is attached to a curved sheet of galvanized steel that reflects sunlight onto the tank (see Figure 18-4). Aluminized Mylar is glued to this

Photo 18-2: Capsule Collector designed by John Golder.

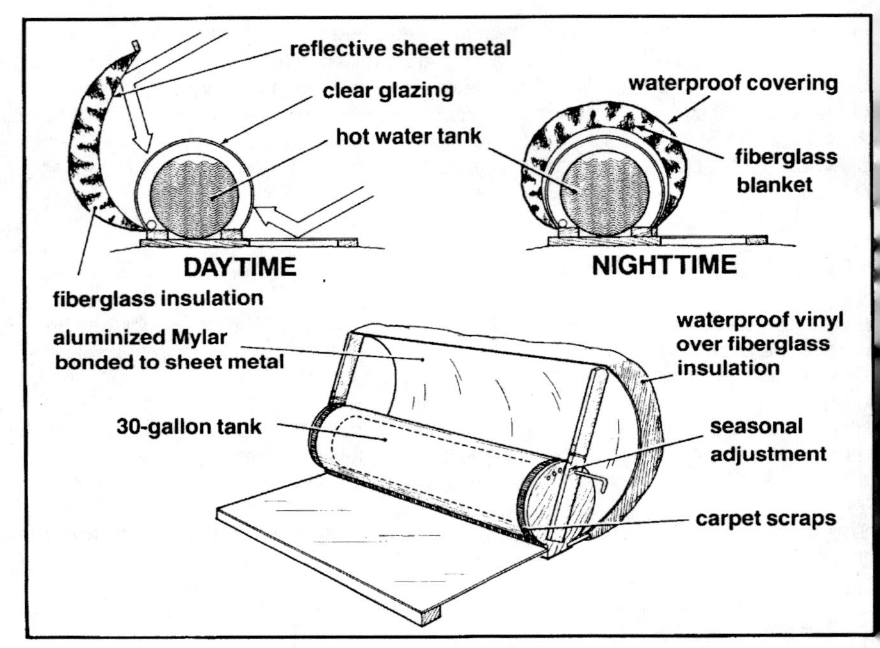

Figure 18-4: The Capsule Collector.

galvanized sheet to increase its reflective properties. Hinged 2-by-4s at each end of the tank hold the reflector in position and guide it as it closes. These arms are shifted seasonally to adjust the tilt of the reflector to winter and summer sun angles. A waterproof vinyl fabric covers the insulation on the outside. Carpet scraps are glued along each end and the bottom of the tank to make a tight seal when the cover is closed. Complete plans for this heater are available from John Golder, P.O. Box 854, Santa Cruz, CA 95061.

This design inspired me to integrate this concept into a second-story split-roof in a home I recently designed for Janis Paquette in Arden, North Carolina (see Figure 18-5). Instead of having the reflector fold around the tank where the insulation is vulnerable to the elements, I chose to make the reflectors stationary. The glazing in this design is two layers of clear, plastic-reinforced fiberglass bent to the curvature shown in the illustration. A 4-inch clearance remains around the tank, except where it is supported on the bottom. A flexible, foam insulation blanket is pulled open and closed by a small cord.

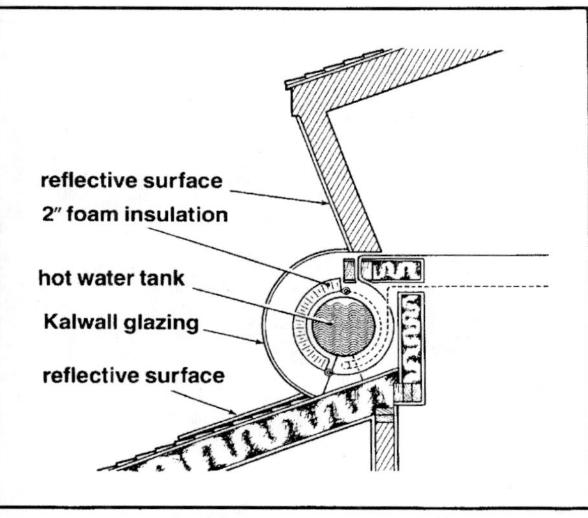

Figure 18-5: A modification of the Capsule Collector design that can be attached to a residence.

This chapter has presented a few options for heating water efficiently without the expense of most conventional, active hot-water systems. New bread box designs are being developed every day and these designs offer effective ways to handle cold climates. At the same time, active solar hot-water systems are now available as packaged units and are reliably installed and serviced by a number of local plumbing companies. The use of active solar hot-water systems is rapidly growing. The Tennessee Valley Authority installed 1,000 active, domestic hot-water heaters on homes in Memphis, Tennessee, and has plans to install 10,000 more in Nashville. Whether bread box water heaters will ever rival the active systems for widespread popularity is yet to be seen. However, for those who want to construct a simple, low-cost, solar water-heating system, the bread box is hard to beat.

Appendix I

Insulated Shade and Shutter Construction

Section 1 Making a Roman Shade*

Assembling the necessary materials for making the roman shade requires accurate measurements of the window. Measure the width (W) from the outer edge of the casing on both sides. Measure the length (L) from the windowsill to the upper edge of the top casing. If there is no sill, measure from the lower edge of the bottom casing or apron. (see Figure AI-1a).

Materials

The shade requires fabrics with insulating and moisture-retaining value. Some comments about the nature of these fabrics and their availability follow the materials list. For more specific recommendations, refer to Table 6-1: Stability of Fibers to Environmental Conditions.

The fabrics are to be cut to the dimensions that follow (see Figure AI-1b). When purchasing materials, allow a few inches of additional fabric length and width to insure that you will be able to cut on the straight and square all corners.

	Width	*Length*
Outer fabric	W + ½ inch	L + 3 inches
Lining fabric	W + ½ inch	L + 3 inches
Fiberfill	W – 1 inch	L + 2½ inches
Vapor barrier	W – 1 inch	L + 2½ inches

*These instructions were condensed from a set of detailed, step-by-step plans prepared by The Center for Community Technology for a number of window insulation systems. For information on available plans and their prices, contact: The Center for Community Technology, 1121 University Avenue, Madison, WI 53715.

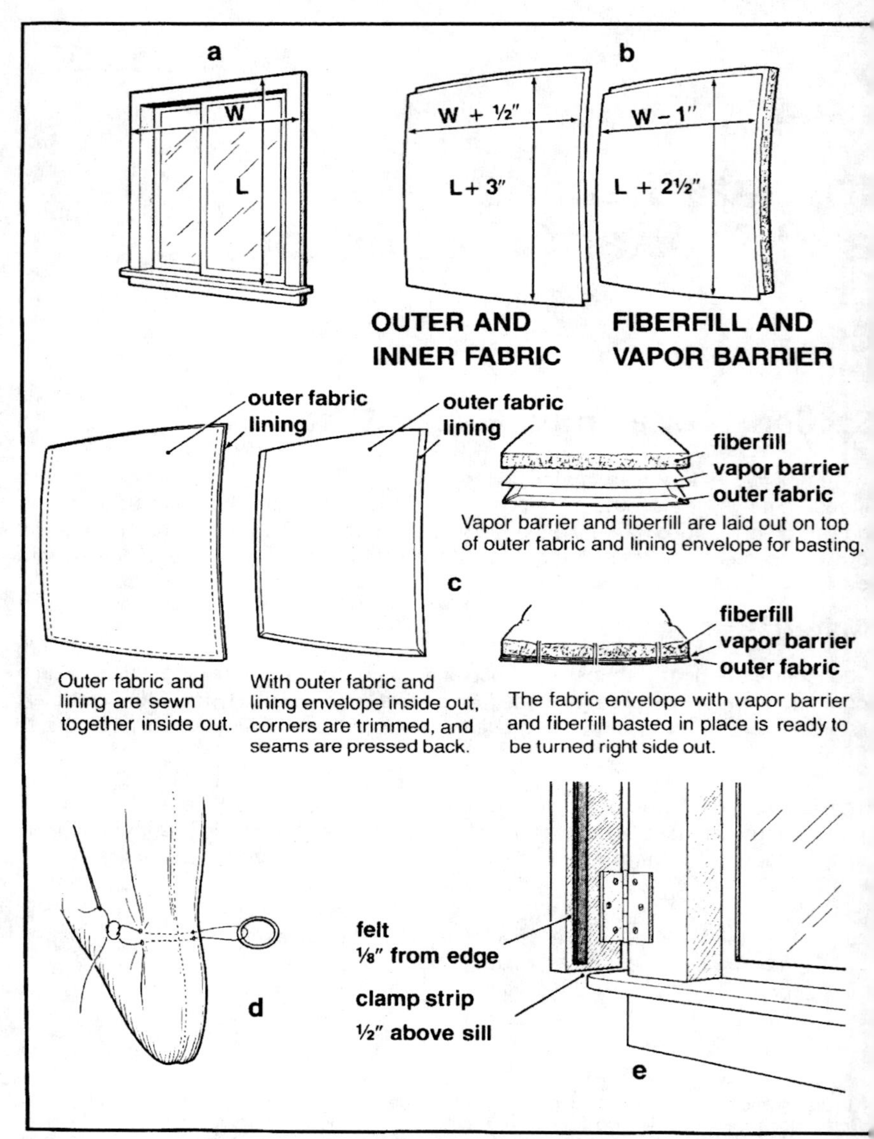

Figure AI-1: Step-by-step construction of a roman shade (here and on facing page).

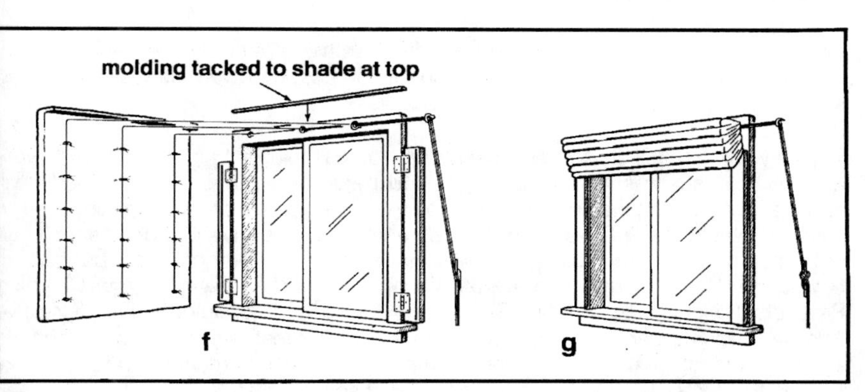

(Figure A I-1 cont.)

½-inch plastic rings—called "bone rings," available in fabric and notions stores
Thread
Heavy thread or crochet cord
One spool 210-pound-test nylon mason's line (twisted, not braided)
Four to six flush-mounted, self-closing cabinet hinges (like Amerock 7929)
1-by-2-by-(L - 3)-inch wooden strips
½-by-(L - 3)-inch felt weather stripping
½-inch-diameter screw eyes
½-inch finishing nails
Flatheaded push tacks, carpet tacks, or ¼-inch staples
Corner or screen molding, quarter-round, or furring strips, cut to width of window
Glue
Cleat

The outer fabric used should be tightly woven and have enough body to hang without sagging. Most standard drapery fabrics are fine, although the tightness of the weave should be checked. Polyester blends are wrinkle-free and recommended for longer wear.

Sheets work well as a lining fabric. Whatever fabric is chosen should be light in color, tightly woven, and resistant to damage by the sun.

Either 1- or 2-mil polyethylene or lightweight, clear vinyl will work well as a vapor barrier. However, most fabric store salespeople cannot tell you the mil thickness of the plastic or vinyl they carry. Painter's drop cloth is a standard ½-mil thickness and is readily available in 9-by-10-foot sheets at most hardware stores. To form ar adequate vapor barrier, use a double thickness.

The polyester fiberfill carried by most fabric stores is usually ½ inch thick, and available by the yard in sheets or prepackaged bags. Three layers of the fiberfill wil form insulation of sufficient thickness to make the roman shade a good energy-saver. However, the R value is less than other polyester fiberfills designed specifically for this purpose. Thinsulate, by 3M Company, is only ¾ inch thick, but has ar R value of 4 (per inch). Write the 3M Company, Building Services and Cleaning Products Division, 3M Center, Saint Paul, MN 55101 for information. (Creative Energy Products packages roman shade kits to fit most windows. These contair all the materials except the outer fabric. For more information, write Creative Energy Products, 1053 Williamson Street, Madison, WI 53703.)

Tools

Carpenter's square
Iron
Soft lead pencil or chalk
Scissors
Heavy sewing needle
Screwdriver—blade must match the screwheads of the hinges and cleat
Sandpaper
Handsaw or circular saw
Hammer

Optional Tools

Sewing machine (preferable, but not necessary; shade can be hand sewn)
Quilting frame
Staple gun
Drill—needed only for working with a hardwood, such as oak
Drill bit—must match the size of screws used to attach hinges and cleat

Construction Procedure

Step 1: Before cutting the fabric, make a tissue paper pattern to the largest dimensions given above. Each corner of the pattern should form a perfect 90-degree angle so the shade will hang correctly on the window. Use a carpenter's square when laying out the material to make sure that all pieces are cut with squared corners. Do not cut on the bias, but on the straight so that the fabric will not stretch Trim the pattern as needed to match the cutting dimensions given above.

If your window is a nonstandard size, it will be necessary to piece together the various layers to make them large enough. After piecing, each layer should be of the given dimensions. The only place where all layers will be sewn together is at the seams on the sides, bottom, and top of the shade, and where the shade is spot-quilted to attach the plastic rings. The largest roman shades of this design ever created by The Center for Community Technology are 12 by 5 feet and 6 by 6 feet (for glass patio doors).

Step 2: Take the lining fabric and outer fabric and put the "right" sides (those that will show on the outside) together (see Figure AI-1c). Sew the outer fabric to the lining down one side, across the bottom, and up the other side, taking a ½-inch seam allowance. Trim the corners and iron, pressing the seams back. Lay out the inside-out "envelope" with the outer fabric up. Lay first the vapor barrier and then the fiberfill layer on top of this. Using only enough stitches to hold these layers in place, baste the vapor barrier and fiberfill through the top layer of the fabric "envelope" only. Do not stitch through the lining. Turn the "envelope" right side out. The vapor barrier and fiberfill will fit snugly at the corners and edges.

To check the measurements and squareness and to mark the top edge, hold the curtain up to the window or measure very carefully while it's lying flat. The shade should drop an inch below the sill to provide a seal at the bottom. If there is no sill, the shade should extend an inch past the bottom edge of the casing.

After marking the placement of the top edge, lay the shade flat on the floor and draw a straight line all the way across the top. Fold in the edges along that line. Stitch along the top, by hand or machine, making a narrow seam, and cut away any excess plastic or fiberfill that protrudes from the top. The shade should now measure 1 inch less than measurement W and be about 2 inches longer than measurement L.

Step 3: The plastic rings to raise the shade are attached to the fabric 6 to 12 inches apart. To determine the spacing, pencil vertical and horizontal lines on the shade lining. Begin the vertical lines 4 inches up from the bottom edge; for an average-size window, space three or four vertical lines evenly across the shade. Begin the horizontal lines 5 to 6 inches from the edge of the shade; for a standard window, pencil in five to seven horizontal lines. Wherever they intersect, you will attach a plastic ring.

Prepare to spot-quilt, attaching the plastic rings to the shade in the process (see Figure AI-1d). Use a quilting frame or work on a rug to keep the shade from sliding around. Thread your needle with heavy thread or crochet cord. If you want the ties to show, work from the room side of the shade; if not, work from the lining side.

If working from the room side, push the needle down through all the layers. Attach a ½-inch plastic ring to the lining of the curtain by stringing the needle through it and then bring the needle back up through the curtain about ¼ inch from the first stitch. Tie the threads loosely together, knot, and trim them down to about ½ inch. Pulling too tightly will crush the fiberfill, and the insulating value will be lost. Quilting reduces the width and length of the shade, and this must be kept to a minimum.

Step 4: Cut the wooden strips about 3 inches shorter than L, the height of the window from the upper edge of the top casing to the sill. Sand and finish. Then attach two flush-mounted, self-closing cabinet hinges on the back of each of the wooden strips. Fix the two hinges at points equidistant from the center of each wooden strip and its ends. If your window is more than 60 inches high, use three hinges per wooden strip (see Figure AI-1e).

Staple, tack, or glue ½-inch-wide felt weather stripping to the backs of the wooden strips, about ⅛ inch from the inside edges. The hinges will be mounted on the face of the window casing near the outside edge. Hold one strip in place at a time and mark the spots for the holes near the edge of the window casing. The bottom of the wooden strip should clear the sill by about ½ inch. If there is no sill, the bottom end of the wooden strip should be even with the bottom edge of the window casing. Make starter holes for the screws with a drill or a nail, and screw the wooden strip to the window casing. Fasten the strip on the other side of the window casing in the same manner.

These strips may now more accurately be called clamp strips, since they will hold the fabric of the roman shade tightly against the window casing, forming an air seal. If you prefer not to drill holes in your walls, attach the cleat to the right clamp strip, inset from the bottom about one-fourth of the total length of the clamp strip. If you prefer to attach the cleat to the wall, make certain there is a supporting wall stud. Mark the location of the cleat, make a starter hole, and screw in the cleat.

Step 5: For pulling the shade up, we recommend two strands of twisted—rather than braided—210-pound-test nylon mason's line. Cut as many lengths of mason's line as there are vertical rows of plastic rings. The length of the shortest line will be equal to twice the height of the window opening and the length of the longest line will be equal to twice the height of the window opening plus the width across the top. Tie a line to the bottom ring in each row and thread it up vertically through the remaining rings in the row (see Figure AI-1f).

Screw eyes about ½ inch in diameter are the top guides for the line which pulls the shade up. If the cleat is fixed to the right wooden clamp strip, the number of eyes installed in the frame at the top of the window opening should match the number of vertical rows of rings. If a cleat is fixed to the wall to the right of the window, an additional screw eye will be necessary to guide the pull cord to the cleat. This screw eye should be placed outside the frame, on the same side as the cleat, where it can be screwed into a wall stud. Mark and drill pilot holes so the screw eyes will line up with the vertical rows of plastic rings, and horizontally with each other. Then screw the eyes in by hand.

Fold the top edge of the shade over the top edge of the head casing and lightly tack or staple it in place. Check to see that the shade is hanging straight along the edges and bottom; make the necessary adjustments if it is not. Cut a piece of corner molding, quarter-round, or furring the width of the shade; finish, if desired, with paint or stain, and nail it in place on the top edge of the head casing to hold the shade securely. Then string the lengths of mason's line through the screw eyes. You should tie the lines together and braid them. Tie them up high enough so that when the curtain is down the knot will be just short of the first screw eye, then braid them and tie again at the bottom (see Figure AI-1g).

Note: To clean the shade, hand wash in warm water with a mild soap or detergent or machine wash on gentle cycle. Soak for 20 to 30 minutes, then press out the water by hand. Do not wring. Repeat until the shade is well rinsed. Air dry or tumble dry at very low temperatures. Dry cleaning is not recommended.

Section 2 Making a Frame and Roll-Down Quilted Shade*

Materials

For the Frame

Planed, shelving grade 1-inch lumber (actual ¾-inch thickness)—
  width and quantity vary by the job.
⅞-inch-diameter wooden dowel rod—1⅝ inches shorter than top window casing
Two or three (6 to 10 inch) pieces of ½-inch-diameter wooden dowel rod
Nylon rope—⅛-inch diameter, twisted rather than braided
Metal ring—1-inch diameter (approximate)
Glue

*This prototype was developed for the Appropriate Community Technology Fair in Washington, D.C., May 1979, by the Sandstone senior citizens as a project incorporating recycled fabrics. Bruce Osen and Jon Averill supplied the design idea and built the frame. Peggy Rossi, Edna Gwinn, and Pauline Foster made the original quilted shade. Jon Averill, who contributed the instructions for the frame, and Peggy Rossi, who described the method of making the shade, would like to hear from anyone who has suggestions for possible insulating and heat-reflective inner layers. Write: Rossi/Averill, General Delivery, Sandstone, WV 25985.

4d finishing nails
1¼-inch wood screws (optional)
Carpet tacks or gun staples
½-inch vinyl weather stripping—approximately the width of the window

For the Shade

Determining the amount of fabric needed to make the shade requires measurements obtainable only after making the frame.
Comments on the nature and availability of the materials needed follow the materials list.
Patchwork or appliqued layer—acts as quilt cover; see "Construction Procedure for the Shade," Step 1, for design requirements
Two layers of polyester fiberfill or quilt batting
Vapor barrier with reflective coating on one side
Lining
Thread
½-inch dowel—approximately the width of the quilt track

Cotton-polyester blend fabrics are recommended for the interior-facing patchwork layer and exterior-facing lining. These fabrics should be closely woven. To line a large, nonstandard-size window, actual quilt backing (which is 90 inches wide) or sheets can be used.

Sandwiched between the quilt cover and the lining are three layers. Two layers insulate and the third layer captures moisture and reflects heat back into the room. Low-density polyester fiberfill sheets work well as insulation. Regular quilt batting may be used instead, but it is less compact and durable. Beachwood polyfill interlining, which is available in some fabric stores and is sold by the yard, is recommended.

The middle layer is a vapor barrier material. Thermos brand all-weather sportsman's blanket, a ground cloth with fiber scrim added for durability, performs well in this role. It is available from most sporting goods stores, but is expensive. Many other plastic fiber-aluminized materials will work well here. (Astrolon was used in a prototype.)

Since most windows are no wider than 36 inches, the fabrics described above for the inner layers are of an adequate width. If you attempt to make a shade for a large nonstandard-size window, each individual inner layer must be cut in sections and pieced to form a layer wide enough to cover the window.

Most hardware stores do not sell dowels longer than 36 inches, the maximum vidth of a standard window. If you are making the quilted shade and frame for a arge, nonstandard window, you will need access to a lathe to cut dowels of sufficient length.

'ools

For the Frame

Radial arm or bench saw—desirable, but a hand circular saw will suffice
Joiner-planer
Rasp
Sandpaper
Drill
Drill bits (1/16, ⅛, ¼, ½, and 1 inch)
Hammer
Screwdriver—blade must match head of wood screws
Staple gun

For the Shade

Sewing machine
Needles
Pins
Scissors
Chalk or soft lead pencil
Cardboard (for quilt pattern shapes)

Construction Procedure for the Frame

Step 1: Measure the width of window side casing. Subtract 1¼ inches. Rip two pieces of planed shelving lumber to this width, making both the same length as the side trim (see Figure AI-2a). Carefully sand the edges, which are to face the quilt to minimize abrasion. Glue and nail one strip to the window side trim, leaving a ¼-inch space on the outside, and a 1-inch space on the inside. Glue and nail the other strip to the opposite side trim, placing it as described above.

Step 2: Cut two boards to match the length and width of the original side trim. After careful sanding, mount with glue and nails on each side of the window opening. Your quilted shade track is now in place.

Step 3: Cut a ¾-by-1-inch strip to bridge the tops of the two new side casings. Rasp and sand the strip to create rounded edges on the inside (this piece will keep the shade somewhat snug against the top casing). Drill starter holes, glue to the top casing, center, and nail into place.

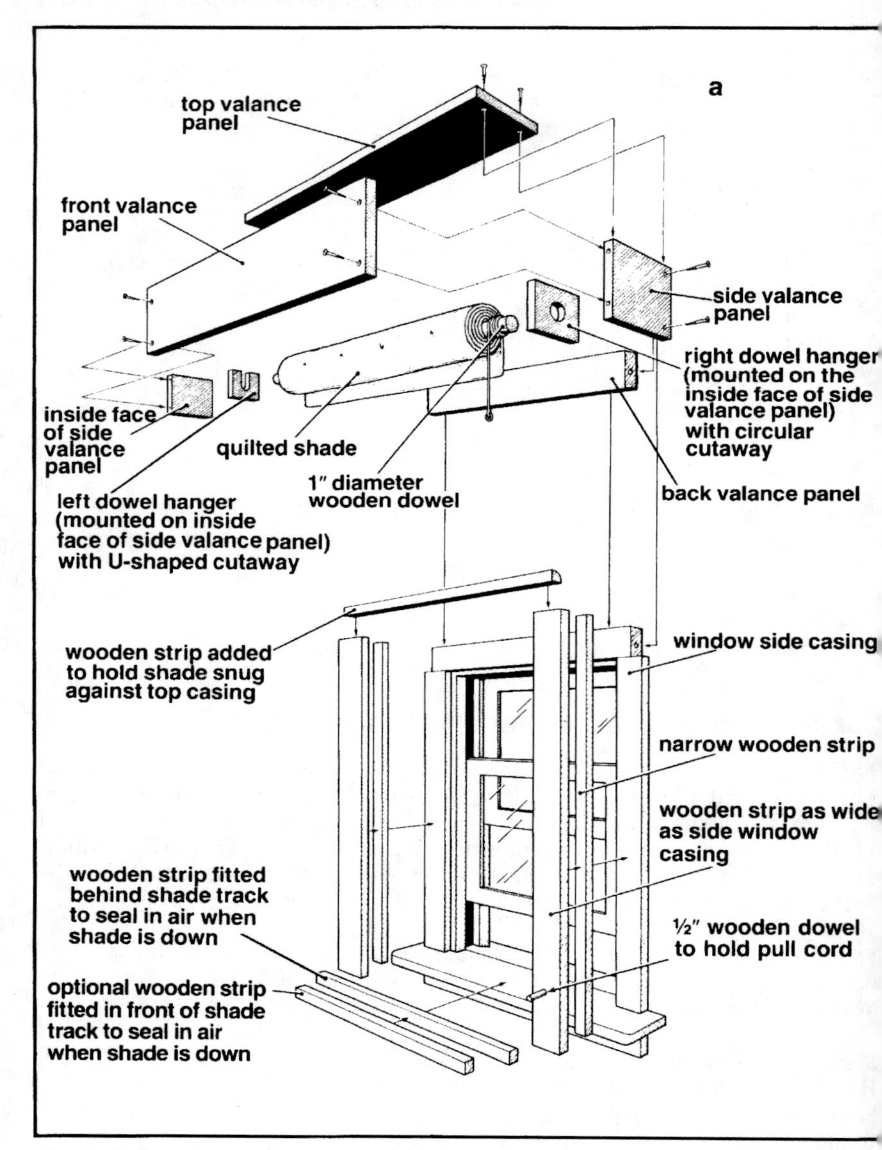

Figure AI-2a: Exploded views of valance and shade track.

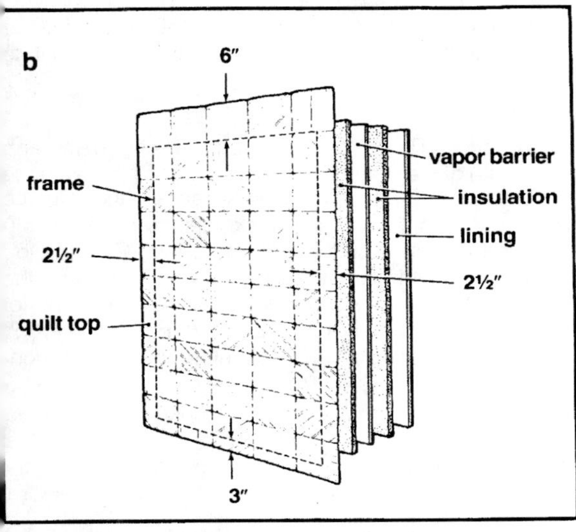

Figure AI-2b: Exploded view of shade fabric.

Step 4: Determine necessary size of the valance. There must be enough room for the quilted shade to roll up inside the wooden frame (see Figure AI-2a). Calculate your dimensions this way: Fold a blanket to the desirable thickness of your shade, which will be in the range of ¼ to ½ inch thick. Roll enough length of blanket around the dowel to match the length of the window opening; measure the diameter. Make sure the valance can accommodate your quilted shade by adding ¼ inch to the diameter of the rolled blanket, and using this figure as your inside valance dimension.

Table AI-2: Guidelines for Shade Measurements (in inches)

Window Opening Height (sill up to bottom of top casing)	Vertical Measurement of Top Window Casing & Back Valance Panel	Diameter of Rolled Blanket	Inside Valance Dimension	Width of Top & Front Valance Panels
54	6¼	6	6¼ by 6¼	7
44	5½	5¼	5½ by 5½	6¼
34	5	4¾	5 by 5	5¾

Step 5: Remove existing window head casing. Cut 1½ inches from one end of this piece and replace it, centering it above the window opening. Cut a piece of shelving lumber the same length as the head casing and rip so the width of the two pieces combined will equal the inside valance dimension. Nail it directly above and flush to the shortened head casing piece.

Step 6: Cut front and top valance panels. Rip to identical width, making them each ¾ inch wider than the combined width of the original head casing and back valance panel. Cut top valance panel to the same length as the back valance panel. Cut the front valance panel 1½ inches longer than the top and back valance panels. Sand the roughness out of the lumber so the quilted shade will not be torn. Glue the top valance panel inside the front panel so that they meet on the perpendicular and the front panel extends ¾ inch past the top panel at both ends. Drill starter holes for the finishing nails ⅜ inch from the top edge of the front valance panel. Nail through the front valance panel into the edge of the top valance panel to which it is glued.

Step 7: Cut two 3-by-3-inch blocks from the shelving lumber to make dowel hangers for the shade. Drill a 1-inch hole in each block at dead center. Then mark ¾ inch in from each end on one edge of one block, and cut down into the block at the ¾-inch marks to the drill hole so that a U-shaped cutaway is made.

Cut valance side panels. They should be square and each side dimension should equal the width of the front valance panel. On the inside face of each side panel draw two lines ¾ inch in from the edges to indicate where the top and back valance panels will overlap.

Center the hangers within the pencil-outlined squares on the inside face of each side valance panel. Drill starter holes in the four corners, glue, and nail in place. Attach the block with the 1-inch round cutaway to the top and front valance panels on the right side, and the block with the U-shaped cutaway on the left side. The rope will work more freely if it is secured to the dowel on the side of the valance with the round cutaway. Consistency makes it possible to change a quilted shade from one window to another.

Step 8: Complete assembly of valance. Drill, glue, and nail to create a permanent structure. Mount with screws and do not use glue if you wish to make the valance detachable.

Step 9: After you have attached the shade to the dowel rod (instructions given in Step 7 of the "Construction Procedure for the Shade"), secure the nylon rope that will raise and lower the shade on the righthand end: Using different drill bits, drill a ⅛-inch hole through the dowel 1 inch from the end, then widen the same hole to ¼ inch on the opposite side of the dowel. Drop your nylon rope through the

⅛-inch hole, knot it, and pull the knot into the dowel so that it does not protrude. Wrap the rope around the dowel a few times (counterclockwise), then tie the metal ring on the end of the rope.

Step 10: Drill two or three ½-inch holes on the right side casing at points where the quilted shade is to be held open. Insert short, sanded dowel rods in these holes. With the quilted shade in place, try raising and lowering the shade. Adjust the rope by looping it around the dowel rod or knotting it at the ring.

Step 11: Mount a ¾-inch wooden strip along the windowsill behind the shade track so that a seal will be made when the shade is closed. Add ½-inch vinyl weather stripping to the wooden strip for a tighter seal.

Step 12: Finally, sand and varnish.

Construction Procedure for the Shade

Step 1: Based on the measurements of the opening in the frame, design the quilt top, pieced or appliqued. If you have no experience in quilting, you should refer to a basic quilting book—*101 Patchwork Patterns* by Ruby S. McKim (New York: Dover, 1962) is recommended. In designing the quilt top, create a 3-inch border on each side and the bottom, and a 6-inch border on the top. If you are appliqueing rather than piecing your quilt top together, do not applique in these areas.

Step 2: Cut two layers of insulating material, one layer of vapor barrier, and one layer of lining. Determine the measurements as follows:

Width—the width of the frame opening + 2 times the width of the shade track + ½ inch to allow for drawing from quilting
Length—distance from top dowel to sill + 2½ inches

Step 3: Assemble five layers with the vapor barrier in the center (see Figure AI-2b) The reflective surface of the vapor barrier should be placed so that it will face into the room. The patchwork layer should overhang the others by approximately 2½ inches on the sides and 3 inches on the bottom, with the remaining extra overhang length at the top. Pin or baste the layers together so they will stay in place while you are machine quilting. It will be necessary to space your basting stitches not only along all four sides, but also in the central area of the shade. Since this is the area in which you will machine quilt, the layers must be attached at enough points so that they do not buckle.

Step 4: Machine quilt the layers together so the pattern compliments the patchwork on the interior-facing layer. (Remember this pattern will also be visible on the back of the shade, so use your imagination. If you feel confident about your quilting abilities, try choosing a thread color to contrast with the backing.)

Step 5: Leaving the top and bottom free, bend the fabric overhanging the sides of the quilt top (the border area) around to the lining. Pin it firmly in place. Fold under ¼ inch, as though turning up a hem, and hand sew the fabric to the lining. Visually this will create a border along the sides of the shade lining. The sides of the quilted shade will now slide smoothly inside the shade track.

Step 6: Wrap the quilt top fabric overhanging the bottom of the quilted layers around to the lining. Since a pocket must be created for the bottom dowel, simply fold over the bottom border 1½ inches, turn the edge under ¼ inch, and hand sew the finished edge to the lining. The pocket created will accommodate a ½-inch-diameter wooden dowel. Looking at the shade lining, you will see a border area now on three sides.

Step 7: Insert the bottom dowel in the quilted shade. Holding the shade up to the dowel in the frame, wrap the fabric overhang at the top of the patchwork layer up over the dowel from the back and around to the front. Mark on the dowel where it should attach. Remove the dowel from the frame by lifting it out of the U-shaped hanger. Turning the top edge of fabric under ½ inch, attach the shade to the dowel with staples or carpet tacks. Set the dowel in the frame, and enjoy the result of your labors.

Section 3 Making Insulated Bifold Shutters*

(The maximum window area to be insulated by these shutters should not exceed 4 by 8 feet.)

Materials

- 1-by-4 lumber with square edges— pine or redwood recommended
- ⅛-inch mahogany door skin—available in 3-foot-by-6-foot-8-inch and 4-by-8-foot sheets; single sheet provides enough skin for 1½ shutters
- Four to six self-closing shutter hinges per window—need 1 inch of hinge vertically for each 12 inches of shutter
- Screws
- Latches
- 4d finishing nails

*This section was prepared with assistance from David Bainbridge and Denny Long of The Passive Solar Institute, 2446 Bucklebury, Davis, CA 95616.

5/8-inch brads
White glue
3½-inch fiberglass insulation
Weather stripping—a soft, compactable nylon foam, 3/8 inch wide by 3/16 inch thick is recommended

Tools

Try square
Hammer
Nail set punch (1/16 inch)
Saw (8 to 10 points per inch)
Tape measure
Sandpaper (#120 grade)
Chisel (¾ inch)
Screwdriver—standard-Phillips, to fit hinge screws

Optional Tools

Carpenter's square
Level
Circular saw with plywood blade
Chalk line
Finish sander
Belt sander
Screw set punch
Utility knife

Construction Procedure

Before you begin to construct a pair of hinged shutters, you must carefully examine your window casing to see if the panel can be made to fit. If you have a canted sash stop, you will need to replace it with a rectangular strip. Hinges should also lie flat and parallel to the wall even if your window has a tapered or curved interior casing (see Figure AI-3a).

If your window does have a tapered or curved casing, there are three ways to make the hinges lie flat and parallel to the wall. You can chisel a flat seat for your hinges if you have had experience with a wood chisel and have a steady hand. You can also shim the hinge level with a wedge-shaped piece of wood, or you can replace the entire curved casing with a flat strip of 1-by-4s. If you have deep windows, you may want to mount the hinges on the jamb, allowing the shutters to swing only 90 degrees.

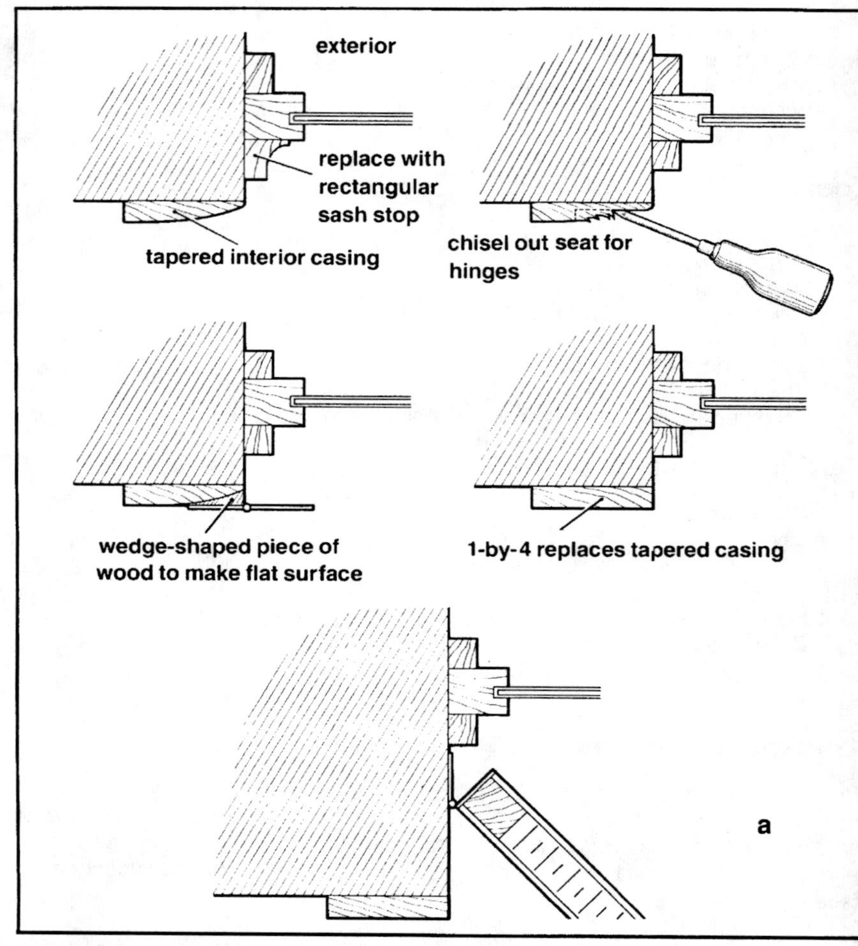

Figure AI-3a: Adjusting window casing for hinged shutter installation.

Step 1: After determining if the sash stops in the window opening will remain or be replaced by square-shaped stops, measure the window opening accurately. Make a drawing of the window with the dimensions indicated (see Figure AI-3b). Determine the dimensions of the shutter by marking a $^{3}/_{16}$-inch margin within the drawing of the window at top, bottom, and on both sides. Indicate the centerfold on the drawing,

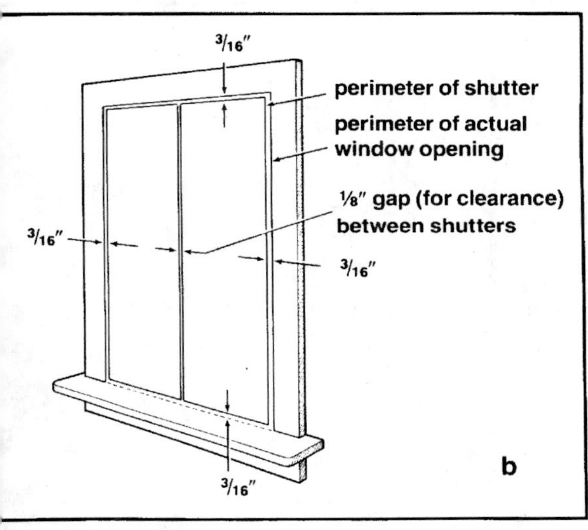

Figure AI-3b: Measuring window for proper fit of shutters.

allowing ⅛-inch clearance between the shutters. Making a cardboard pattern can be helpful.

Step 2: Check the sill and sides, ensuring they are level and in plumb, and make note on the measurement drawing. When things are out of square, build the shutter to the smallest dimension unless there is a significant discrepancy (more than 1/16 to ⅛ inch). If the difference is larger than this, remodel or rebuild the window opening so that shutters will hang properly. A shim or other minor adjustment will often correct the problem.

Step 3: Cut materials. Cut the skin material carefully to avoid splintering, according to the measurements from Step 1. If a circular saw blade is used, do not set the blade any deeper than necessary to cut the material. A plywood blade is best. Cut the frame side pieces to match the side lengths of the skin material. Cut the top-frame piece and bottom-frame piece to fit inside the side pieces. Be certain that they match the skin materials, top and bottom, in case the window is out of square. Cut the insulation to fit inside the frame.

Step 4: Assemble the frame (see Figure AI-3c). Set the 4d finishing nails in the edge pieces (if splitting occurs, blunt the nail points by hitting gently with a hammer against

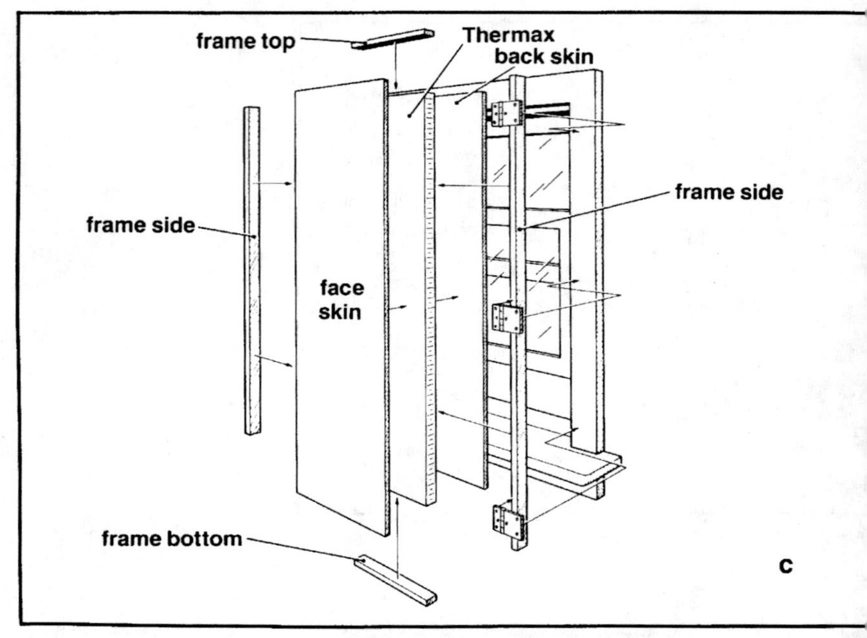

Figure AI-3c: Exploded view of one-half of a bifolding shutter.

a hard surface), apply glue to the inside piece on the end grain, and nail together Leave the nailhead out to set later and wipe up any glue spills. Do all sides.

Note: If the opening is larger than 1 by 2 feet, there should be some reinforcement where the hardware will be attached. Add wooden blocks as backing for the hinges and latch. A shutter 2 by 3 feet or larger will need a reinforced frame. Add a wooden block across the width of the shutter frame, midway from the ends, to brace it.

Step 5: Put a bead of glue all around the frame. Lay the skin over the frame. Nail ⅝-inch brads every 6 inches down one edge, then line up and tack the other corners Finish nailing the brads every 6 inches around the remaining edges of the shutter

Step 6: Turn frame over and lay in insulation. Nail on the other skin.

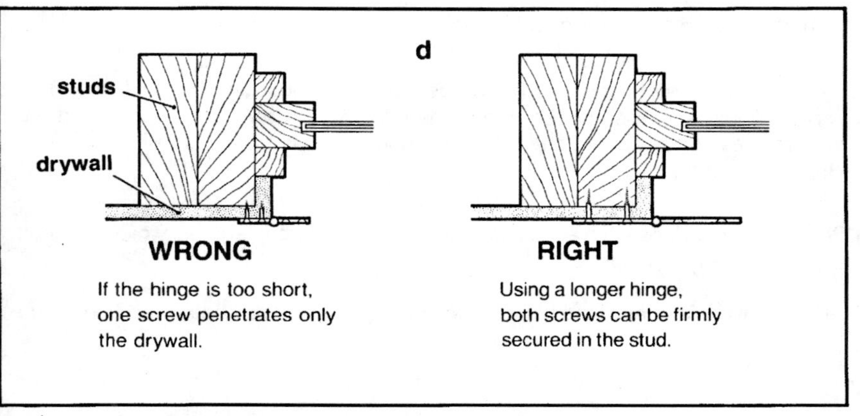

Figure AI-3d: Proper placement of hinges.

Step 7: Allow glue to dry, then set the nails with the nail set punch and putty the nail holes.

Step 8: Sand the faces and edges with #120 sandpaper after putty has dried.

Step 9: Figure hinge placement on the shutter (a top and bottom inset of one-quarter to one-third of the total shutter length is adequate). Mark the hinge placement. Then mortise the hinge-mounting spot with a chisel until the hinge will set flush with the wood. Use a screw set punch to begin the screwholes, then put in the screws.

Step 10: Stain or paint the shutters.

Step 11: To mount, set the shutter in the window frame. Use 12- to 18-inch shakes as wedges underneath the shutter to hold it in proper alignment. Mark the location of the hinges on the window frame, punch the starter holes, and drive in the screws.

When you mount the shutter in the window be sure that the hinge screws are anchored into a solid piece of window casing so that the weight of the shutter will be well supported (see Figure AI-3d). Reinforce any loose casing by securing to the framing behind it with additional finishing nails or wood screws. If your window has no trim but only drywall which wraps around the corner (a common detail in newer apartments), using shutter hinges will provide a surface long enough that the hinge can be anchored into the wall stud behind it. As a rule of thumb, you should have 1 inch of hinge vertically for each 12 inches of shutter height.

Step 12: Make certain the shutter swings freely. Cut wooden strips to fit around the inside perimeter of the shutter. Attach them to the window casing. Sand and finish the surfaces that will be visible from the interior of the room. These will act as a stop strip so the shutters close flush.

Step 13: Carefully weather-strip the surface of the stop strip against which the shutter closes to form an airtight seal.

Add hardware latches to the back of the shutter so that it may be locked from inside.

Appendix II

The Economics of Window Insulation

If you are purchasing or constructing a window insulation system, the amount of money the system returns may determine the size of the initial investment. The expected yearly savings compared to the cost of the system determines the *payback* period for the system (the length of time required to recover your investment). Below is a method of determining the savings. You can record in Table AII-6 the annual savings and payback by this method.

Before you can compute the savings provided by a window insulation system, you must estimate the R value of the system you will be installing. If you are purchasing a movable insulation product, using the manufacturer's stated R value is one way to determine the dollar savings the system will provide. However, don't blindly accept the value they give you. Ask who made the tests. If they were done by a reliable, independent testing lab, the stated R value usually has more merit than if it is from a test done by the manufacturer. In either case, the product was probably tested under ideal conditions which do not exist in your home.

If you cannot find a reliably tested R value for the system you are planning to install, you can make a guess of what effective thermal resistance you can expect from the system. To be safe, you can make a high or optimistic guess and a low or conservative one, find the percentage of heat loss reduction in Table AII-1 and then find the fuel savings for both the optimistic and conservative value. The difference between increasing an R value from 4 to 5 is not as significant, in terms of the percentage of heat loss reduction, as increasing R from 1 to 2. In going from R-1 to 2 the rate of heat loss is halved, but it is reduced by only 20% in going from R-4 to 5. It is necessary to double R to reduce the rate of heat loss by 50%.

Listed below are some guidelines for determining the R value of an untested system for *rough* approximations only:

Curtains and Drapes	Sheer curtains provide practically no protection at all. The standard, heavy drape which touches the floor may provide an R of 2. For special thermal curtains with a fiber-fill material or foil layers, R values range between 2 and 4.
Roman Shades	These shades generally have similar R values to the curtains, which are between 2 and 4.
Roller Shades	Regular, single-layer vinyl window shades without edge seals have been tested to have R values of around 0.5 or less. Some thermal shades with reflective Mylar layers and edge tracks have R values of over 10. The R values of shades vary tremendously and can only be determined by testing each one. If you do not have a manufacturer's R value, go ahead and assume an R value, now that you can make an educated guess.
Rigid Shutters	Rigid shutters are more predictable than shades or curtains if their edge seals are secure. The true R value for one of these shutters usually falls between 75 and 95% of the computed R value for a panel with anywhere from good to nearly perfect edge seals. Thus, the true R value can be found by adding the R values of the materials in the shutter and multiplying the sum by an efficiency factor of 0.75 to 0.95, depending on the tightness of the edge seal.

The Dollar Heat Savings per Winter

The dollar savings from heating for any window insulation system you install is a product of the three factors in the equation below:

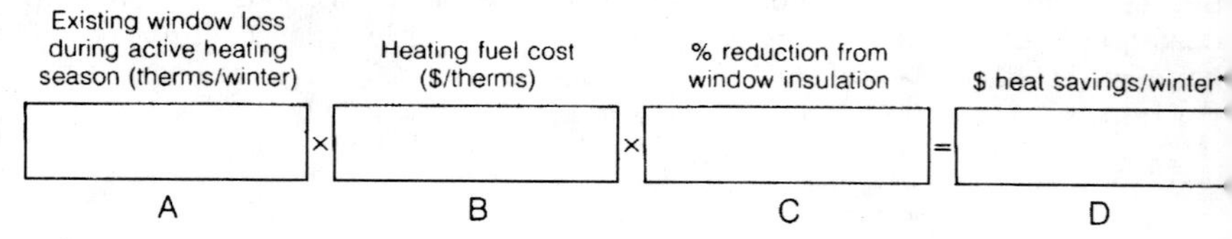

*If a system with exterior solar reflectors is used, the additional usable solar gain should be estimated and added to this figure.

A. Existing Window Loss =

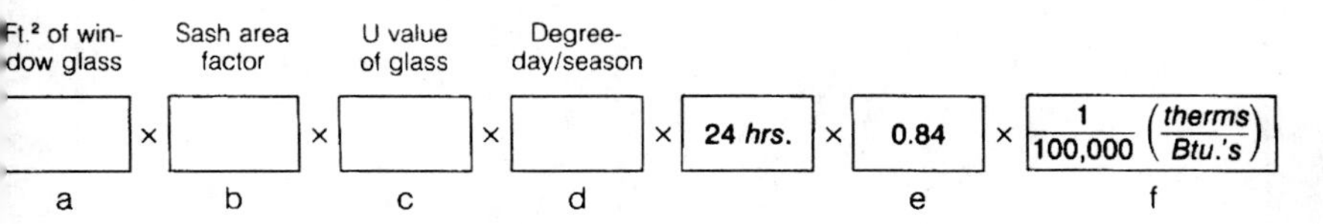

a. To find the square footage of your window opening with a tape measure:

$$\textit{square footage} = \frac{L \times W\ (inches)}{144}$$

b. Sash area factor (for calculating heat loss): If sashes are aluminum, enter 1. If sashes are wooden, enter 0.9. Compared to the glazing, the wooden sashes reduce heat loss, while aluminum sashes will have about the same loss as glass.

c. U value (heat conductance) of glass:*

Single glazing	1.13
Double glazing, sealed, ¼-inch space between panels	0.65
Single glazing with storm window	0.56
Triple pane, sealed, ¼-inch air spaces	0.47
Double glazing, sealed, with storm window	0.32

d. Degree-days per heating season is a number which defines how cold the heating season is in a given locale. A map of the United States showing the degree-days per season is shown in Figure AII-1. A call to your local oil or utility company should provide you with a specific number for your locale. Degree-day-per-season fluctuations due to local factors, particularly changes in altitude, are not shown accurately on the U.S. map. Tables of degree-days are also listed city by city in many solar heating books.

e. The number 0.84 is an adjustment factor to the degree-days for the portion of the heating season in which purchased fuels are burned. During early fall and late spring, the heat from lights, people, appliances, and solar

*_ASHRAE Handbook 1977 Fundamentals & Product Directory_. See Appendix V, Section 1 for book description and complete bibliographical information.

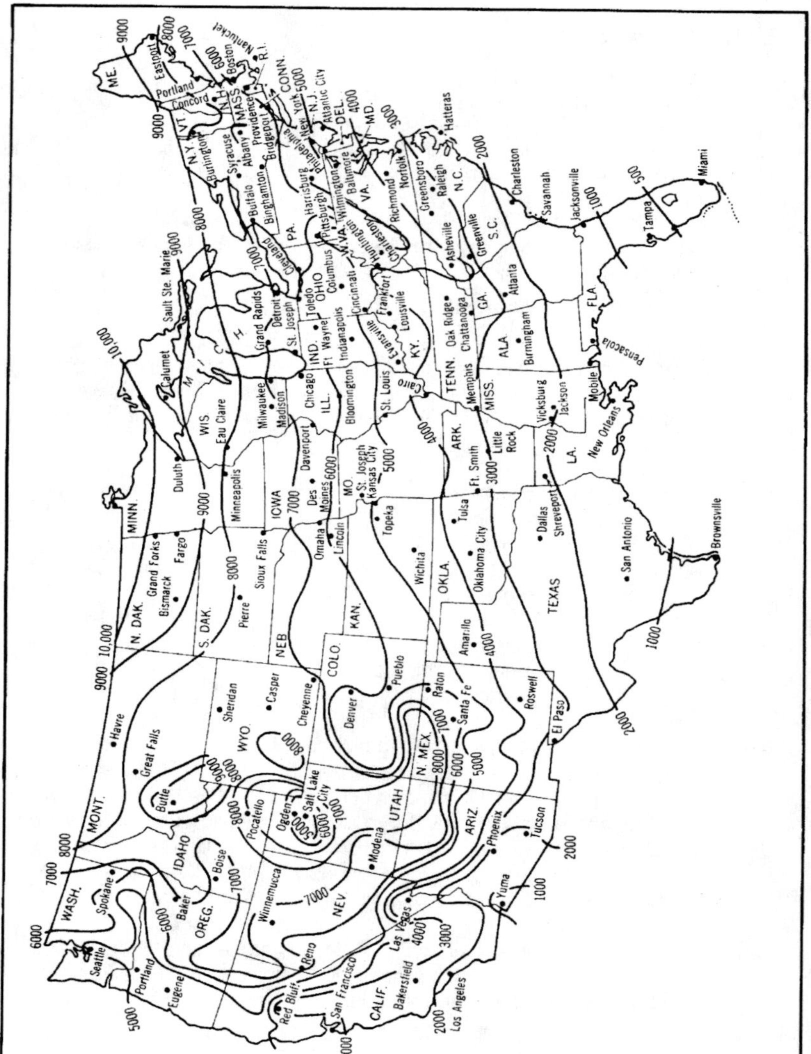

Figure AII-1: ***Degree-day map of the United States (reprinted by permission from William J. McGuinness and Benjamin Stein, Mechanical and Electrical Equipment for Buildings, 5th ed., Asheville, N.C.: John Wiley & Sons, 1971)***

gain through windows, is usually enough to offset losses without the furnace running. For passive solar homes or ones with larger than usual south-facing windows, this factor should be 0.6 to 0.8, depending on how much the movable insulation is used or needed each winter.

f. One therm for our purposes here is 100,000 Btu.'s. One therm is the energy equivalent of 100 cubic feet of natural gas, which is also equal to 0.0025 cords of hardwood or 0.71 gallons of oil.

B. Heating Fuel Cost—The costs of heating fuels vary a great deal. If your heat source is the electric baseboard heater, movable insulation will generate the greatest dollar savings. If your are burning lower-cost fuels, the annual savings for a system will be less and the payback period longer. Table AII-2 lists heating and electric air-conditioning costs per therm of heat delivered.

C. Percentage of Heating Load Reduction from Window Insulation—This value can be found in Table AII-1, according to an estimated or assumed R value of the window insulation.

D. Heat Savings per Winter—The product of the three above items (D) can be entered into Table AII-6 to calculate the payback period.

The Savings in Cooling Cost per Summer

The sunlight entering windows is reduced by window insulation or by shading devices and therefore heat doesn't have to be extracted by electric air conditioners. Although summer comfort will be increased in all homes where solar gains are blocked, the dollar savings should only be computed for those homes with electric (compressor coil type) air conditioners (A/C). These savings are determined by the following equation:

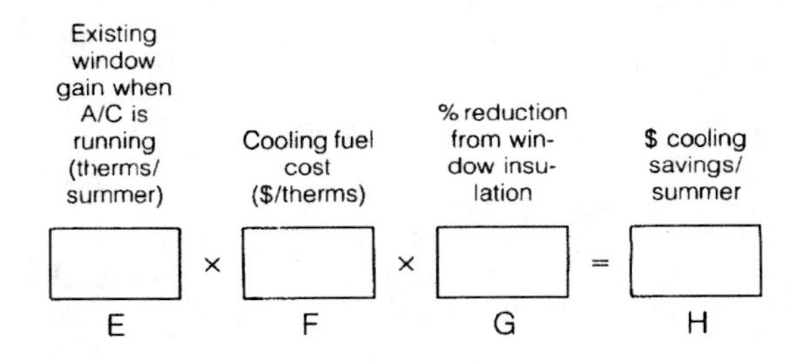

E. Existing Window Solar Gain =

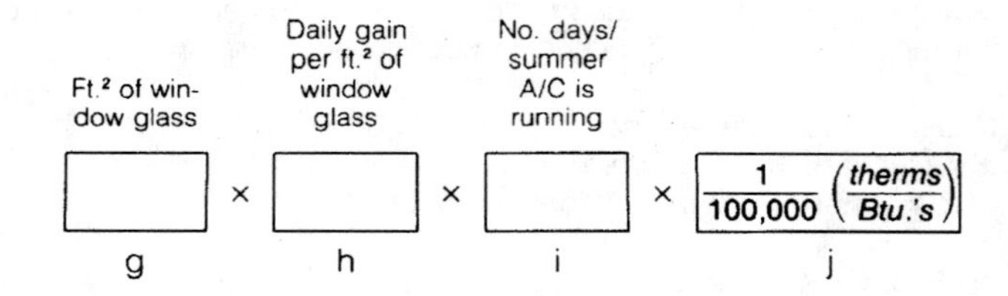

g. This is only the total square footage of window glass and does not include window sash areas. You may want to remeasure your glass areas for the figure or simply multiply 0.9 times your overall sash area. This 0.9 is called the sash area factor (for calculating heat gain).

h. This is the daily gain for the window in a given location and orientation. Table AII-3 gives average daily heat gain values, June through August for 45 cities in the United States. Windows which face north, northeast, and northwest are assumed to have no solar heat gain.

If a window is partially shaded during the summer by vegetation, overhangs, or other buildings, estimate how much of a day's direct sunshine this window receives and reduce the daily gain accordingly. For the period of time when the window is shaded, use the value from Table AII-4. *Example:* For one-third of the day the window is in dense shade, the other two-thirds of the day the window receives direct sun.

h = ⅔ × daily solar heat gain per ft.² (from Table AII-3) + ⅓ × 0.2 (value for dense shade from Table AII-4) × daily solar heat gain per ft.² (taken from Table AII-3).

i. Estimate the number of days in each year that you use your air conditioner.

j. (Same as "f.")

F. Cooling Fuel Cost—This value can be found in Table AII-2.

G. The Percentage of Solar Gain Reduction—Shading devices are most effective outside the home, but any device which blocks incoming sunlight and/or reflects it back to the outside is helpful. Table AII-5 lists the percentage of solar gain reduction for several shades and shutters.

H. Dollar Cooling Savings per Season—This value can also be entered right into Table AII-6.

The System Cost and Payback Time Period

In Table AII-6 you can list various windows in your home and determine the length of time it will take for a system to pay for itself. Two examples are worked out on the top lines of the table.

Table AII-1: Percentage of Heating Load Saved with Movable Insulation

R Value of Movable Insulation in Addition to Window	Percentage Saved* (with single glazing)	(with double glazing)
1	32	22
2	42	31
3	46	38
4	49	41
5	51	44
6	52	46
7	53	48
8	54	48
9	55	50
10	55	50
11	56	51
12	56	52

*Based on ΔT = 35 degrees with movable insulation over the window 14 hours per day.

Table AII-2: Fuel Cost per Therm (100,000 Btu's) of Heat Delivered

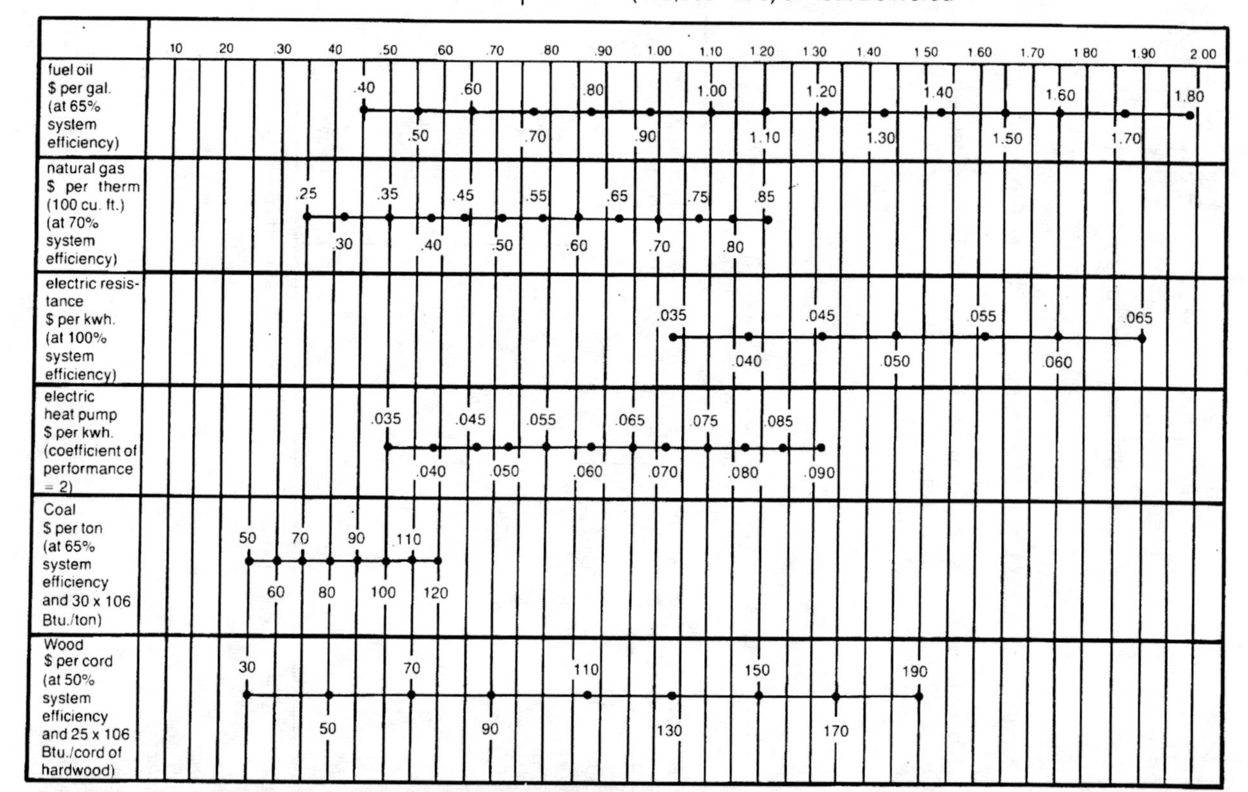

Table AII-3: Average Daily Heat Gain through Double Glazing during the Summer*

(Values are in Btu. ft.²-day for vertical south, southeast, southwest, east, west, and for south-facing glazing tilted at 30 degrees and 45 degrees. For single glazing multiply values by 1.213.)

City	South	Southeast/ Southwest	East West	South 30°	South 45°
Albuquerque, N. Mex.	618	832	906	2,287	1,978
Astoria, Oreg.	645	716	695	1,742	1,548
Atlanta, Ga.	516	663	719	2,219	1,909
Bismarck, N. Dak.	746	839	811	2,014	1,790
Boise, Idaho	746	875	867	2,238	2,014
Boston, Mass.	571	656	659	1,770	1,599
Charleston, W. Va.	493	632	686	2,120	1,824
Cleveland, Ohio	623	735	743	1,994	1,807
Columbus, Ohio	568	680	701	1,874	1,704
Davis, Calif.	677	855	894	2,437	2,166
Dodge City, Kans.	640	807	848	2,368	2,048
East Lansing, Mich.	588	675	676	1,823	1,646
El Paso, Tex.	540	813	864	2,430	2,052
Fort Worth, Tex.	549	743	820	2,416	2,086
Fresno, Calif.	641	831	885	2,500	2,179
Gainesville, Fla.	457	611	683	2,148	1,782
Glasgow, Ky.	798	890	849	1,995	1,835
Grand Junction, Colo.	674	846	882	2,359	2,089
Greensboro, N.C.	541	676	718	1,943	1,701
Indianapolis, Ind.	590	705	723	2,065	1,829
Lake Charles, La.	474	643	722	2,180	1,849
Las Vegas, Nev.	634	839	904	2,473	2,156
Lemont, Ill.	615	719	724	1,968	1,784
Lincoln, Nebr.	612	722	732	1,958	1,775
Little Rock, Ark.	537	683	733	2,202	1,933
Los Angeles, Calif.	563	744	811	2,308	2,027
Madison, Wis.	628	725	724	1,884	1,696
Medford, Oreg.	746	891	892	2,313	2,089
Nashville, Tenn.	551	690	733	2,149	1,873
Newport, R.I.	589	689	698	1,818	1,647

*Figures derived from *Hourly Solar Radiation Data for Vertical and Horizontal Surfaces on Average Days in the United States and Canada,* by the following methods. For vertical glazing, the average solar heat gain equals the sum of the average daily insolation of June, July, and August, divided by 3; multiplied by 0.68 for double glazing. The average daily solar heat gain for tilted, south-facing glazing equals the three-month average daily gain for vertical, south-facing glazing times a tilt correction factor. This correction factor is determined by dividing the July 30 degrees figure or 45 degrees clear-day insolation by the 90 degrees clear-day insolation for July.

(Table AII-3 cont.)

City	South	Southeast/ Southwest	East/ West	South 30°	South 45°
New York, N.Y.	568	669	683	1,874	1,704
Oak Ridge, Tenn.	534	664	705	2,083	1,816
Oklahoma City, Okla.	561	724	780	2,076	1,795
Portland, Maine	646	743	737	1,938	1,744
Rapid City, S. Dak.	707	814	803	1,980	1,768
Saint Cloud, Minn.	679	765	747	1,901	1,698
San Antonio, Tex.	561	702	801	2,637	2,188
Sault Sainte Marie, Mich.	693	779	757	1,871	1,663
Schenectady, N.Y.	536	608	607	1,662	1,501
Seattle, Wash.	625	685	660	1,625	1,500
Spokane, Wash.	806	901	862	2,096	1,934
State College, Pa.	583	687	700	1,924	1,749
Tampa, Fla.	441	616	708	2,125	1,766
Tucson, Ariz.	537	743	831	2,309	1,987
Washington, D.C.	561	675	699	1,964	1,739

Table AII-4: A Factor for Existing Exterior Shading*

Light shade from trees	0.55
Dense shade from trees	0.2
Building providing shade	0.2
Continuous overhang	0.2

*Derived from shading coefficient values found in *Energy Conservation Building Code Workbook* Davis, California.

Table AII-5: Percentage of Solar Gain Reduction*

Inside, dark roller or roman shade, drawn	0.19
Inside, medium roller or roman shade, drawn	0.38
Inside, white roller or roman shade, drawn	0.49
Inside, dark venetian blind, drawn	0.25
Inside, medium venetian blind, drawn	0.35
Inside, white venetian blind, drawn	0.44
Inside, venetian blind, reflective aluminum	0.45
Outside, venetian blind, light colored	0.85
Outside, venetian blind, awning type, white	0.85
Dark, open-weave fabric curtain	0.09
Medium, open-weave fabric curtain	0.11
Light, open-weave fabric curtain	0.14
Dark, semi-open weave fabric curtain	0.12
Medium, semi-open weave fabric curtain	0.16
Light, semi-open weave fabric curtain	0.21
Dark, closed-weave fabric curtain	0.14
Medium, closed-weave fabric curtain	0.21
Light, closed-weave fabric curtain	0.26
Clear, domed skylight with translucent diffuser	0.39
Translucent, domed skylight	0.43
Outside canvas awning, dark or medium	0.75
Kaiser shade screen, 30-degree profile angle	0.85
Regular Koolshade, 40-degree profile angle	0.82
Reflecting shade, sealed edges	0.8
Insulated shade, sealed edges	0.9
Inside, rigid insulated shutter or panel, sealed edges	0.95
Outside, rigid insulated shutter or panel, sealed edges	1.
Pop-in panel, sealed edges	0.95
Glass-hugging panel	0.8

A value derived from the shading coefficient for usage only in the economic calculations in this appendix. Factors based on shaded coefficients found in *Energy Conservation Building Code Workbook*, Davis, California; *ASHRAE 1977 Handbook Fundamentals & Product Directory*, and other sources.

Table AII-6: Tally Sheet for the Economic Payback of Window Insulation Added to Your Home

Window Designation				% Savings					Movable Insulation		
(location)	(single or double glazing)	Sash Material & Opening Size	Orientation	(heating)	(cooling)	$ Heat Savings/Window/Winter (calculated earlier as D)	$ Cooling Savings/Summer (calculated earlier as H)	Combined $ Savings/Window/Year (D + H)	(type)	($ cost/window)	Payback Period (D + H)
Example 1	double (storm)	wood 15 ft.²	north	46	———	3.12	———	3.12	shutter	30	9.6 yrs.
Example 2	single	aluminum 15 ft.²	southeast	51	95	3.23	5.20	8.43	shutter	30	3.6 yrs.

Example 1

What is the payback for a 15 ft.² shutter system costing $2 per square foot in Boston where fuel oil is being used?

Step A—Calculate window loss:

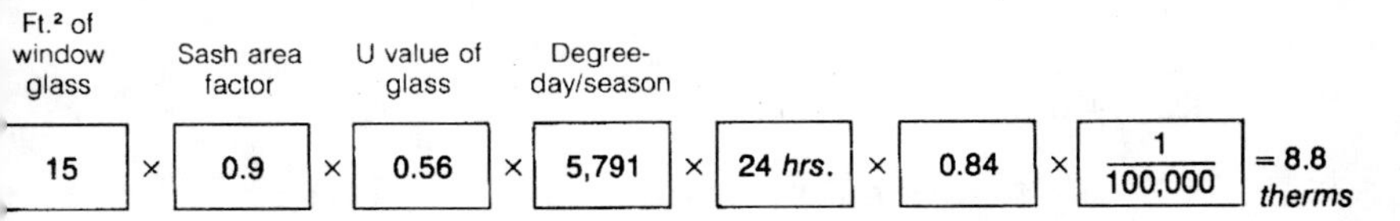

Step B—Look up cost per therm of fuel oil in Table AII-2:

$.70 per gallon, therefore it is $.77 per therm

Step C—The total R value of the movable insulation is R-6.
Look up % reduction for double glazing with R-6 in Table AII-1:

46% = 0.46

Step D—A × B × C = savings per winter:

8.8 × 0.77 × 0.46 = $3.12

Since this is a north window, a cooling value is not calculated. Next, multiply the cost of the movable insulation system per square foot times the number of square feet:

$2 × 15 = $30

Divide the yearly savings into the cost of the system:

$$\frac{30}{3.12} = 9.6 \textit{ years for payback}$$

Example 2

What is the payback for R-5 interior hinged shutter costing $2 per square foot on a southeast, aluminum sashed, single-glazed window in Atlanta where a heat pump is utilized for heating and cooling and where the window is lightly shaded by vegetation one-third of the day. The cooling cycle of the heat pump is used for approximately 90 days each summer.

Step A—Calculate window loss:

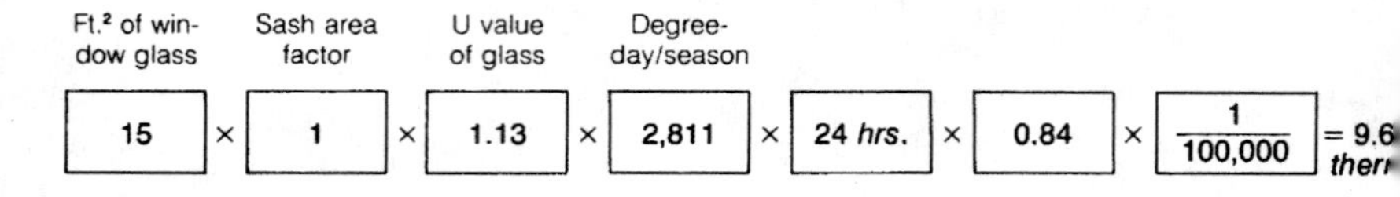

Step B—Look up cost per therm of heat pump fuel in Table AII-2 with electricity:

$.045 *per kilowatt hour, therefore it is* $.66 *per therm*

Step C—Look up % reduction for movable insulation in Table AII-1:

51% = 0.51

Step D—A × B × C = savings per winter:

9.6 × 0.66 × 0.51 = $3.23

Step E—Calculate the existing solar gain: The existing gain equals the square footage of glass times the daily solar gain per square foot, times the number of days the cooling cycle of the heat pump is used each summer, divided by 100,000 Btu./therm.

The square footage of glass = 15 (ft.² of sash) × 0.9 (sash area factor for calculating heat gain) = 13.5

The daily gain per square foot of glass is more difficult to calculate. Table AII-3 lists a gain of 663 Btu./ft.² for southeast-facing double glazing in Atlanta. Because this is single glazing, the gain will be 1.213 times greater. Two-thirds or 0.67% of the day the window is unshaded, but during the other third of the day, the window is lightly shaded receiving only 0.55 the amount of solar radiation as listed in Table AII-4. Therefore, the average daily gain on each square foot of glass:

663 × 1.213 × [0.67 + (0.33 × 0.55)] = 685

Plugging in these values, the existing solar gain can be computed as follows:

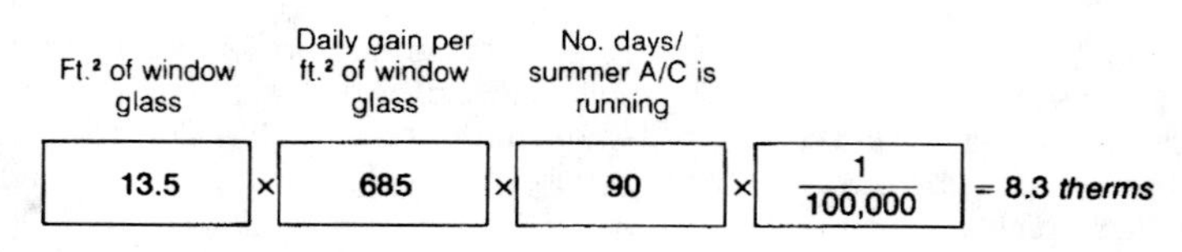

Step F—Check the cost per therm for the cooling cycle of the heat pump in Table AII-2:

$.45 *per kilowatt,* therefore it is $.66 *per therm*

Step G—Look up % reduction for interior hinged shutter with R-5 in Table AII-5:

95% = 0.95

Step H—E × F × G = summer savings:

8.3 × 0.66 × 0.95 = $5.20

To calculate how many years it will take until the movable insulation system pays back the initial investment, add winter savings $3.23 and summer savings $5.20 = $8.43 total savings. Cost of system 15 ft.² × $2 per ft.² = $30:

$$\frac{\$30}{\$8.43} = 3.6 \textit{ years for payback}$$

Appendix III

Movable Insulation Products, Hardware, and Components

The following sources for commercially available movable insulation systems and related hardware are up-to-date as of this book's publication, but the field is rapidly expanding as manufacturers respond to new and increased consumer demand for energy conservation products. The following periodicals will help to keep you abreast of new developments in materials and complete movable insulation systems:

Alternative Sources of Energy
Route 2
Milaca, NM 56353

New Shelter
Rodale Press
33 East Minor Street
Emmaus, PA 18049

Popular Science
Times Mirror Magazines, Inc.
380 Madison Avenue
New York, NY 10017

Solar Age
Church Hill
Harrisville, NH 03450

Section 1 Ready-Made Movable Insulation Products

Product Trade Name	*Manufacturer or Distributor*
ATC Window Quilt	Appropriate Technology Corporation P.O. Box 975 Brattleboro, VT 05301
Beadwall Skylid	Zomeworks Corporation P.O. Box 712 Albuquerque, NM 87103
Brattleboro Design Group Shade (under development)	Brattleboro Design Group P.O. Box 235 Brattleboro, VT 05301
Greenshield Systems (greenhouse blankets)	Automatic Devices Company 2121 South Twelfth Street Allentown, PA 18103
Insealshaid	Arc-Tic-Seal Systems, Inc. P.O. Box 428 Butler, WI 53007
Insul Shutter	Insul Shutter, Inc. P.O. Box 338 Silt, CO 81652
IS High R Shade	The Insulating Shade Company, Inc. P.O. Box 282 Branford, CT 06405
Mecho Shade System	Joel Berman Associates, Inc. 102 Prince Street New York, NY 10012
Metrovox Roll Shutter	Metrovox, Inc. 1308 Gresham Road Silver Spring, MD 20904
NRG Shade	Sun Control Products, Inc. 431 Fourth Avenue, SE Rochester, MN 55901

Appendix III/Products, Hardware, and Components

Product Trade Name	*Manufacturer or Distributor*
Panel Drape	Shelter, Inc. P.O. Box 108 Oakland, MD 21550
Pease Rolling Shutter	The Pease Company Ever-Straight Division 7100 Dixie Highway Fairfield, OH 45023
Plastic-View Shade	Plastic-View Transparent Shade, Inc. P.O. Box 25 Van Nuys, CA 91408
Roldoor	One Design, Inc. Mountain Falls Route Winchester, VA 22601
Rolladen Shutter	American German Industries, Inc. 14601 North Scottsdale Road Scottsdale, AZ 85260
Serrande Shutter	Serrande of Italy P.O. Box 1034 West Sacramento, CA 95691
Shutter Shield	C. D. Davidson & Associates, Inc. P.O. Box 1293 Pontiac, MI 48057
Sol-R-Veil	Sol-R-Veil, Inc. 60 West Eighteenth Street New York, NY 10011
Sunflake Window	Sunflake 625 Goddard Avenue P.O. Box 676 Ignacio, CO 81137
Sun Quilt Thermal Gate	Sun Quilt Corporation P.O. Box 374 Newport, NH 03773

Product Trade Name	*Manufacturer or Distributor*
Sun Saver Kit	Homesworth Corporation P.O. Box 565 Department MT-1 Brunswick, ME 04011
Thermafold Shutters	Shutters, Inc. 110 East Fifth Street Hastings, MN 55033
Thermo-Shade	Solar Energy Components, Inc. 212 Welsh Pool Road Lionville, PA 19353
TTC Self-Inflating Curtain Wall	The Thermal Technology Corporation P.O. Box 130 Snowmass, CO 81654
Warm-In Drapery Liner	Conservation Concepts, Ltd. P.O. Box 376 Stratton Mountain, VT 05155
Wind-n-Sun Shield	Wind-n-Sun Shield, Inc. P.O. Box 1434 E.Y. Melbourne, FL 32935
Window Blanket	Window Blanket Company, Inc. Route 1, Box 83 Lenoir City, TN 37771

Section 2 Storm Windows and Kits

Product Type or Trade Name	*Manufacturer or Distributor*
Frames for Plastic Film Acrylic Frame (under development)	Thermo Tech Corporation 410 Pine Street Burlington, VT 05401

Temp-Rite	SUN Catalog Solar Usage Now, Inc. Box 306 Bascom, OH 44809
Plastic Storm Window Kits	Available locally at hardware and home supply stores.
	Minute Man Storm Windows Minute Man Anchors, Inc. 305 West Walker Street East Flat Rock, NC 28726
	Perkasie Industries Corporation 50 East Spruce Street Perkasie, PA 18944
	Plaskolite, Inc. 1770 Joyce Avenue Box 1497 Columbus, OH 43216
	Reynolds Metals Company E. T. Stevak P.O. Box 27003 Richmond, VA 23261
Polyester Glazing Film Flexiguard	3M Company Special Enterprises 3M Center 223-2-02 Saint Paul, MN 55101
Polyethylene or Clear Vinyl	Available locally at hardware and home supply stores.

Section 3 Sealants and Weather Strippings

Product Type or Trade Name	*Manufacturer or Distributor*
Latex, Butyl, and Silicone Caulkings	Available locally at paint, building supply, and hardware stores.
Polypropylene Spring Strip 3M Type #2743	Energy Control Products Industrial Tape Division 3M Company 3M Center 220-8E Saint Paul, MN 55101
Urethane Foam Sealant Great Stuff	Insta-Foam Products, Inc. Joliet, IL 60435
Various Plastic Weather Strippings Polyflex, Fin-Seal, and Vinyl-Clad Foam	Schlegel Corporation P.O. Box 23113 Rochester, NY 14692 Attention: CWO Department
Various Weather-Stripping Products	Available locally at hardware stores. Manufactured by: Stanley Hardware Division of The Stanley Works New Britain, CT 06050

Section 4 Solar Control Films

Product Type or Trade Name	*Manufacturer or Distributor*
Dun-Ray	Dunmore Corporation Newtown Industrial Commons Pennsylvania Trail Newtown, PA 18940
Heat Mirror Films (under development)	Thermo Film Corporation 385 Sherman Avenue #3 Palo Alto, CA 94306

Product Type or Trade Name	*Manufacturer or Distributor*
Kool Vue	Solar Screen 53-11 105th Street Corona, NY 11368
Llumar	Martin Processing, Inc. P.O. Box 5068 Martinsville, VA 24112
Nunsun Sumsun	Standard Packaging Corporation National Metallizing Division Cranbury, NJ 08512
Plastic-View Film	Plastic-View Transparent Shades P.O. Box 25 Van Nuys, CA 91408
P-19 Film	3M Company 3M Center Saint Paul, MN 55101
Reflecto-Shield	Madico 64 Industrial Parkway Woburn, MA 01801
Sungard	Metallized Products 224 Terminal Drive S Saint Petersburg, FL 33712

Section 5 Aluminized Reflector Films

Product Type or Trade Name	*Manufacturer or Distributor*
King-Lux Aluminum Reflector Sheet for Solar Application	Mark Sherwood Solar Products Division Kingston Industries 205 Lexington Avenue New York, NY 10016

Product Type or Trade Name	*Manufacturer or Distributor*
Llumar	Martin Processing, Inc. P.O. Box 5068 Martinsville, VA 24112
Reflective Films	SUN Catalog Solar Usage Now, Inc. Box 306 Bascom, OH 44809
Self-Adhesive Mylar Type Film	Parsec, Inc. P.O. Box 38534 Dallas, TX 75328

Section 6 Insulation Boards

Product Type or Trade Name, and Description	*Distributor*
Beadboard—a white insulation board containing polystyrene beads that have been expanded like popcorn and pressed together. This board is inferior in quality to extruded polystyrene or isocyanurate foam, but is usually a bargain in cost.	Available at most building supply outlets.
Cardboard—can be laminated to make shutters.	Often available at no cost at appliance and furniture stores.
Extruded Polystyrene—sold under the trade name Styrofoam. This foam board is far superior to beadboard in thermal performance and wear resistance. It can be recognized by its characteristic sky blue color.	Available at most building supply outlets.

Product Type or Trade Name, and Description	*Distributor*
Isocyanurate Foam Board—a foil-faced, high R value, strong, lightweight, brown foam insulation board for shutters. Don't be fooled by foil-faced beadboard, an inferior product sold under trade names similar to foil-faced isocyanurate board at about the same cost.	Available at most building supply outlets under the trade names: Thermax, Thermofax, High R, R-Max, and Technifoam.
Thermoply Board—a dense cardboard aluminized on one side.	Available at many building supply outlets.
Wood-Fiber Exterior Wall Sheathing—a low-cost material with moderate insulating value that can also be used as a tackboard.	Available at just about any lumberyard.

Section 7 Fibrous Insulation

Product Type or Trade Name	*Manufacturer or Distributor*
Fiberglass Batting—can be used inside shutters, not recommended for curtains or shades.	Available at building supply outlets in various thicknesses as wall and ceiling insulation batts.
Hollofill II—a polyester insulating fiberfill of short, hollow fibers made by Du Pont.	Frostline Kits Department C Frostline Circle Denver, CO 80241
PolarGuard—a polyester fiberfill made by Celanese Fibers, used in many sleeping bags and outdoor clothing.	Available in large quantities from: Celanese Fibers Marketing Company 1211 Avenue of the Americas New York, NY 10036

Product Type or Trade Name	*Manufacturer or Distributor*
Thinsulate—a very high-quality fiberfill material recently developed by 3M Company used in some cold-weather, outdoor clothing.	Write for more information on how to obtain: 3M Company Building Services and Cleaning Products Division 3M Center Saint Paul, MN 55101

Section 8 Reflective Fabrics for Shades and Curtains

Product Type or Trade Name	*Manufacturer or Distributor*
Aluminized Polyester Films—numerous aluminized Mylar-polyester films are listed under solar control films. Thin layers of these films sometimes can be used as reflective fabrics in shades and curtains.	See solar control films in Section 4.
Aluminized Reinforced Polyethylene Film (Astrolon)—a line of Astrolon fabrics is available; the most widely distributed is the Thermos all-weather sportsman's blanket. The cheaper grades of this material burn very rapidly. Be sure to order Astrolon III or Astrolon VIII; they are fire-resistant.	Manufactured by: Kingly Sealy Thermos Company 37 East Street Winchester, MA 01890 Distributed by: Shelter Institute 38 Centre Street Bath, ME 04530
Foil-Laminated, Reinforced Fabrics (Foylon)—aluminum foil is laminated to layers of nylon or vinyl cloth by Duracote Corporation to make Foylon their line of aluminized fabric. The Foylon series includes:	Manufactured and distributed by: Duracote Corporation 350 North Diamond Street Ravenna, OH 44266

Product Type or Trade Name	*Manufacturer or Distributor*
Foylon 7001—a foil on a lightweight nylon scrim for drapery liners.	
Foylon Durashade 4413—a white, vinyl fabric on one side with aluminum foil on the other side. Designed for use in shades.	
Foylon 7137—black vinyl on one side and foil on the other for use in greenhouse covers.	

Section 9 Adhesives

Product Type or Trade Name	*Manufacturer or Distributor*
Aliphatic Resin Glues—look like white glue tinted yellow. These glues form much stronger bonds than the white glues and are more impervious to moisture.	Available locally at hardware stores under brand names Titebond, Wilhold, and others.
Panel Adhesives—a wide range of adhesives that comes in tubes to be applied with a caulking gun. Directions on the tube describe the intended uses of each adhesive.	Available locally at hardware stores
Resorcinol—a red, waterproof resin for exterior wood laminations.	Available locally at hardware stores
Scotch Grip Plastic Adhesive 4693—excellent for gluing foil and Mylar films to flat surfaces.	3M Company 3M Center Saint Paul, MN 55101
White Glue—a very common glue. White glues work best on porous materials such as cardboard. They do not hold up well under high humidity.	Available locally at hardware stores One common brand is Elmer's Glue-All.

Section 10 Vapor Barriers and Waterproof Fabrics

Product Type or Trade Name	*Manufacturer or Distributor*
Duratuff—a vinyl-coated nylon material	Duracote Corporation 350 North Diamond Street Ravenna, OH 44266
Polyethylene or Vinyl	Available locally at hardware and home supply stores.
Sealed Air Solar Pool Blanket	Sealed Air Corporation 2015 Saybrook Avenue Commerce, CA 90040
Vinyl-Coated Canvas	Available locally at awning fabricators.

Section 11 Hardware for Shutters

Product Type or Trade Name	*Manufacturer or Distributor*
Nightwall Magnetic Clips—used to attach pop-in shutters to glass	Zomeworks Corporation P.O. Box 712 Albuquerque, NM 87103
3M Plastiform Magnetic Tape—self-adhesive flexible tape, available in 1/16-inch and 1/32-inch thicknesses	Zomeworks Corporation P.O. Box 712 Albuquerque, NM 87103 Brookstone 127 Vose Farm Road Peterborough, NH 03458
Window, Closet Door, and Cabinet Hardware	Available locally at hardware stores and cabinet shops. Catalogs available from: Blaine Window Hardware, Inc. 1919 Blaine Drive, RD4 Hagerstown, MD 21740

Product Type or Trade Name	*Manufacturer or Distributor*
	Lawrence Brothers, Inc. Sterling, IL 61081
	Stanley Hardware Division of The Stanley Works New Britain, CT 06050

Section 12 Sun-Shading Screens

Product Type or Trade Name	*Manufacturer or Distributor*
KoolShade	KoolShade Corporation 722 Genevieve Street P.O. Box 210 Solona Beach, CA 92075
Rolscreen Pella Slimshade (window with venetian blinds between glazing)	Rolscreen Company Pella, IA 50219
Shade Screens	Kaiser Aluminum 300 Lakeshore Drive Oakland, CA 94643
	Phifer Wire Products, Inc. P.O. Box 1700 Tuscaloosa, AL 35401
	Vimco Corporation P.O. Box 212 Laurel, VA 23060
	Sears Home Improvement Catalog
	Chicopee Manufacturing Company P.O. Box 47520 Atlanta, GA 30362
	J.C. Penney Catalog

Product Type or Trade Name	*Manufacturer or Distributor*
Somfy Motors for Solar Shades	Somfy Systems Division of Capano and Pons, Inc. 268 Terminal Avenue W Clark, NJ 07066
Symtrac	Symtrac, Inc. 8243 North Cristana Drive Skokie, IL 60076

Section 13 Miscellaneous

Product Type or Trade Name	*Manufacturer or Distributor*
Quaker Window Channels (for window repair)	Quaker Manufacturing Sharon Hill, PA 19079

Appendix IV

Further Technical Information

Section 1 Air Infiltration Losses at Windows

Although air infiltration losses are difficult to predict accurately, modern testing methods have helped to refine the science to some extent. In one modern testing method, the building is vacated of all occupants and then is filled with a special gas. For several hours instruments record the rate the outside air mixes with and displaces this gas. The air infiltration through non-weather-stripped windows in older homes has sometimes been found to account for over 50% of the total air infiltration heat losses. On the other hand, window air infiltration losses in new homes with modern windows are surprisingly low, often less than 25% of total infiltration losses.

Several methods are available for computing air infiltration losses in a home. The simplest method assigns a somewhat arbitrary air-change-per-hour value to the home, based on the type of wall construction, windows, entrances, etc. Whether ¾ air changes per hour or 1½ air changes per hour are assumed makes a tremendous difference in the computed rate of heat loss. A more accurate but involved method involves estimating the sizes of the cracks around windows and estimating the wind speed outside the home or the pressure difference between the inside and outside. It is very difficult to determine the size of all the cracks around your windows with great accuracy, since many of those cracks are hidden to the eye. All methods of calculating air infiltration losses leave a lot to be desired.

Table AIV-1 shows the heat loss per linear foot of crack for several types of windows according to the outside wind speed. The air enters through some windows in a home and exits through others. Therefore, when using this table, only half the windows in the home should be counted as leaking air to the outside.

Trees, shrubs, and the orientation of a wall to the wind all determine the effect outside air movement has on window air infiltration. As a result, current ASHRAE methods

Table AIV-1: Air Leakage Around Window Sashes in Cubic Feet of Air per Linear Foot of Crack per Hour

Window Type	Remarks	Wind Velocity (miles per hour) 5	10	15	20	25	30
Double-Hung Wooden Sash	Around frame in wooden frame construction	2	6	11	17	23	30
	Average window, non-weather-stripped, 1/16" crack and 3/64" clearance. Includes leakage around frame.	7	21	39	59	80	104
	Same fit, weather-stripped	4	13	24	36	49	63
	Poorly fitted window, non-weather-stripped, 3/32" crack and 3/32" clearance. Includes leakage around frame.	27	69	111	154	199	249
	Same fit, weather-stripped	6	19	34	51	71	92
Double-Hung Metal Sash	Non-weather-stripped	20	47	74	104	137	170
	Weather-stripped	6	19	32	46	60	76
Rolled Section Steel Sash	Industrial; pivoted, 1/16" crack	52	108	176	244	304	372
	Architectural; projected, 1/32" crack	15	36	62	86	112	139
	Architectural; projected, 3/64" crack	20	52	88	116	152	182
	Residential; casement, 1/64" crack	6	18	33	47	60	74
	Residential; casement, 1/32" crack	14	32	52	76	100	128

are based on pressure differences between inside and outside, rather than on outside wind speeds. (See *ASHRAE 1977 Fundamentals & Product Directory*, Chapter 21). Table AIV-1 is derived from the older ASHRAE method because it is easier for the lay person to use. The following is an example of this method:

A house has 10 poorly fitted, double-hung windows which measure 2½ by 5 feet, and 6 average-fit, double-hung windows which are 2 by 4 feet. None of the windows are weather-stripped. What will be the hourly heat loss from those windows if the temperature inside is 70°F. and the air outside is at 20°F. with a 15 mile-per-hour wind? What would the heat loss be if these windows were weather-stripped?

Step 1—Estimate the length of the crack:
Poorly fit windows—17.4 feet/window × 10 windows = 175 feet
Average-fit windows—14 feet/window × 6 windows = 84 feet

Step 2—Find hourly air leakage:

Non-weather-stripped:

$$\frac{(175 \times 111) + (84 \times 39)}{2} = 11{,}350\ cu.\ ft./hr.$$

Weather-stripped:

$$\frac{(175 \times 34) + (84 \times 24)}{2} = 3{,}983\ cu.\ ft./hr.$$

Step 3—Multiply the air infiltration by the temperature difference between insid and outside and multiply by a factor of 0.018.

Non-weather-stripped:

$$11{,}350\ cu.\ ft./hr. \times 50°F. \times 0.18\ Btu./cu.\ ft.–°F. = 10{,}215\ Btu./hr.$$

Weather-stripped:

$$3{,}983\ cu.\ ft./hr. \times 50°F. \times 0.018\ Btu./cu.\ ft.–°F. = 3{,}585\ Btu./hr.$$

Therefore, the heat loss after the windows are weather-stripped is about one-thir of that without weather stripping.

Section 2 Radiant Heat Transfer and Transparent Glazing Materials

All objects radiate heat as a 4th power function of their temperature abov absolute zero. Warm objects therefore radiate more heat than cool ones. Th temperature of an object also determines the wavelengths of radiation it emits. Hc objects give off shorter wavelengths than cold ones.

Figure AIV-1 shows wavelength distribution curves for the radiant energy receive from the sun and for room-temperature radiation. The vertical dotted lines on thi graph define the three bands of wavelengths mentioned in chapter 2: the visible the near-visible infrared band, and the room-temperature infrared band. Visibl light is in the middle of the radiant spectrum. Ultraviolet light, X rays, an gamma rays have shorter wavelengths than visible light and can penetrate man forms of matter that are opaque to visible light. On the long wave end of th spectrum are infrared radiation, microwave radiation, and radio frequency waves.

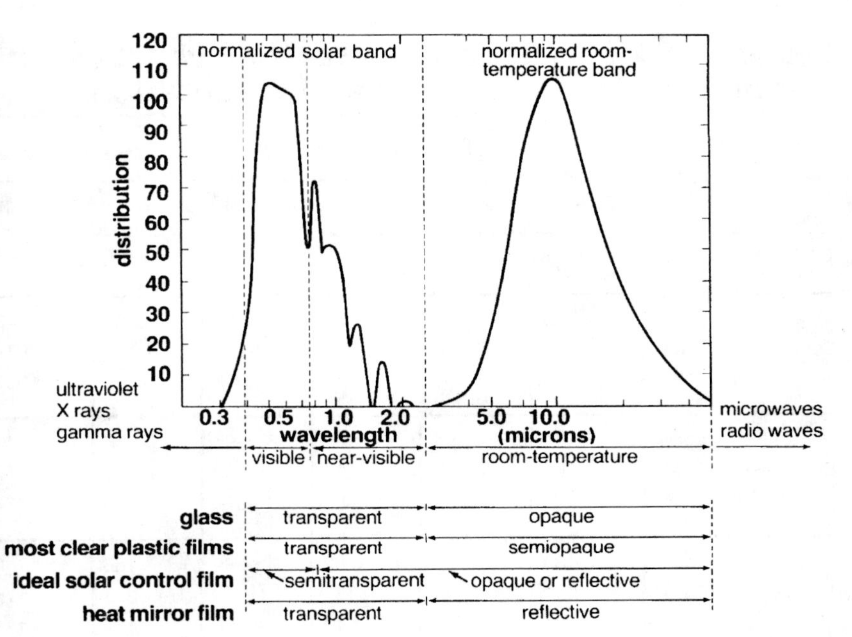

Figure AIV-1: Wavelength distribution of solar and room-temperature radiations. (Adapted from Windows for Energy-Efficient Buildings, *January 1979. Vol. 1 No. 1. U.S. Department of Energy.)*

Glass is an excellent glazing material for the capture of solar heat since it is transparent to the visible and near-visible bands but is opaque to room-temperature infrared radiation.

Plastic films can also achieve transparency to the solar band but are not usually as opaque to the room-temperature radiation as glass. Thus they aren't quite as effective at trapping heat.

Solar control films are tinted with metallic substrates to reduce the amount of solar heat entering a building. These films still allow useful daylight to enter. The ideal solar control film is reflective or opaque to the near-visible infrared band, but allows a useful portion of the visible band to enter.

Heat mirror films are designed to transmit the full solar spectrum but at the same time reflect back into the room any infrared radiation emitted by room-temperature objects.

Section 3 Net Gains and Losses for Five Cities with Various Glazing Assemblies and Orientations*

ALBUQUERQUE

		Single	Double	Triple	Single MI†	Double MI	Triple MI
OCTOBER (218 DD‡)							
South	Loss	189	94	62	95	56	40
	Gain	1,437	1,185	972	1,437	1,185	972
	Net	+1,248	+1,091	+910	+1,342	+1,129	+932
East/West	Loss	189	94	62	95	56	40
	Gain	769	631	517	769	631	517
	Net	+580	+537	+455	+674	+575	+477
North	Loss	189	94	62	95	56	40
	Gain	260	213	175	260	213	175
	Net	+71	+119	+113	+165	+157	+135
NOVEMBER (630 DD)							
South	Loss	564	282	186	285	162	116
	Gain	1,432	1,182	975	1,432	1,182	975
	Net	+868	+900	+789	+1,147	+1,020	+859
East/West	Loss	564	282	186	285	162	116
	Gain	582	480	396	582	480	396
	Net	+18	+198	+210	+297	+318	+280

*Figures (calculated in Btu./ft.² /day and total Btu./ft.² /season) are based on average isolation value in *Hourly Solar Radiation Data for Vertical and Horizontal Surfaces on Average Days in the Unite States and Canada*. Shading coefficients are 0.825 for single glazing, 0.68 for double glazing, 0.56 f triple glazing. Movable insulation calculated with R-5 and the system is in place 14 hours per day. Th heat loss is calculated using Formula 3 in the following section "Calculations for Window Heat Losse with Movable Insulation."

†MI is an abbreviation for movable insulation.

‡DD is an abbreviation for degree-day.

		Single	Double	Triple	Single MI	Double MI	Triple MI
NOVEMBER (630 DD)							
North	Loss	564	282	186	285	162	116
	Gain	200	165	136	200	165	136
	Net	−364	−117	−50	−85	+3	+20
DECEMBER (899 DD)							
South	Loss	780	390	257	394	223	160
	Gain	1,509	1,245	1,027	1,509	1,245	1,027
	Net	+729	+855	+770	+1,115	+1,022	+867
East/West	Loss	780	390	257	394	223	160
	Gain	549	453	374	549	453	374
	Net	−231	+63	+117	+155	+230	+214
North	Loss	780	390	257	394	223	160
	Gain	177	146	120	177	146	120
	Net	−603	−244	−137	−217	−77	−40
JANUARY (970 DD)							
South	Loss	841	420	278	425	242	174
	Gain	1,497	1,234	1,016	1,497	1,234	1,016
	Net	+656	+814	+738	+1,072	+992	+842
East/West	Loss	841	420	278	425	242	174
	Gain	605	498	410	605	498	410
	Net	−236	+78	+132	+180	+256	+236
North	Loss	841	420	278	425	242	174
	Gain	201	166	137	201	166	137
	Net	−640	−254	−141	−224	−76	−37

		Single	Double	Triple	Single MI	Double MI	Triple MI
FEBRUARY (714 DD)							
South	Loss	685	343	226	346	197	142
	Gain	1,395	1,151	950	1,395	1,151	950
	Net	+710	+808	+724	+1,049	+954	+808
East/West	Loss	685	343	226	346	197	142
	Gain	770	635	524	770	635	524
	Net	+85	+292	+298	+424	+438	+382
North	Loss	685	343	226	346	197	142
	Gain	267	221	182	267	221	182
	Net	−418	−122	−44	−79	+24	+40
MARCH (589 DD)							
South	Loss	511	255	169	258	147	106
	Gain	1,250	1,031	851	1,250	1,031	851
	Net	+739	+776	+682	+992	+884	+745
East/West	Loss	511	255	169	258	147	106
	Gain	765	631	551	765	631	551
	Net	+254	+376	+382	+507	+484	+445
North	Loss	511	255	169	258	147	106
	Gain	323	266	220	323	266	220
	Net	−188	+11	+51	+65	+119	+114
APRIL (289 DD)							
South	Loss	259	129	85	131	75	54
	Gain	882	728	600	882	728	600
	Net	+623	+599	+515	+751	+653	+546

		Single	Double	Triple	Single MI	Double MI	Triple MI
APRIL (289 DD)							
East/West	Loss	259	129	85	131	75	54
	Gain	936	772	637	936	772	637
	Net	+677	+643	+552	+805	+697	+583
North	Loss	259	129	85	131	75	54
	Gain	413	341	281	413	341	281
	Net	+154	+212	+196	+282	+266	+227
SEASON (4,309 DD)							
South	Net	+169,142	+177,210	+155,492	+226,463	+201,739	+169,740
East/West	Net	+34,607	+66,080	+64,870	+91,928	+90,609	+79,118
North	Net	−60,164	−11,974	−386	−2,843	+12,555	+13,862

ATLANTA

		Single	Double	Triple	Single MI	Double MI	Triple MI
NOVEMBER (387 DD)							
South	Loss	347	173	114	175	100	72
	Gain	951	785	647	951	785	647
	Net	+604	+612	+533	+776	+685	+575
East/West	Loss	347	173	114	175	100	72
	Gain	444	366	302	444	366	302
	Net	+97	+193	+188	+269	+266	+230
North	Loss	347	173	114	175	100	72
	Gain	209	172	142	209	172	142
	Net	−138	−1	+28	+34	+72	+70

		Single	Double	Triple	Single MI	Double MI	Triple MI
DECEMBER (611 DD)							
South	Loss	530	265	175	268	153	110
	Gain	866	715	590	866	715	590
	Net	+336	+450	+415	+598	+562	+480
East/West	Loss	530	265	175	268	153	110
	Gain	379	312	257	379	312	257
	Net	−151	+47	+82	+111	+159	+147
North	Loss	530	265	175	268	153	110
	Gain	182	150	124	182	150	124
	Net	−348	−115	−51	−86	−3	+14
JANUARY (632 DD)							
South	Loss	548	274	181	277	158	114
	Gain	912	753	621	912	753	621
	Net	+364	+479	+440	+635	+595	+507
East/West	Loss	548	274	181	277	158	114
	Gain	433	357	295	433	357	295
	Net	−115	+83	+114	+156	+199	+181
North	Loss	548	274	181	277	158	114
	Gain	209	172	142	209	172	142
	Net	−339	−102	−39	−68	+14	+28
FEBRUARY (515 DD)							
South	Loss	494	247	163	250	142	102
	Gain	905	747	616	905	747	616
	Net	+411	+500	+453	+655	+605	+514

		Single	Double	Triple	Single MI	Double MI	Triple MI
FEBRUARY (515 DD)							
East/West	Loss	494	247	163	250	142	102
	Gain	563	465	384	563	465	384
	Net	+69	+218	+221	+313	+323	+282
North	Loss	494	247	163	250	142	102
	Gain	273	225	186	273	225	186
	Net	−221	−22	+23	+23	+83	+84
MARCH (392 DD)							
South	Loss	340	170	112	172	98	70
	Gain	856	706	583	856	706	583
	Net	+516	+536	+471	+684	+608	+513
East/West	Loss	340	170	112	172	98	70
	Gain	699	576	476	699	576	476
	Net	+359	+406	+364	+527	+478	+406
North	Loss	340	170	112	172	98	70
	Gain	338	279	230	338	279	230
	Net	−2	+109	+118	+166	+181	+160
SEASON (2,537 DD)							
South	Net	+67,324	+77,775	+69,780	+101,047	+92,205	+78,142
East/West	Net	+7,725	+28,510	+29,188	+41,448	+42,940	+37,550
North	Net	−31,687	−3,994	+2,352	+2,036	+10,436	+10,714

BOSTON

		Single	Double	Triple	Single MI	Double MI	Triple MI
OCTOBER (315 DD)							
South	Loss	273	137	90	138	79	57
	Gain	807	666	549	807	666	549
	Net	+534	+529	+459	+669	+587	+492
East/West	Loss	273	137	90	138	79	57
	Gain	428	353	291	428	353	291
	Net	+155	+216	+201	+290	+274	+234
North	Loss	273	137	90	138	79	57
	Gain	210	174	143	210	174	143
	Net	−63	+37	+53	+72	+95	+86
NOVEMBER (618 DD)							
South	Loss	554	227	183	280	159	115
	Gain	587	484	399	587	484	399
	Net	+33	+207	+216	+307	+325	+284
East/West	Loss	554	277	183	280	159	115
	Gain	257	212	175	257	212	175
	Net	−297	−65	−8	−23	+53	+60
North	Loss	554	277	183	280	159	115
	Gain	136	112	93	136	112	93
	Net	−418	−165	−90	−144	−47	−22

		Single	Double	Triple	Single MI	Double MI	Triple MI
ECEMBER (988 DD)							
outh	Loss	865	433	286	437	249	179
	Gain	625	515	425	625	515	425
	Net	−240	+82	+139	+188	+266	+246
ast/West	Loss	865	433	286	437	249	179
	Gain	236	195	161	236	195	161
	Net	−629	−238	−125	−201	−54	−18
orth	Loss	865	433	286	437	249	179
	Gain	118	97	80	118	97	80
	Net	−747	−336	−206	−319	−152	−99
ANUARY (1,113 DD)							
outh	Loss	965	483	318	488	278	200
	Gain	682	563	464	682	563	464
	Net	−283	+80	+146	+194	+285	+264
ast/West	Loss	965	483	318	488	278	200
	Gain	290	239	197	290	239	197
	Net	−675	−244	−121	−198	−39	−3
orth	Loss	965	483	318	488	278	200
	Gain	143	118	97	143	118	97
	Net	−822	−365	−221	−345	−160	−103
EBRUARY (1,002 DD)							
South	Loss	962	481	317	486	277	199
	Gain	753	621	513	753	621	513
	Net	−209	+140	+196	+267	+344	+314

		Single	Double	Triple	Single MI	Double MI	Triple MI
FEBRUARY (1,002 DD)							
East/West	Loss	962	481	317	486	277	199
	Gain	414	342	282	414	342	282
	Net	−548	−139	−35	−72	+65	+83
North	Loss	962	481	317	486	277	199
	Gain	209	172	142	209	172	142
	Net	−753	−309	−175	−277	−105	−57
MARCH (849 DD)							
South	Loss	736	368	243	372	212	153
	Gain	772	637	526	772	637	526
	Net	+36	+269	+283	+400	425	+373
East/West	Loss	736	368	243	372	212	153
	Gain	559	461	381	559	461	381
	Net	−177	+93	+138	+187	+249	+228
North	Loss	736	368	243	372	212	153
	Gain	285	235	194	285	235	194
	Net	−451	−133	−49	−87	+23	+41
APRIL (534 DD)							
South	Loss	478	239	158	242	138	99
	Gain	678	559	462	678	559	462
	Net	+200	+320	+304	+436	+421	+363
East/West	Loss	478	239	158	242	138	99
	Gain	669	552	455	669	552	455
	Net	+191	+313	+297	+427	+414	+356

		Single	Double	Triple	Single MI	Double MI	Triple MI
APRIL (534 DD)							
North	Loss	478	239	158	242	138	99
	Gain	369	304	251	369	304	251
	Net	−109	+65	+93	+127	+166	+152
MAY (236 DD)							
South	Loss	205	102	68	103	59	42
	Gain	686	566	467	686	566	467
	Net	+481	+464	+399	+583	+507	+425
East/West	Loss	205	102	68	103	59	42
	Gain	826	681	562	826	681	562
	Net	+621	+579	+494	+723	+622	+520
North	Loss	205	102	68	103	59	42
	Gain	483	398	328	483	398	328
	Net	+278	+296	+260	+380	+339	+286
SEASON (5,665 DD)							
South	Net	+17,506	+63,874	+65,294	+92,820	+96,182	+84,002
East/West	Net	−40,379	+16,134	+25,887	+34,935	+48,442	+44,595
North	Net	−92,849	−27,183	−15,443	−17,535	+5,125	+8,845

MADISON

		Single	Double	Triple	Single MI	Double MI	Triple MI
OCTOBER (419 DD)							
South	Loss	363	182	120	184	105	75
	Gain	941	777	641	941	777	641
	Net	+578	+595	+521	+757	+672	+566
East/West	Loss	363	182	120	184	105	75
	Gain	475	392	323	475	392	323
	Net	+112	+210	+203	+291	+287	+248
North	Loss	363	182	120	184	105	75
	Gain	215	178	147	215	178	147
	Net	−148	−4	+27	+31	+73	+72
NOVEMBER (864 DD)							
South	Loss	774	387	255	378	216	155
	Gain	639	527	435	639	527	435
	Net	−135	+140	+180	+261	+311	+280
East/West	Loss	774	387	255	378	216	155
	Gain	268	221	182	268	221	182
	Net	−506	−166	−73	−110	+5	+27
North	Loss	774	387	255	378	216	155
	Gain	135	112	92	135	112	92
	Net	−639	−275	−163	−243	−104	−63
DECEMBER (1,287 DD)							
South	Loss	1,116	558	368	564	321	231
	Gain	771	636	525	771	636	525
	Net	−345	+78	+157	+207	+315	+294

		Single	Double	Triple	Single MI	Double MI	Triple MI
DECEMBER (1,287 DD)							
East/West	Loss	1,116	558	368	564	321	231
	Gain	268	221	182	268	221	182
	Net	−848	−337	−186	−296	−100	−49
North	Loss	1,116	558	368	564	321	231
	Gain	120	99	82	120	99	82
	Net	−996	−459	−286	−444	−222	−149
JANUARY (1,417 DD)							
South	Loss	1,229	614	405	621	354	255
	Gain	868	716	591	868	716	591
	Net	−361	+102	+186	+247	+362	+336
East/West	Loss	1,229	614	405	621	354	255
	Gain	341	281	232	341	281	232
	Net	−888	−333	−173	−280	−73	−23
North	Loss	1,229	614	405	621	354	255
	Gain	148	123	101	148	123	101
	Net	−1,081	−491	−304	−473	−231	−154
FEBRUARY (1,207 DD)							
South	Loss	1,159	579	382	585	334	240
	Gain	874	721	595	874	721	595
	Net	−285	+142	+213	+289	+387	+355
East/West	Loss	1,159	579	382	585	334	240
	Gain	458	378	312	458	378	312
	Net	−701	−201	−70	−127	+44	+72

		Single	Double	Triple	Single MI	Double MI	Triple MI
FEBRUARY (1,207 DD)							
North	Loss	1,159	579	382	585	334	24
	Gain	214	177	146	214	177	14
	Net	−945	−402	−236	−371	−157	−9
MARCH (1,011 DD)							
South	Loss	877	438	289	443	252	18
	Gain	939	775	639	939	775	63
	Net	+62	+337	+350	+496	+523	+45
East/West	Loss	877	438	289	443	252	18
	Gain	653	538	444	653	538	44
	Net	−224	+100	+155	+210	+286	+26
North	Loss	877	438	289	443	252	18
	Gain	297	245	202	297	245	20
	Net	−580	−193	−87	−146	−7	+2
APRIL (573 DD)							
South	Loss	513	257	769	259	148	10
	Gain	742	612	505	742	612	50
	Net	+229	+355	−264	+483	+464	+39
East/West	Loss	513	257	169	259	148	10
	Gain	722	596	491	722	596	49
	Net	+209	+339	+322	+463	+448	+38
North	Loss	513	257	169	259	148	10
	Gain	380	314	259	380	314	25
	Net	−133	+57	+90	+121	+166	+15

		Single	Double	Triple	Single MI	Double MI	Triple MI
MAY (266 DD)							
South	Loss	231	115	76	117	66	48
	Gain	690	569	469	690	569	469
	Net	+459	+454	+393	+573	+503	+421
East/West	Loss	231	115	76	117	66	48
	Gain	818	675	557	818	675	557
	Net	+587	+560	+481	+701	+609	+509
North	Loss	231	115	76	117	66	48
	Gain	479	395	326	479	395	326
	Net	+248	+280	+250	+362	+329	+278
SEASON (7,044 DD)							
South	Net	+7,023	+67,372	+53,261	+101,092	+107,711	+94,604
East/West	Net	−67,629	+5,762	+20,390	+35,616	+46,101	+43,733
North	Net	−128,887	−44,673	−21,198	−34,818	−4,334	+2,145

SEATTLE

		Single	Double	Triple	Single MI	Double MI	Triple MI
OCTOBER (329 DD)							
South	Loss	285	143	94	144	82	159
	Gain	606	500	413	606	500	413
	Net	+321	+357	+319	+462	+418	+254
East/West	Loss	285	143	94	144	82	59
	Gain	309	255	211	309	255	211
	Net	+24	+112	+117	+165	+173	+152

		Single	Double	Triple	Single MI	Double MI	Triple MI
OCTOBER (329 DD)							
North	Loss	285	143	94	144	82	59
	Gain	167	137	113	167	137	113
	Net	−118	−6	+19	+23	+55	+54
NOVEMBER (540 DD)							
South	Loss	484	242	160	244	139	100
	Gain	340	280	231	340	280	231
	Net	−144	+38	+71	+96	+141	+131
East/West	Loss	484	242	160	244	139	100
	Gain	154	127	105	154	127	105
	Net	−330	−115	−55	−90	−12	+5
North	Loss	484	242	160	244	139	100
	Gain	98	81	67	98	81	67
	Net	−386	−161	−93	−146	−58	−33
DECEMBER (679 DD)							
South	Loss	589	294	194	297	170	122
	Gain	276	228	188	276	228	188
	Net	−313	−66	−6	−21	+58	+66
East/West	Loss	589	294	194	297	170	122
	Gain	116	95	79	116	95	79
	Net	−473	−199	−115	−181	−75	−43
North	Loss	589	294	194	297	170	122
	Gain	76	63	52	76	63	52
	Net	−513	−231	−142	−221	−107	−70

		Single	Double	Triple	Single MI	Double MI	Triple MI
JANUARY (753 DD)							
	Loss	653	326	215	330	188	135
South	Gain	299	247	204	299	247	204
	Net	−354	−79	−11	−31	+59	+69
	Loss	653	326	215	330	188	135
East/West	Gain	145	120	99	145	120	99
	Net	−508	−206	−116	−185	−68	−36
	Loss	653	326	215	330	188	135
North	Gain	97	80	66	97	80	66
	Net	−556	−246	−149	−233	−108	−69
FEBRUARY (602 DD)							
	Loss	578	289	191	292	166	120
South	Gain	498	411	339	498	411	339
	Net	−80	+122	+148	+206	+245	+219
	Loss	578	289	191	292	166	120
East/West	Gain	275	227	187	275	227	187
	Net	−303	−62	−4	−17	+61	+67
	Loss	578	289	191	292	166	120
North	Gain	163	134	111	163	134	111
	Net	−415	−155	−80	−129	−32	−9
MARCH (558 DD)							
	Loss	484	242	160	244	139	100
South	Gain	747	617	509	747	617	509
	Net	+263	+375	+349	+503	+478	+409

		Single	Double	Triple	Single MI	Double MI	Triple MI
MARCH (558 DD)							
East/West	Loss	484	242	160	244	139	100
	Gain	505	417	344	505	417	344
	Net	+21	+175	+184	+261	+278	+244
North	Loss	484	242	160	244	139	100
	Gain	256	211	174	256	211	174
	Net	−228	−31	+14	+12	+72	+74
APRIL (396 DD)							
South	Loss	355	177	117	179	102	74
	Gain	776	640	528	776	640	528
	Net	+421	+463	+411	+597	+538	+454
East/West	Loss	355	177	117	179	102	74
	Gain	706	583	481	706	583	481
	Net	+351	+406	+364	+527	+481	+407
North	Loss	355	177	117	179	102	74
	Gain	365	301	248	365	301	248
	Net	+10	+124	+131	+186	+199	+174
MAY (246 DD)							
South	Loss	213	107	70	108	61	44
	Gain	727	600	495	727	600	495
	Net	+514	+493	+425	+619	+539	+451
East/West	Loss	213	107	70	108	61	44
	Gain	806	665	549	806	665	549
	Net	+593	+558	+479	+698	+604	+505

		Single	Double	Triple	Single MI	Double MI	Triple MI
AY (246 DD)							
	Loss	213	107	70	108	61	44
orth	Gain	470	388	320	470	388	320
	Net	+257	+281	+250	+362	+327	+276
EASON (4,103 DD)							
outh	Net	+19,431	+51,926	+51,960	+74,050	+75,342	+62,401
ast/West	Net	−18,487	+20,634	+26,177	+36,132	+44,050	+39,718
orth	Net	−58,798	−12,673	−1,348	−4,179	+10,743	+12,193

ection 4 Calculations for Window Heat Losses with Movable Insulation

he simplest method for calculating the heat loss (Q) through a window with movable nsulation is with the equation:

(1)
$$Q = (65 - T_o) \times [(t_{open} \times U_g) + (t_{closed} \times U_{gi})]$$
$$(units = Btu./ft.^2 - day)$$

Where T_o = average daily temperature outside (inside temperature at 65°F.)

t_{open} = daily time period (hours) the window insulation is open, or not in place

U_g = U value of window glass (Btu./ft.²/hr. − °F.)

$t_{closed} = 24 - t_{open}$ = time period the window insulation is closed or in place

U_{gi} = U value of glass + window insulation (Btu./ft.²/hr. − °F.)

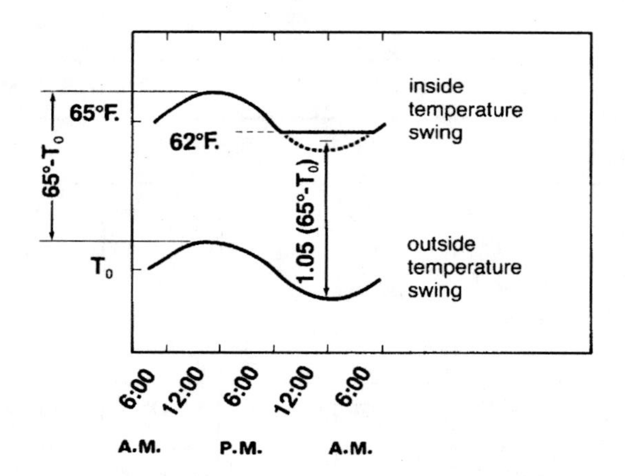

***Figure AIV-2:** Temperature difference between inside and outside over a typical 24-hour period in an average sun-tempered residence.*

If the window insulation is opened at 7:30 A.M. and closed at 5:30 P.M. (a common assumption), t_{open} = 10 hours and t_{closed} = 14 hours. Equation (1) thus becomes:

$$Q = (65 - T_o) \times [(10\,U_g) + (14\,U_{gi})] \quad (Btu./ft.^2 - day) \tag{2}$$

Equation 2 does not generally take into account that it is colder outside when the shutter is closed than when it is open. However, the temperature difference is not as great as one might expect since the temperatures inside a dwelling usually rise during the daytime, offsetting any rise in the outside temperature when the shutter is open. If a 62°F. nighttime temperature is maintained inside a dwelling, the daily inside and outside temperature curves will be similar to those shown in Figure AIV-2. During the night the upper curve levels off at 62°F., the minimum interior temperature, while the outside temperatures continue to decline. Therefore 65°F. minus the mean daily temperature appears to be a close approximation of the daytime indoor-outdoor temperature difference (Δt), but an average of several U.S. cities found Δt with a 62°F. nighttime setback to be larger than $(65° - T_o)$ by about 5%. Adding this increased 5% loss to equation (2):

$$Q = (65 - T_o) \times [(10\,U_g) + (1.05 \times 14\,U_{gi})]$$

(3)
$$Q = (65 - T_o) \times (10\ U_g + 14.7\ U_{gi})$$
(*Btu./ft.*^{2} *– day*)

Another useful approach to building thermal analysis is to compute the window heat loss in Btu./degree-days. The Btu./month or Btu./heating season load can then be derived simply by multiplying this value times the degree-days/month or degree-days/heating season of a given climate zone. To find this value for a window or series of windows:

(4)
$$Q = A \times (10\ U_g + 14.7\ U_{gi})$$
(*Btu./day* × °*F.*)

or

(5)
$$Q = A \times \frac{(10\ U_g + 14.7\ U_{gi})}{24}$$
(*Btu./hr.* – °*F.*) *average*

(6)
$$Q = A \times (65 - T_o) \times (10\ U_g + 14.7\ U_{gi})$$
(*Btu.–day*)

Where A = the area of the window

If, for example, a region has an average 5,000 degree-day heating season, then the total Btu. load from equation 4 is:

(7)
$$Q = A \times (10\ U_g + 14.7\ U_{gi}) \times 5{,}000$$
(*Btu./heating season*)

Appendix V

Design Sources

Section 1 Recommended Reading

Books, Pamphlets, and Periodicals

The Complete Greenhouse Book
by Peter Clegg and Derry Watkins
Garden Way Publishing Company
Charlotte, VT 05445
(1978)

An in-depth guide to greenhouse design and construction, this publication contain a wealth of information on solar design principles in greenhouses and present several options for movable insulation. Unfortunately, most of these designs are cor ceptual. However, on the whole, this book is well planned and offers an exceller presentation of the solar greenhouse.

The First Passive Solar Home Awards
by the Office of Policy Development and Research
U.S. Government Printing Office
Stock No. 023-000-00517-4
Washington, DC 20402
(1979)

Following a passive-solar home-design competition, sponsored by the U.S. Depar ment of Housing and Urban Development, a team of writers and designers wer drawn together at the National Solar Heating and Cooling Center in Rockville, Mary land, to compile a summary of the most important, passive solar home design and details. Written primarily for the builder, this book contains a wealth of passiv solar design information, including several, innovative movable insulation systems ue to the limited space, only a few details of movable insulation systems were ncluded, but additional designs can be explored by contacting the designers who re listed in this book.

From the Walls In
by Charles Wing
Little, Brown and Company
Boston, MA 02114
(1979)

his is a very unusual book which thoroughly addresses many of the issues in etrofitting a home. The most important section of the book, entitled "Buttoning Up," lescribes how to insulate an existing home. There is an excellent discussion on vays to avoid moisture problems in walls and ceilings and an informative section n caulking and weather-stripping windows. Several window assemblies are examned in terms of net heat losses or gains per heating season. These figures are ased on a lengthy heating season, disregarding the reduced heating needs in arly fall and late spring. These statistics are probably a little optimistic, particuarly on east and west windows.

The Fuel Savers
by Dan Scully, Don Prowler, and Bruce Anderson
Brick House Publishing Company
Church Hill
Harrisville, NH 03450

This booklet presents many ways that homes can be retrofitted with south-facing vindows and low-cost solar devices to provide supplemental heating. A major portion lescribes the addition of thermal shutters and curtains, solar windows, and an nterior heat-storage mass. The window insulation designs are not detailed, but several excellent ideas are presented.

New Inventions in Low-Cost Solar Heating
by William A. Shurcliff
Brick House Publishing Company
Church Hill
Harrisville, NH 03450
(1979)

William Shurcliff does it again. Instead of the traditional approaches to passive solar heating, he offers a challenging package of new ideas. Of particular interest are his movable insulation variations for discontinuous Trombe walls, and his approaches to solar water heating. These are only some of many creative designs, tried and

untried, in this refreshing book which provides the basics for a number of passiv
solar possibilities.

The Passive Solar Energy Book
by Edward Mazria
Rodale Press
33 East Minor Street
Emmaus, PA 18049
(1979)

This excellent manual on passive solar design includes a chapter on movabl insulation, one on reflectors, and another on shading devices. The chapter on mov able insulation examines window insulation primarily for south-facing windows The technical data and tables presented in it are indispensable. Some of the infor mation presented in *Movable Insulation* was arrived at in part from the method an data that forms the basis of Mazria's book.

Reader's Digest Complete Do-It-Yourself Manual
The Reader's Digest Association
Pleasantville, NY 10570
(1973)

This book is one of the best guides ever printed on general home repair. Illustra tions, diagrams, and photographs show step-by-step methods of home repair. Thi book makes it easy even for the do-it-yourselfer with little home repair experience There is an excellent section on window repair.

Regional Guidelines for Building Passive Energy-Conserving Homes
by the Office of Policy Development and Research
U.S. Government Printing Office
Stock No. 023-000-00481-0
Washington, DC 20402
(1978)

Breaking the country down into 13 climatic regions, this publication makes a stron case for a return to regionally defined, climatically responsive house designs. Man of the illustrations provide conceptual designs for movable insulation. It is interest ing to compare the map of climatic regions in this book with maps of bioregions.

Solar Control and Shading Devices
by Aladar and Victor Olgyay
Princeton University Press
Princeton, NJ 08540
(1976)

No bibliography on movable insulation would be complete without this classic reference. This book contains valuable advice on sun angles and shading devices, and emphasizes exterior shading for warm climates.

The Solar Greenhouse Book
edited by James C. McCullagh
Rodale Press
33 East Minor Street
Emmaus, PA 18049
(1978)

A survey of attached solar greenhouses across the United States, this book presents several greenhouses which capitalize on movable insulation systems. The movable insulation is experimental and generally there is room for refinement, but unlike so many idea books, this one presents things that have been actually tried. On the broader subject of attached greenhouses, this book contains a great deal of useful information.

Thermal Shutters and Shades
by William A. Shurcliff
19 Appleton Street
Cambridge, MA 02138
(1977)

This book is the first comprehensive survey on window insulation for the home. Most of it is a systematic inventory of designs for movable insulation, ranging from commercial systems to home-fabricated designs invented by the author. There are over 100 schematic designs for thermal shutters and shades and a great deal of additional information, including designs which affect the acceptability and performance of a system. The economics of each system is also addressed.

This book has been expanded and revised by Brick House Publishing Company, 3 Main Street, Andover, MA 01816, and should be available soon under the same title. Judging from the quality of William Shurcliff's previous work, this book should be an excellent companion to *Movable Insulation*.

Village Homes' Solar House Designs
by David Bainbridge, Judy Corbett, and John Hofacre
Rodale Press
33 East Minor Street
Emmaus, PA 18049
(1979)

Forty-three "energy-conscious" house designs are presented in this short book. The graphics are clear and the photographs informative. The designs used in the homes feature direct-gain windows, including skylights and clerestories; heat-storing masonry floors and walls; water-filled culverts and drums; bread box water heaters exterior shading screens and trellises, and several kinds of thermal shutters. The window and other energy-control components should be useful in home designs in many parts of the United States, not just in Davis, California, where Village Home is located.

Window Design Strategies to Conserve Energy
by Robert S. Hastings and Richard W. Crenshaw
Written for The National Bureau of Standards
U.S. Government Printing Office
Stock No. 003-003-01794-9
Washington, DC 20402

This publication successfully presents a systematic survey of approaches to window energy conservation, including windscreens and sunshades, exterior appendages frame and glazing improvements, and interior accessories. The diagrams, photographs, and tables in this book clearly present a great deal of useful information about window designs. Product names are missing, as they are in most government publications, and more attention could also have been directed toward reducing heat losses with movable insulation.

Windows for Energy-Efficient Buildings
Energy Efficient Windows Program
Building 90, Room 3111
E.O. Lawrence Laboratory
1 Cyclotron Road
Berkeley, CA 94720

This periodical is supported by the U.S. Department of Energy and designed to bring together inventors, manufacturers, new products, and research to the attention of architects and engineers. It provides grass-roots communication with those involved in the complex structure of federal, state, and local energy programs.

Solar Conference Proceedings

Three solar conferences and their proceedings provided valuable information for this book. The 2d National Passive Solar Conference had several good papers on movable insulation, and a couple of papers on movable insulation were given at the 3rd National Passive Solar Conference. *Solar Glazing: 1979 Topical Conference* includes some excellent information on enhanced glazing assemblies.

Proceedings of the 2d National Passive Solar Conference
Mid-Atlantic Solar Energy Association
2233 Gray's Ferry Avenue
Philadelphia, PA 19146

Proceedings of Solar Glazing: 1979 Topical Conference
Mid-Atlantic Solar Energy Association
2233 Gray's Ferry Avenue
Philadelphia, PA 19146

Proceedings of the 3rd National Passive Solar Conference
American Technological University
P.O. Box 1416
Killeen, TX 76541

Technical Sources

The following list contains several technical sources for the study of window energy transactions and movable insulation.

ASHRAE 1977 Handbook Fundamentals & Product Directory
American Society of Heating, Refrigerating and Air-Conditioning Engineers, Inc. (ASHRAE)
United Engineering Center
345 East Forty-Seventh Street
New York, NY 10017

Energy Conservation Building Code Workbook
City of Davis
Department of Building Inspection
Davis, CA 95616
(1976)

Energy Conservation and Window Systems
by Samuel Berman and Seth Silverstein
National Technical Information Service
U.S. Department of Commerce
Springfield, VA 22161
(1975)

Hourly Solar Radiation Data for Vertical and Horizontal Surfaces on Average Days in the United States and Canada
by T. Kusuda and K. Ishii
U.S. Government Printing Office
Stock No. 003-003-01698-5
Washington, DC 20402
(1977)

"Roller Shade System Effectiveness in Space Heating Energy Conservation"
by Maureen Grosso and David Buchanan
ASHRAE Transactions 1979, Vol. 85, Part I
(1979)

"Window Shades and Energy Conservation"
Illinois Institute of Technology
Department of Mechanics, Mechanical and Aerospace Engineering
Chicago, IL 60616
(1974)

Section 2 Design Credits

Chapter 5

Steve Baer has been working for several years with pop-in shutters which are held to the glass with magnetic clips. Bill Shurcliff suggests a number of other designs for pop-in shutters. I have constructed a number of pop-in shutters and many of the suggestions in Chapter 5 (including the sun-collecting pop-in shutter) are from my own experience. Finally, the Muppet Shutter is a special design developed by The Center for Community Technology.

Chapter 6

I have constructed several PolarGuard-filled curtains and experimented with magnetic edge seals. The wooden baseboard hopper in Figure 6-1, the bottom-spring clamp strip in Figure 6-4c, the canvas valance top seal in 6-6d are also my designs. Clare Moorhead has recently done more extensive work with magnetic edge seals, and her design work is noted in the text. Bill Shurcliff made several suggestions on the edge seals of curtains, and A. B. LaVigne has contributed several ideas on reflective curtain linings to this chapter.

The roman shade for centuries has been a practical way to draw shades open and closed. Recently, this system has been adapted into a thermally effective design by Abby Marier of *Alternative Sources of Energy* magazine, David Wright, a California architect, and Nancy Korda and Susan Kummer of The Center for Community Technology.

Chapter 7

The modifications of existing shades, shown in Figure 7-1, were designed by me. Numerous other parties have contributed to other designs in this chapter and are listed; many are mentioned in the text. Those not mentioned there include: Ron Shore of Thermal Technology Corporation, and The Insulating Shade Company, Inc. for larger reflective shades; A. B. LaVigne, John Snebly (designer of the ATC Window Quilt), Peggy Rossi et al., Sandstone, West Virginia, for fiberfill shades; spring-loaded clamp strips were suggested by Nancy Korda of The Center for Community Technology; and the C-shaped sail track was suggested by Appropriate Technology Corporation (ATC); the wooden pulley mechanism was suggested by Peggy Rossi et al.; and the floating roller design is the Shelter Institute's suggestion. Other design sources are mentioned in the text.

Chapter 8

The shutters examined in this chapter combine information from numerous sources, including Bill Shurcliff, David Bainbridge, Denny Long, Dr. Luis H. Summers, Peter Dobrovolny, Steve Merdler, Charlie Wing, Living Systems, and The Center for Community Technology. Other design sources are mentioned in the text.

Chapter 9

The loose-fill insulation was suggested by John Kubricky in *Alternative Sources of Energy* magazine, No. 31. The between-glazing shade was designed by Bill Shurcliff, Norman B. Saunders, and myself. The Beadwall was developed by David Harrison and Steve Baer. The RIB Design was designed by Hanna Shapiro and Paul Barnes. The seasonal, hinged foam board was designed by me on suggestion by Bill Shurcliff.

Chapter 10

Steve Baer is the original source of the bottom-hinged shutter design which appears on several homes in this book. Bill Shurcliff suggested several other designs and design guidelines for this chapter. The guidelines for the optimum angle of bottom-hinged shutters are Ed Mazria's work. The reflectors, angles, and suggestions for optimum solar gains from side-hinged shutters were derived from my work, as were the block and turnbuckles to remove warping and shutter supports. Suggestions for reflector surfaces were made by Lawrence Lindsey of the Princeto Energy Group. The two sliding shutters were designed by Clinton Sheerr and Tot Environmental Action.

Chapter 11

Victor V. Olgyay is the original source for most of the material in this chapte Beyond his work, most of the designs in this chapter are either manufactured item or common knowledge.

Chapter 12

The rolladen shutter has been used for centuries. The adaptation of the garag door for movable insulation has been explored by Tim Maloney. The suggestio to use doors for seasonal shelter enclosure was made by Fuller Moore.

Chapter 13

Mike Corbett and David Norton developed the skylight shutter used in the Villag Homes in Davis, California. The Skylid was developed by Steve Baer of Zome works. The top-hinged shutter which I designed was prompted by an earlier desig of Steve Baer. The louvered, skylight shutter in Figures 13-4 and 13-5, was designe by Steve Merdler and redesigned by myself.

Chapter 14

Steve Merdler suggested a couple of the designs in this chapter. Additional credit should be given to John Hofacre, Doug Kelbaugh, and Susan Nichols.

Chapter 15

The magnetically sealed shade was designed by Fuller Moore. The fiberfill shad was designed by Andy Shapiro at the National Center for Appropriate Technolog The hinged-shutter designs were by Steve Baer, Reed Maes, and Charlie Win The overhead blanket designs were developed by the author, as was the hinge adaption of the pop-in shutter.

Chapter 16

The exterior greenhouse blanket was originally devised by Kathryn Taylor and Edit Gregg. The hinged cold frame was suggested by numerous sources. The Sola Frame was designed by Leandre Poisson, and the bottom-hinged shutters use at Marlboro College were designed by John Hayes. The bifolding shutters wer developed by Bob Forer, Bob Nelson, and Scott Hicks and the accordion-foldin shutter was suggested by Peter Clegg.

Chapter 17

The Roldoor system was developed by Tim Maloney of One Design, Inc. The curtain wall system was developed by Ron Shore of The Thermal Technology Corporation, and my work for the Aycock residence was inspired in part by this design. The louvered, segmented Trombe wall was suggested by the work of Jim Bier, and the sliding panels to insulate these walls were suggested by Bill Shurcliff. The curtain to insulate the wall segments was designed by myself. The venetian blind collector is from the work of Peter Calthorpe and the earlier work of Jeffery Cook. The concept for the solar hot-air collector-shutter design came from the work of Nick Nicholson. The hydraulic shutters for the roof pond house were designed by Living Systems, and also should be credited to the earlier work of Harold Hay.

Chapter 18

The solar collector blanket design was devised by me. The batch-type (or bread box) solar water heater has been in existence for many years. It has recently become popular once again thanks to the efforts of Steve Baer. Many other bread box designs are emerging from numerous sources including the Bainbridge-Corbett-Hofacre designs at Village Homes in Davis, California, and the works of John Burton, John Golder, and Bill Shurcliff.

Section 3 Designers and Research Groups

Don Abrams
Southern Solar Energy Center
61 Perimeter Park
Atlanta, GA 30341

Brattleboro Design Group
P.O. Box 235
Brattleboro, VT 05301

The Center for Community Technology
1121 University Avenue
Madison, WI 53715

Cornerstones
54 Cumberland Street
Brunswick, ME 04011

Gwen Cukierski
407 Citizens Bank Building
Ithaca, NY 14850

Josann Duane
Exploratory Research Laboratory
Owens-Corning Technical Center
Box 415
Granville, OH 43023

EGGF Research
Box 394B, RFD 1
Kingston, NY 12401

Dale B. Gerdeman
1403 Fifth Street
Las Vegas, NM 87701

Appendix V/Design Sources

John Golder
P.O. Box 854
Santa Cruz, CA 95061

Denise A. Guerin, Interior Designer
School of Education
and Allied Professions
150 McGuffey Hall
Miami University
Oxford, OH 45056

A. B. LaVigne, Research Engineer
8025 Thirtieth Avenue, NE
Seattle, WA 98115

Richard S. Levine
University of Kentucky
College of Architecture
Lexington, KY 40506

Living Systems
Route 1, Box 170
Winters, CA 95654

Denny Long
Route 1, Box 158
Woodland, CA 95695

Steve Merdler
300 Calle Sierpe
Santa Fe, NM 87501

Fuller Moore
Department of Architecture
Miami University
Oxford, OH 45056

Clare F. Moorhead
Conservation Concepts, Ltd.
P.O. Box 376
Stratton Mountain, VT 05155

National Center
for Appropriate Technology
3040 Continental Drive
P.O. Box 3838
Butte, MO 59701

Oak Ridge National Laboratory
Energy Division
P.O. Box X
Oak Ridge, TN 37830

One Design, Inc.
Mountain Falls Route
Winchester, VA 22601

The Passive Solar Institute
2446 Bucklebury
Davis, CA 95616

Princeton Energy Group
729 Alexander Road
Princeton, NJ 08540

Barbara Putnam
P.O. Box 29
Harrisville, NH 03450

David E. Russell
Consulting Engineer
110 Riverside Avenue
Jacksonville, FL 32202

Sandstone Senior Citizens Center
Sandstone, WV 25985
Attention: Peggy Rossi

Stephen Selkowitz
Energy and Environment Division
E.O. Lawrence Laboratory
University of California
Berkeley, CA 94220

Shelter Institute
38 Centre Street
Bath, ME 04530

William A. Shurcliff
19 Appleton Street
Cambridge, MA 02138

Doug Taff
Parallax, Inc.
P.O. Box 180
Hinesburg, VT 05301

Thermal Technology Corporation
P.O. Box 130
Snowmass, CO 81654

Thermo Tech Corporation
410 Pine Street
Burlington, VT 05401

Total Environmental Action, Inc.
Church Hill
Harrisville, NH 03450

Zomeworks Corporation
P.O. Box 712
Albuquerque, NM 87103

Section 4 Plans

Plan Descriptions	*Designer*
Bread Box Water Heater	Zomeworks Corporation P.O. Box 712 Albuquerque, NM 87103
Capsule Collector (bread box water heater)	John Golder P.O. Box 854 Santa Cruz, CA 95061
Horizontal and Vertical, Two-Tank, Passive Solar Water Heaters (bread box type)	John Burton Integral Design 4438 G Street Sacramento, CA 95819
Insulating Shades	Rodale Plans Department 33 East Minor Street Emmaus, PA 18049
Shutters	Domestic Technology Institute, Inc. Box 2043 Evergreen, CO 80439

Appendix V/Design Sources

Plan Descriptions	*Designer*
Shutters and Drapery Liners	Solpub Company Box 2351 Gaithersburg, MD 20760
Window Coverings	The Center for Community Technology 1121 University Avenue Madison, WI 53715
Window Shade Plans	Rainbow Energy Works 190 East Seventh Street Arcata, CA 95521
Wing Shutters	Charlie Wing c/o Cornerstones 54 Cumberland Street Brunswick, ME 04011

hoto Credits

oel Berman Associates, Inc.	7-5, 7-6
rattleboro Design Group	7-7
he Center for Community Technology	5-3, 5-4, 6-8, 7-4b, 8-2, 8-5a
arl Doney	5-5, 6-3, 6-4, 6-5, 11-5, 11-6, 13-1, 13-2b, 13-3, 13-4, 14-1, 14-3, 15-4, 18-1
uracote Corporation	6-2
enneth Hagard	14-2
ohn P. Hamel	5-1, 5-2, 8-5b
cott Hicks	16-3
oolShade Corporation	11-1
ohn Kubricky, *Alternative Sources of Energy*, No. 31	9-1
Villiam K. Langdon	4-1, 6-1, 6-6, 6-7, 6-9, 7-1, 7-3, 7-4a, 8-1a, 8-1b, 8-3, 8-4, 9-2, 9-3, 10-1, 10-2, 10-3, 11-2, 11-4, 11-7, 12-1, 12-2, 13-2a, 13-5, 13-6, 13-7, 14-5, 15-3, 16-1, 17-2
. B. LaVigne	7-2
James C. McCullagh	15-1, 16-2
Stephen Merdler	14-4
Oak Ridge National Laboratory	9-4a, 9-4b
Reynolds Metals Company	11-3
Robert Rodale	1-1
Jim Samsel	18-2
Clinton Sheerr	10-5
Shelter, Inc., Pre-built Homes	8-6
Solar Age magazine	10-4, 10-6
Sun Quilt Corporation	15-2
Thermal Technology Corporation	17-1

Index

Index

E

F

Index

G

H

O

P